PCR Troubleshooting and Optimization

The Essential Guide

Edited by

Suzanne Kennedy

MO BIO Laboratories, Inc.,
Carlsbad, CA, USA

and

Nick Oswald

BitesizeBio
Edinburgh, UK

Caister Academic Press

Caister Academic Press
Norfolk, UK

www.caister.com

British Library Cataloguing-in-Publication Data
A catalogue record for this book is available from the British Library

ISBN: 978-1-904455-72-1

Cover image adapted from Figure 1.4

Printed and bound in Great Britain

Contents

Contributors

Christopher Adams
Life Technologies
Carlsbad, CA
USA

christopher.adams@lifetech.com

Vasiliki Anest
Life Technologies
Carlsbad, CA
USA

vasiliki.anest@lifetech.com

N. Reginald Beer
Lawrence Livermore National Laboratory
Livermore, CA
USA

beer2@llnl.gov

Caifu Chen
Genomic Assays R&D
Molecular Biology Systems Division
Life Technologies Corporation
Foster City, CA
USA

caifu.chen@lifetech.com

Miroslav Dudas
Life Technologies
Carlsbad, CA
USA

miro.dudas@lifetech.com

Jared S. Farrar
Department of Pathology
University of Utah Health Sciences Center
Salt Lake City, UT
USA

j.farrar@utah.edu

Jack M. Gallup
Department of Veterinary Pathology
College of Veterinary Medicine
Iowa State University
Ames, IA
USA

eag@iastate.edu

Cameron N. Gundry
Idaho Technology, Inc.
Salt Lake City, UT
USA

cameron_gundry@idahotech.com

Jan Hellemans
Ghent University Center for Medical Genetics
Ghent;
Biogazelle
Zwijnaarde
Belgium

jan.hellemans@UGent.be
jan.hellemans@biogazelle.com

Sandrine Javorschi-Miller
Illumina Inc.
San Diego, CA
USA

smiller@illumina.com

Gerwyn Jones
Thermo Fisher Scientific
Epsom
Surrey
UK

gerwyn.jones@thermofisher.com

Peter Jozsi
Epigenie LLC
Delmar, CA
USA

pete@epigenie.com

Ian Kavanagh
Thermo Fisher Scientific
Epsom
Surrey
UK

ian.kavanagh@thermofisher.com

Mark Landers
Life Technologies
Carlsbad, CA
USA

mark.landers@lifetech.com

John H. Leamon
Ion Torrent Systems
Guilford, CT
USA

jleamon@iontorrents.com

John F. Mackay
dnature diagnostics and research Ltd
Gisborne
New Zealand

john@dnature.co.nz

Gavin Meredith
Life Technologies
Carlsbad, CA
USA

gavin.meredith@lifetech.com

Saima N. Nayab
Thermo Fisher Scientific
Epsom
Surrey
UK

saima.nayab@thermofisher.com

Ivan Delgado Orlic
Mouse Genotype
Carlsbad, CA
USA

ivan@mousegenotype.com

Michael W. Pfaffl
Physiology Weihenstephan
Technical University Munich
Freising
Germany

michael.pfaffl@wzw.tum.de

Matthew D. Poulson
Idaho Technology, Inc.
Salt Lake City, UT
USA

matthew_poulson@idahotech.com

Martina Reiter
BioEPS GmbH
Freising
Germany

martina.reiter@bioeps.com

Greg Shipley
UT Health Science Center – Houston
Houston, TX
USA

Gregory.L.Shipley@uth.tmc.edu

Jo Vandesompele
Ghent University Center for Medical Genetics
Ghent;
Biogazelle
Zwijnaarde
Belgium

joke.vandesompele@UGent.be
jo.vandesompele@biogazelle.com

Jonathan Wang
PCR Systems R & D
Life Technologies Corporation
Foster City, CA
USA

jonathan.wang@lifetech.com

Carl T. Wittwer
Department of Pathology
University of Utah School of Medicine;
ARUP Institute for Clinical and Experimental Pathology
Salt Lake City, UT
USA

carl.wittwer@path.utah.edu

Current Books of Interest

Viruses and Interferon: Current Research	2011
Essentials of Veterinary Parasitology	2011
Hepatitis C: Antiviral Drug Discovery and Development	2011
Streptomyces: Molecular Biology and Biotechnology	2011
Alphaherpesviruses: Molecular Virology	2011
Recent Advances in Plant Virology	2011
Vaccine Design: Innovative Approaches and Novel Strategies	2011
Salmonella: From Genome to Function	2011
PCR Troubleshooting and Optimization: The Essential Guide	2011
Insect Virology	2010
Environmental Microbiology: Current Technology and Water Applications	2011
Sensory Mechanisms in Bacteria	2010
Bifidobacteria: Genomics and Molecular Aspects	2010
Molecular Phylogeny of Microorganisms	2010
Nanotechnology in Water Treatment Applications	2010
Iron Uptake and Homeostasis in Microorganisms	2010
Caliciviruses: Molecular and Cellular Virology	2010
Epstein-Barr Virus: Latency and Transformation	2010
Anaerobic Parasitic Protozoa: Genomics and Molecular Biology	2010
Lentiviruses and Macrophages: Molecular and Cellular Interactions	2010
Microbial Population Genetics	2010
Borrelia: Molecular and Cellular Biology	2010
Influenza: Molecular Virology	2010
RNA Interference and Viruses: Current Innovations and Future Trends	2010
Retroviruses: Molecular Biology, Genomics and Pathogenesis	2010
Metagenomics: Theory, Methods and Applications	2010
Aspergillus: Molecular Biology and Genomics	2010
Environmental Molecular Microbiology	2010
Neisseria: Molecular Mechanisms of Pathogenesis	2010
Frontiers in Dengue Virus Research	2010
ABC Transporters in Microorganisms	2009
Pili and Flagella: Current Research and Future Trends	2009
Lab-on-a-Chip Technology: Biomolecular Separation and Analysis	2009
Lab-on-a-Chip Technology: Fabrication and Microfluidics	2009
Bacterial Polysaccharides: Current Innovations and Future Trends	2009
Microbial Toxins: Current Research and Future Trends	2009
Acanthamoeba	2009
Bacterial Secreted Proteins	2009
Lactobacillus Molecular Biology: From Genomics to Probiotics	2009
Mycobacterium: Genomics and Molecular Biology	2009
Real-Time PCR: Current Technology and Applications	2009

Magic in Solution: an Introduction and Brief History of PCR

1

Carl T. Wittwer and Jared S. Farrar

Abstract

The polymerase chain reaction (PCR) has become a fundamental tool in molecular research and clinical testing. The origins of PCR and its early evolution are described, including adaptation to RNA, thermostable polymerases, automation, improvements in specificity and rapid temperature cycling. Perhaps the most significant advance is real-time PCR, combining both amplification and detection into one instrument as a superior solution for nucleic acid quantification. Real-time PCR is enabled by monitoring the reaction with double stranded DNA dyes or specific probes, including hydrolysis, hybridization and conformation-sensitive probes. Early real-time instruments are compared. PCR product and probe melting analysis continues to improve in resolution, allowing greater sequence detail for genotyping and variant scanning. Microfluidic platforms and digital PCR are destined to find more applications in the future.

Origins of PCR

With 3 billion bases in the human genome, it is not easy to find and analyse the small sequence regions that confirm a genetic disorder, identify an oncogenic change or detect microbial infection. The polymerase chain reaction (PCR) provides this focus. Since its development 25 years ago, it has become the most important tool for working with nucleic acids in molecular biology and clinical diagnostics. It deserves such central recognition because of its simplicity.

Before PCR, molecular methods were multistepped, laborious and time-consuming. To amplify DNA, it had to be cloned into plasmids, the plasmids inserted into bacteria, the bacteria grown in culture, the bacteria harvested, the plasmids isolated from the bacteria, and the DNA inserts separated from the plasmid DNA. Southern blotting required multiple steps of restriction enzyme digestion, electrophoresis, blotting onto membranes, hybridization with radioactively labelled oligonucleotide probes and development on X-ray film. These early techniques required large amounts of DNA and strong technical expertise for consistent results.

Many of the components necessary for PCR were available before its invention in the 1980s by Kary Mullis and colleagues at the Cetus Corporation (Mullis *et al.*, 1986; Saiki *et al.*, 1985). The synthesis of DNA oligonucleotides had been perfected and strides towards its automation were under way (Agarwal *et al.*, 1970). The Sanger method for sequencing,

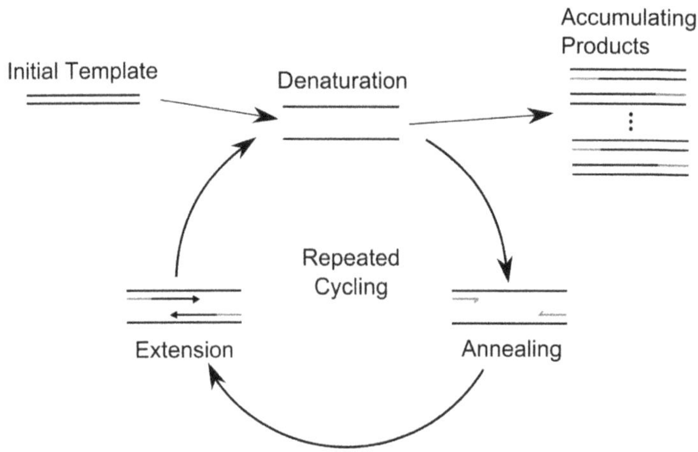

Figure 1.1 The PCR cycle. Initial template DNA is denatured by heat. The reaction is then cooled to anneal two oligonucleotide primers, each annealing to opposite strands with their 3′-ends pointed inward toward each other. A polymerase then extends each primed template to double the amount of targeted DNA. The cycle is repeated 20–40 times through successive steps of denaturation, annealing and extension. Eventually if reagents are not limiting, the concentration of accumulating products becomes high enough that double strand product formation is favoured over primer annealing, and the reaction plateaus.

which won the Nobel Prize in 1980, demonstrated DNA polymerase extension of single oligonucleotide primers annealed to their complementary strands (Sanger *et al.*, 1977). PCR combined aspects of these developments into a two primer mediated amplification of a specific DNA sequence.

PCR requires template DNA, a DNA polymerase, two sequence specific primers and deoxyribonucleoside triphosphates (dNTPs). These components are cycled in solution through three temperatures corresponding to denaturation, annealing and extension (Fig. 1.1). Denaturation separates the double stranded template DNA into single strands which then hybridize to specific primers during the annealing step. Extension allows the polymerase to incorporate dNTPs into the extending 3′ end of the primer complementary to the template DNA. After 30 temperature cycles, the DNA template bounded by the primers is amplified a million- to a billion-fold. Instead of relying on a complex *in vivo* process such as cloning, the replication of DNA was performed *in vitro*. PCR greatly reduced the number of steps required to generate appreciable quantities of DNA necessary for many downstream applications. The acceptance of PCR in the scientific community was relatively swift, with an independent research group using the technique within a year (Verlaan-de Vries *et al.*, 1986). PCR has since revolutionized molecular biology and clinical diagnostics, earning Kary Mullis the Nobel Prize in 1993.

Early PCR evolution

PCR is simple and elegant. It is remarkably robust and tolerates the addition of many diverse reagents such as electrophoresis dyes (Wittwer and Garling, 1991). PCR has spawned innumerable modifications and many related techniques. However, if it were not for a series of enhancements over the years, it might have remained a laboratory curiosity. Reverse transcriptase PCR, thermostable polymerases, automation, improved specificity, and rapid temperature cycling were early developments in the evolution of PCR. Dissemination of PCR was hindered by restrictive patent policies, particularly licence fees for clinical laboratories that extracted 12–20% of clinical testing revenues. However, even under this liability, PCR often proved to be the best solution as a flexible and robust method for a variety of applications, resulting in many laboratory-developed tests.

Reverse transcriptase PCR

The discovery of an 'RNA-dependent DNA polymerase', or reverse transcriptase, was initially reported by two independent groups (Temin and Mizutani, 1970). The suggestion of such an enzyme was new and went against the widely accepted 'Central Dogma' propounded by Watson and Crick of DNA being transcribed into RNA and then translated into protein. It was therefore met with mixed feelings in the scientific community. However, the concept gained wider acceptance after its independent confirmation (Green *et al.*, 1970; Spiegelman *et al.*, 1970a–c), and in 1975 Temin and Baltimore were jointly awarded the Nobel Prize for their discovery.

It did not take long after the discovery of PCR for its power to be used to study transcription levels in various tissues (Becker-Andre and Hahlbrock, 1989; Wang *et al.*, 1989). The amount of RNA required for PCR was much less than that required for other assays such as northern blotting, nuclease protection assays or *in situ* hybridization. PCR's ability to amplify low copy number transcripts made it immediately applicable for detection of low copy mRNAs and mRNA from a small number of cells or a small amount of tissue sample.

Traditional reverse transcriptase PCR is a useful technique for quantification of mRNA transcription. However, in terms of relative or absolute quantification, traditional reverse transcriptase PCR (as opposed to real-time RT-PCR) falls short because of the equalizing effect of the PCR plateau. Small differences in the initial concentration of templates are not easily discerned by simple end-point staining. PCR with a competitive template is one solution, with the relative amounts of target and competitor often distinguished by size, but the method requires either prior knowledge of the approximate template concentration, or multiple reactions at different concentrations. Today, real-time, reverse transcriptase, quantitative PCR (qPCR) with SYBR Green I is the most common method for transcript quantification because of simplicity and cost (Morrison *et al.*, 1998).

Thermostable polymerases

The earliest reports of PCR used a modified version of DNA polymerase I from *E. coli* known as the Klenow fragment (Klenow and Henningsen, 1970; Mullis and Faloona, 1987; Saiki *et al.*, 1985). Though this enzyme produces double stranded DNA from primed single stranded template at 37°C, it is inactivated at the high temperatures necessary for the denaturation of genomic DNA. The first PCRs were performed in three water baths set to three temperatures for denaturation, annealing and extension. Reactions were manually transferred from one water bath to the next in the cyclic manner essential for PCR. Owing to the thermolabile nature of the Klenow fragment, fresh enzyme had to be added to the reaction mixture after each denaturation step.

A potential solution to the problem of heat-induced enzyme denaturation was discovered long before the creation of PCR. Thomas D. Brock and Hudson Freeze, microbiologists working at the hot springs of Yellowstone National Park, discovered a hyperthermophilic bacterium, *Thermus aquaticus*, which thrived at high temperatures (Brock and Freeze, 1969). In 1976, Chien and colleagues, at the University of Cincinnati, reported on the isolation and purification of a DNA polymerase from *Thermus aquaticus*, noting its optimum enzymatic activity of 80°C.

It was not until 1988, 12 years after its isolation, that the first report of PCR using *Thermus aquaticus* (*Taq*) polymerase was published (Saiki *et al.*, 1988). Before *Taq* polymerase, PCR was time-consuming and inefficient because the enzyme had to be replenished after each denaturation, requiring in aggregate a large amount of enzyme. With a thermostable polymerase, no addition steps were necessary because the enzyme retained activity each cycle. Furthermore, higher annealing and extension temperatures could be used, increasing the efficiency, sensitivity and speed of the reaction. Efficiency was increased because annealing and extension were less affected by secondary structures at higher temperatures. Sensitivity was increased because higher primer annealing temperatures resulted in more stringent hybridization. Finally, the smaller spread between annealing and denaturation temperatures, and the high extension rate of *Taq* polymerase between 70°C and 80°C provided the potential for rapid cycling.

Automation

Before thermostable polymerases were used in PCR, thermal cyclers were unwieldy instruments with integrated fluidics to add fresh enzyme after each denaturation. *Taq* polymerase greatly reduced the engineering complexity of thermal cyclers, requiring only temperature cycling but not liquid handling. It did not take long before a variety of thermal cycling solutions appeared. Instruments progressed rapidly from laboratory oddities to mainstream commodities. Some early homemade examples changed the temperature of stationary reactions with flowing water or robotically transferred samples between constant temperature water baths (Foulkes *et al.*, 1988; Weier and Gray, 1988). However, water has some drawbacks. Owing to its large thermal mass a great amount of energy and time is required to heat or cool water to a specific temperature. In contrast, air has a very low thermal mass and was used in some systems (Wittwer *et al.*, 1989). Many thermal cyclers now use Peltier elements and metal blocks for heating and cooling (Collasius *et al.*, 1989; Wittbrodt and Erhardt, 1989).

Today, PCR hardware and reagents are commonplace in research and diagnostic laboratories. The instruments have evolved to fill a variety of batch size and time-to-result needs. Thermal cycling concerns now focus on issues of speed, temperature uniformity, sample volume and increased throughput. Many thermal cycling solutions, heat-stable polymerases, and commercial PCR mastermixes that include all components except primers and template DNA are available commercially.

The next big step in PCR automation was connecting the amplification and detection stages to control PCR product contamination. Laboratories can be plagued by false-positive results if products from a prior reaction find their way into a future reaction with the same primers. This contamination is usually controlled by separating pre- and post-amplification processes and careful attention to reaction preparation (Kwok and Higuchi, 1989). Another solution is to automate both amplification and detection in a closed-vessel system, eliminating PCR product exposure to the environment (Findlay *et al.*, 1993). The best solution is to amplify and analyse at the same time by real-time PCR and/or melting analysis.

Specificity

PCR specificity under ideal conditions is extraordinary. However, the human genome is very large and primers may bind not only to their intended target but also to other areas of the genome. Furthermore, the primers in PCR are at high concentrations, so even minor self or cross complementation may initiate primer dimers. These side reactions can create so-called 'non-specific' products other than the desired product. A number of methods to prevent what Kary Mullis termed 'cold oligodeoxyribonuclear fusion' (Mullis, 1991) have been developed to prevent primer extension at low (room) temperatures where polymerase activity, although greatly reduced, is still capable of extending primers.

The first 'hot-start' techniques relied on adding an essential reaction component (such as the polymerase) only after the reaction had reached high temperatures to favour specific primer annealing (D'Aquila *et al.*, 1991). This required opening the reaction container and increased the possibility of PCR contamination. To circumvent this problem, waxes and greases were used to physically partition reagents with a barrier that would melt at high temperature, mixing the essential reagent(s) with the other reaction components (Chou *et al.*, 1992).

Instead of physical separation, polymerase activity can be inhibited at room temperature. For example, monoclonal antibodies against the active site of the polymerase can inhibit the enzyme until they denature at high temperature (Kellogg *et al.*, 1994). Alternatively, the polymerase active site can be chemically modified by heat-labile covalent modifications that break down and activate the enzyme at high temperature (Birch, 1996). Instead of inactivating the polymerase, oligonucleotide primers or dNTPs can be modified at their 3′-end with similar heat-labile linkages (Koukhareva *et al.*, 2008; Lebedev *et al.*, 2008).

Many different reagents are now available to augment PCR specificity, but they are usually only necessary when the template copy number is low. Nevertheless, such reagents are an easy way to increase the robustness of PCR, sometimes making optimization unnecessary. If a hot start method is required, the best method depends on the circumstances. For example, an antibody-mediated hot-start is more useful in rapid PCR because chemically modified polymerases typically require 15–30 min for activation, longer than an entire rapid-cycle PCR protocol.

Rapid temperature cycling

When PCR first became popular in the late 1980s the process was slow. A typical automated protocol was 1 min for denaturation at 94°C, 2 min for annealing at 55°C, and 3 min for extension at 72°C. With temperature transitions included, 8 min cycles were typical requiring 4 h for 30 cycles. No one knew how much time was necessary for each stage. Cycle times were slow because of the large volumes (100 µl), containers (microfuge tubes), and heaters (metal blocks) common at the time. Using samples in capillary tubes and temperature control with circulating air, 10 min PCR (30 cycles of 20 s each) soon demonstrated that these physical limitations were not biochemical (Wittwer *et al.*, 1990). Rapid-cycle protocols use momentary or '0' s holds at the denaturation and annealing temperatures, showing that denaturation and annealing occur very quickly if the temperature is accurately controlled. The time and temperature requirements for each PCR stage were then established (Wittwer and Garling, 1991) using rapid-cycle instrumentation (Fig. 1.2).

Terms such as 'rapid' or 'fast' are relative and vague. A 1 h PCR is fast compared to 4 hours, but slow compared to 10 min. Furthermore, faster PCR is possible if you start with a higher template concentration or use fewer cycles. It is better to define the time required for each cycle and rapid-cycle PCR has been defined as 30 cycles in 10–30 min (Wittwer *et al.*, 1994) so that each cycle is 20–60 s each. The actual time of each cycle is longer than the sum of the times programmed for denaturation, annealing and extension. Indeed, during rapid PCR the temperature is usually changing. This challenges the 'equilibrium paradigm' of PCR, where three reactions (denaturation, annealing and extension) occur at three temperatures over three time periods each cycle (Fig. 1.3, left). Although intuitive, the equilibrium paradigm does not fit well with physical reality. Instantaneous temperature changes do not occur and reactions occur over a range of temperatures at different rates. More accurate is a kinetic paradigm for PCR where reaction rates and the temperature are always changing (Fig. 1.3, right). Under the equilibrium paradigm, a cycle is defined by 3 temperatures each held for a time period, whereas the kinetic paradigm requires transition rates and target temperatures. Table 1.1 shows typical rates and target temperatures found by rapid-cycle PCR (Wittwer *et al.*, 2009).

Paradigms are not right or wrong and should be judged by their usefulness. The equilibrium paradigm is simple to understand and lends itself well to the engineering mindset. The kinetic paradigm is more relevant to biochemistry, rapid PCR and melting curve analysis. Although most commercial instruments still follow equilibrium protocols, rapid protocols are a nice match for microsystems, where small volumes and rapid PCR are natural

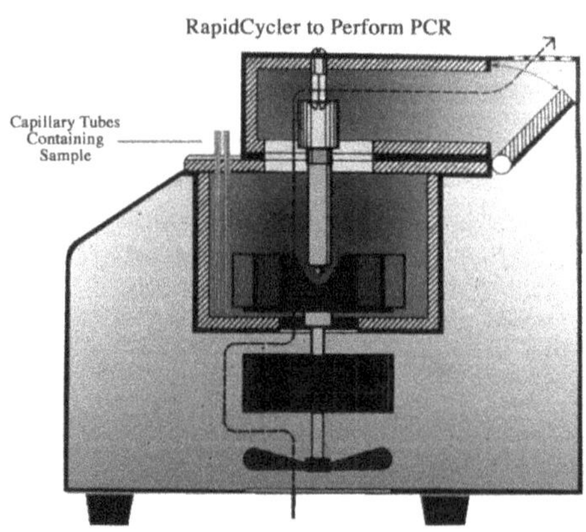

Figure 1.2 Cross-sectional diagram of a rapid, air-controlled thermal cycler. An Idaho Technology ATC 1605 thermal cycler from 1990 is shown. PCR was performed in glass capillaries which were placed in a cylindrical air chamber. A central halogen lamp functioned as a heater and an internal fan ensured even temperature distribution within the chamber. For cooling, room temperature air was drawn through the chamber by opening a door (top right). The use of air as both a heating and cooling medium allowed for rapid heat transfer to the capillaries, resulting in 30-cycle amplification times of 10–30 min.

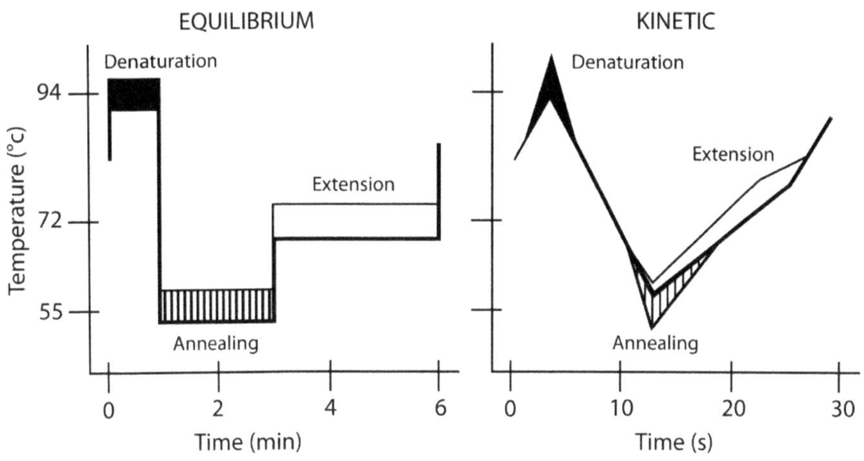

Figure 1.3 Equilibrium and kinetic paradigms of PCR. PCR can be described by equilibrium (left) or kinetic (right) models. Both paradigms highlight three events which occur during the PCR cycle (denaturation, annealing and extension). The more conventional equilibrium paradigm focuses on a specific temperature at which a reaction occurs and its time period (e.g. extension at 72°C for 15 s). The transitions between specific temperatures are generally ignored. The kinetic paradigm, however, differs in that the temperature of the reaction is never static. Denaturation, annealing and extension occur over a range of temperatures and can occur simultaneously. Reaction rates depend on temperature but are not limited to a single temperature. (Reprinted by permission of Cambridge University Press from Wittwer, C.T., Rasmussen, R.P., and Ririe, K.M. (2009). Rapid PCR and melting analysis. In: The PCR Revolution: Basic Technologies and Applications, S.A. Bustin, ed.)

Table 1.1 Optimal rates and target temperatures during rapid cycle PCR under the kinetic paradigm. (Reprinted by permission of Cambridge University Press from Wittwer, C.T., Rasmussen, R.P., and Ririe, K.M. (2009). Rapid PCR and melting analysis. In: The PCR Revolution: Basic Technologies and Applications, S.A. Bustin, ed.)

	Approach rate (°C/s)	Target temperature (°C)
Denaturation	10–30	Product T_m+3
Annealing	10–30	Primer T_m – 5
Extension	1–10 (usually 2–5)	65–80 (usually 70–74)

(Zhang and Xing, 2007). Rapid-cycle PCR is used in real-time instruments such as the Roche carousel LightCycler® and Cepheid's SmartCycler®. Other companies now promote 'Fast' protocols on more conventional thermal cyclers, apparently in a bid for independent branding and to promote reagents that are presumably required. Few instruments based on microtitre plates and heat blocks can approach rapid-cycling speeds and rapid PCR does not require special reagents.

Real-time PCR

Real-time PCR not only automates both amplification and detection, but integrates them so that they occur concurrently. Time, temperature and fluorescence are monitored during PCR in real-time instruments. The earliest report of continuous monitoring of PCR and acquiring fluorescence at each cycle was by Higuchi *et al.* (1992, 1993) utilizing ethidium bromide, a double-stranded DNA (dsDNA) specific dye. This allowed for a truly homogenous or 'closed-tube' assay in which product amplification was combined with detection. The most important application of real-time PCR is quantification of the initial template, known as quantitative PCR or qPCR.

Quantitative PCR

PCR was destined to be a quantitative technique. By both theory and practice, a well-optimized PCR doubles the amount of product each cycle for many cycles. Early attempts to harness the quantifying power of PCR were limited by dependence on end-point analysis of the products generated, either by removal of an aliquot of the reaction at predetermined cycle numbers (PCR cycle titration; Ozawa *et al.*, 1990) or serial dilution PCR (Becker *et al.*, 1996). Additional attempts were made to measure PCR products in the log phase of the reaction (Kellogg *et al.*, 1990; Pang *et al.*, 1990) or include a competitive internal control in the reaction (Piatak *et al.*, 1993). These methods were time-consuming and labour intensive, often using agarose gels to quantify the amount of PCR product and from this determine an initial template concentration.

Real-time PCR greatly simplified quantification. By monitoring fluorescence once each cycle, fluorescence as a surrogate of PCR product amount can be plotted against cycle number. No longer is there a need to physically sample a reaction at multiple cycles or guess when PCR is exponential. By acquiring data at all cycles, exponential data can be selected in retrospect. The exponential region is identified by plotting fluorescence on a log plot and the earliest cycle 'significantly above background' chosen to correlate with the initial template amount. Such quantification cycles (Cqs) are usually determined by either a fluorescence threshold or by the maximum second derivative. In either case, these fractional cycle numbers are inversely related to the log of the initial template concentration. Technical aspects of qPCR and performance guidelines have recently been published (Bustin *et al.*, 2009).

Fluorescent indicators

Real-time PCR requires monitoring the reaction during amplification. Fluorescence is a convenient method of interrogation that only requires a clear optical path for excitation and emission. Double-stranded DNA (dsDNA) dyes and fluorescently labelled probes are both commonly used. dsDNA dyes directly measure the amount of double-stranded product produced. Probes used in real-time PCR function indirectly through fluorescence resonance energy transfer (FRET) or fluorescence quenching. Initially proposed by Theodor Förster in the late 1940s (Förster, 1948, 1965), it was not until the 1980s that FRET was applied to DNA (Cardullo *et al.*, 1988; Morrison *et al.*, 1989). However, real-time monitoring with probes was only achieved several years later after dsDNA dyes were established in real-time PCR. One advantage of probes over dsDNA dyes is multiplexing by colour with different fluorescent dyes. Nevertheless, this advantage comes at a cost in instrumentation and analysis complexity. Furthermore, multiplex analysis with dsDNA dyes is possible by melting temperature separation of products and/or probes.

dsDNA dyes

dsDNA dyes are commonplace in the molecular biology laboratory. Although ethidium bromide was first used in real-time PCR, SYBR® Green I is by far the most common dye in real-time PCR today. Introduced along with the LightCycler, it is more fluorescent than ethidium bromide and is easily excited at the same wavelength as fluorescein (Wittwer *et al.*, 1997a). Most real-time PCR is performed with dsDNA dyes for reasons of cost and convenience. Any PCR can be monitored with SYBR Green I. However, because dsDNA dyes are generic, there is a risk of non-specific detection of alternative PCR products. This risk can be partly eliminated by acquiring fluorescence at a temperature where only the desired product is double-stranded (Morrison *et al.*, 1998). Melting analysis can also differentiate between specific and non-specific products (Ririe *et al.*, 1997).

Hydrolysis probes

In 1991, Holland and colleagues at the Cetus Corporation used the 5′ to 3′ exonuclease activity of *Taq* polymerase to detect amplification products post-PCR (Holland *et al.*, 1991). An oligonucleotide probe complementary to the PCR product was used with a non-extendable 3′-end and a radioactively labelled 5′-end. During amplification the polymerase degraded the probe, releasing the radioactive label as smaller fragments of the probe. However, a post-PCR radiograph was required in order to visualize the degraded probe. By replacing the radioactive label with two fluorescent labels in a FRET relationship, successful allele discrimination (Lee *et al.*, 1993) and later real-time monitoring (Heid *et al.*, 1996) were achieved. These dual-labelled fluorescent probes were hydrolyzed by the 5′ to 3′ exonuclease activity of *Taq* during PCR, separating the fluorescent labels with a loss of FRET to generate fluorescence. Specificity was enhanced over dsDNA dyes because complementation to three independent oligonucleotides (two primers and one probe) was necessary for probe hydrolysis and signal generation. Hydrolysis probes (also known by the trademark TaqMan®, among others) are the most commonly used probes today. Their popularity was advanced by simplified design (Livak *et al.*, 1995) and a strong commercial effort to provide synthesis services. Signal generation is produced by probe hydrolysis and is irreversible and cumulative.

Hybridization probes

In contrast to hydrolysis probes, the fluorescence from hybridization probes is reversible and depends only on probe hybridization. The first hybridization probes used in real-time PCR were dual hybridization probes consisting of two oligonucleotides, one labelled at the 3′-end, the other at the 5′-end (Wittwer *et al.*, 1997a). Upon hybridization to their complementary sequences and fluorescent excitation, FRET increases. Signal generation with dual hybridization probes requires annealing of four oligonucleotides (two primers and two probes), suggesting even better specificity than hydrolysis probes. Later, single hybridization probe designs were developed, including FRET between an internally labelled primer and a single-labelled probe (Bernard *et al.*, 1998b; Lay and Wittwer, 1997) and deoxyguanosine quenching of a single-labelled probe (Crockett and Wittwer, 2001). In contrast to hydrolysis probes that are consumed during amplification, the fluorescence of hybridization probes is reversible, enabling melting analysis. The first FDA-approved genetic tests in the USA (*F5* and *F2* single base variants) used dual hybridization probes and melting analysis for genotyping.

Conformation-sensitive probes

After the introduction of hydrolysis and hybridization probes, several other probe designs were adapted to, or created for, real-time PCR. Some of these probes increase in fluorescence when their conformation changes with hybridization and may function by hybridization and/or hydrolysis mechanisms depending on the reaction conditions. Hairpin probes, or 'molecular beacons' (Tyagi and Kramer, 1996) can be monitored in real-time (Tyagi *et al.*, 1998). These probes have a central sequence complementary to the DNA target and flanking ends complementary to each other. This configuration creates a hairpin at low temperatures. At higher temperatures in the presence of target, the probe hybridizes preferentially to the target. One end of the probe is labelled with a fluorophore and the other with a quencher so that when hybridized to the target, fluorescence increases. Hairpin probes use quenchers that release transferred energy as heat rather than light.

Hairpins can also be attached to the 5′-end of a PCR primer to generate self-probing amplicons during PCR (Thelwell *et al.*, 2000; Whitcombe *et al.*, 1999). A fluorophore/quencher pair in the hairpin stem is linked to the primer with a blocking agent that prevents PCR read-through (Newton *et al.*, 1993). The loop of the hairpin is complementary to the extension product of the primer so that once extension occurs, intramolecular hybridization separates the fluorophore/quencher pair and a larger hairpin is formed. This intramolecular hybridization of self-probing amplicons is faster than the intermolecular hybridization of other probes. Hairpin primers can also be made without a blocking agent, allowing PCR readthrough and incorporation of the stem–loop into the product (Nazarenko *et al.*, 1997). During PCR, the quencher/reporter pair is separated and fluorescence increases (Solinas *et al.*, 2001).

Melting analysis

Melting curve analysis is a powerful and practical extension of real-time PCR. While real-time PCR focuses on collecting fluorescence at a single temperature each PCR cycle, melting analysis monitors fluorescence over time as the temperature is changing (Ririe *et al.*, 1997). Melting analysis fits nicely into the kinetic paradigm of PCR (Fig. 1.3). Duplexes melt as the temperature increases, and the hybridization of both PCR products and probes can be monitored. Similar to 'old' (slow) PCR being considered an equilibrium process, 'old' (dot blot) hybridizations were performed at a single temperature. Dynamic monitoring of the entire melting curve as the temperature changes defines the entire melting transition, not just a single point.

Melting curve analysis was first integrated with real-time PCR on the LightCycler (Wittwer *et al.*, 1997b). No separations or reagent additions were required and melting analysis was fast (typically 2–15 min). The dye SYBR Green I conveniently provided quantification during PCR and melting analysis after PCR (Ririe *et al.*, 1997; Wittwer *et al.*, 1997a). The melting temperature of a DNA duplex is determined in large part by its sequence, G/C content and length. Specific PCR products can be easily distinguished from non-specific PCR products. In many cases melting analysis eliminates the need for post-PCR processing such as gel electrophoresis.

Genotyping by melting analysis was first demonstrated with a single hybridization probe and FRET to monitor probe melting (Lay and Wittwer, 1997). Different single-base variants produced different probe stabilities, which were revealed by melting analysis. Later, dual hybridization probes were used for genotyping (Bernard *et al.*, 1998a) and both colour and temperature multiplexing exploited (Bernard and Wittwer, 2000). The use of a single fluorescein-labelled probe instead of two probes was a further simplification (Crockett and Wittwer, 2001). Genotyping by melting without labelled probes was first shown with allele-specific PCR and SYBR Green I (Germer and Higuchi, 1999). Three primers were used, one with a GC-tail to discriminate alleles by melting temperature. Genotyping without GC-tails or labelled probes became possible with the availability of saturation dyes that detect heteroduplexes. These methods are detailed later in the section on high-resolution melting analysis.

Real-time instrumentation

The first two commercial real-time PCR platforms were the ABI 7700 and the LightCycler. The LightCycler was initially developed through a small business NIH grant(Lyon and Wittwer, 2009). The prototype, constructed at the University of Utah, integrated rapid temperature cycling with fluorescent monitoring adapted from a flow cytometer (Fig. 1.4). Idaho Technology converted the prototype to a 24-sample instrument (Wittwer *et al.*, 1997b) with a small footprint and simplified optics for commercial sale (Fig. 1.5). In 1997, the system was licensed to Boehringer Mannheim which was subsequently acquired by Roche that same year. A 32-sample LightCycler was released by Roche in 1998 integrating rapid-cycling, SYBR Green I, dual hybridization probes and melting curve analysis.

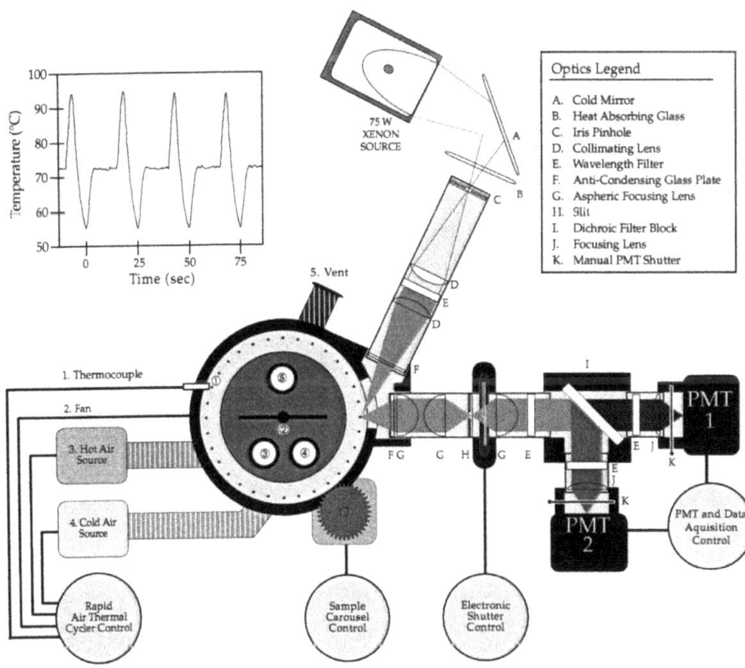

Figure 1.4 Schematic diagram of the academic LightCycler prototype. Built during 1995 at the University of Utah, the LightCycler prototype was NIH-funded. It combined the rapid temperature cycling of the RapidCycler (Fig. 1.2) with the fluorescence excitation and emission of a flow cytometer.

Figure 1.5 The Idaho Technology LC 24 was the first commercial carousel LightCycler, first sold in 1996. While the prototype used discrete excitation and emission paths, xenon lamp excitation and PMT detection, the commercial product used epifluorescent interrogation of the capillary tip, LEDs and photodiodes.

The ABI 7700 was a large, plate-based 96-well instrument focused on hydrolysis probes (Bassam *et al.*, 1996). The 7700 used a 488 nm laser and fibre optics, in contrast to the light emitting diodes and epifluorescence optics of the LightCycler. These first two real-time instruments are shown side-by-side in Fig. 1.6. Today there are many product offerings in the arena of real-time instrumentation. Competition has driven down the costs of instruments and reagents. Foundational developments in the evolution of PCR and real-time PCR are listed in Table 1.2.

Current and future trends

By measuring time, temperature, and fluorescence throughout PCR, real-time 3-dimensional spirals can be acquired and plotted (Fig. 1.7). Software on commercial instruments usually only present selected data. For example, qPCR experiments only acquire fluorescence at one temperature each cycle. Typical melting analysis only acquires fluorescence from one melting curve at the end of amplification. Much more data is available during PCR, and it is likely that this additional data will find further use in the years to come.

Homogeneous monitoring of PCR is the method of choice for gene expression quantification and closed-tube genotyping. As a 'gold standard', it has evolved from early conception to present-day reality. Future improvements will be focused on reducing cost and complexity (high-resolution melting), decreasing reaction volumes (microfluidic PCR) and increasing throughput and sensitivity (digital PCR). These approaches will allow homogeneous monitoring of PCR to continue its evolution as a useful tool for many years to come.

Figure 1.6 The Roche carousel LightCycler (right) and the ABI 7700 (left) were the first two real-time instruments marketed worldwide. Besides the obvious difference in instrument size, the batch size (32 vs. 96), sample format (capillary vs. plate), excitation (laser vs. LED), detection (CCD vs. photodiode), optical path (fibre optic vs. epifluorescence) and cycle speed (rapid-cycle vs. standard) were as different as the groups that created them. The LightCycler enabled 15–30 min PCR with real-time display of fluorescence during amplification, allowing users to watch amplification curves grow in real-time. The 7700 cycled more samples at once in plate format, but the results could not be observed during amplification and were analysed only after completion of the run. The LightCycler was introduced with SYBR Green I and dual hybridization probes while the 7700 used hydrolysis probes. The carousel LightCycler became a favourite for rapid development and analysis where turn-around-times were important while the 7700 was adopted by many large labs where processing convenience and batch size were paramount.

Table 1.2 Early PCR innovations

Innovation	Year	Reference
Invention of the PCR	1982–1985	Saiki *et al.* (1985)
Use of a thermostable polymerase	1988	Saiki *et al.* (1988)
Rapid cycle PCR	1990	Wittwer *et al.* (1990)
Hydrolysis probes (radiolabelled)	1991	Holland *et al.* (1991)
Ethidium bromide real-time PCR	1992–1993	Higuchi *et al.* (1993)
Closed-tube allele discrimination	1993	Lee *et al.* (1993)
qPCR with hydrolysis probes	1996	Heid *et al.* (1996)
SYBR green I and hybridization probe real-time PCR	1997	Wittwer *et al.* (1997a)
Product differentiation by melting analysis	1997	Ririe *et al.* (1997)

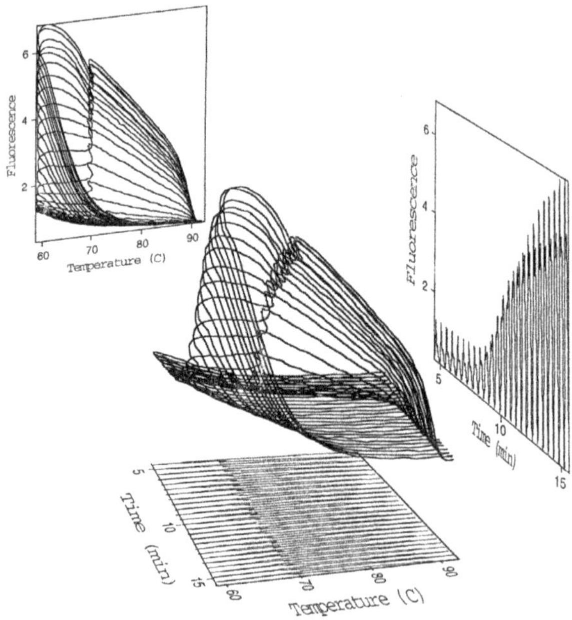

Figure 1.7 Continuous monitoring of rapid cycle PCR. A 110 bp β-globin fragment was amplified from human genomic DNA by cycling between 60°C and 90°C over 15 min in the presence of SYBR Green I. In the centre is a three-dimensional plot of temperature, time and fluorescence. The bottom graph traces temperature versus time over the 15 min of thermal cycling. The far left graph plots fluorescence over temperature and monitors the hybridization state of the PCR product. The sharp decrease in fluorescence on the right of the plot corresponds to the PCR product's melting temperature. The graph on the right is a plot of fluorescence versus time and depicts the increase in fluorescence as the reaction product accumulates with increasing cycle number. (Reprinted by permission of Cambridge University Press from, Wittwer, C.T., Rasmussen, R.P., and Ririe, K.M. (2009). Rapid PCR and melting analysis. In: The PCR Revolution: Basic Technologies and Applications, S.A. Bustin, ed.)

High-resolution melting analysis

High-resolution melting was first reported using a 5′-labelled primer for generation of fluorescence (Gundry *et al.*, 2003). Together with the development of high-resolution melting instrumentation, high-quality, reproducible melting curves for variant scanning and genotyping became possible. High-resolution instruments were necessary because standard real-time PCR instruments did not possess the precision necessary for distinguishing small differences between melting curves (Herrmann *et al.*, 2007). However, the requirement for labelled primers and the limitation that variants had to be in the same melting domain as the primer were disadvantages.

Labelled primers became unnecessary with the development of saturating dsDNA dyes that could detect heteroduplexes throughout an amplicon (Wittwer *et al.*, 2003). Single base genotyping within small amplicons required only two PCR primers and became the simplest method of genotyping (Liew *et al.*, 2004). Because heterozygotes were easily identified by a change in melting curve shape, variant scanning was also enabled. Unlike other scanning techniques, high-resolution melting does not require any physical processing or separations. It has been applied to cancer, many human genetic disorders and has been recently reviewed (Farrar *et al.*, 2009; Vossen *et al.*, 2009; Wittwer *et al.*, 2009). In clinical diagnostics, greater sequence detail can be obtained if necessary with unlabelled probes(Zhou *et al.*, 2004) or snapback primers (Zhou *et al.*, 2008).

Unlabelled probe genotyping requires only two standard primers and an unlabelled probe complementary to the sequence of interest. Unlabelled probes are generally designed to be 20–35 bases long and can match either the wild-type or variant sequence (Zhou *et al.*, 2004). Snapback primers incorporate the unlabelled probe as a 5′-extension on one of the primers (Zhou *et al.*, 2008). At low temperature the probe element snaps back onto its complementary sequence to form an intramolecular hairpin. With both unlabelled probes and snapback primers, different genotypes result in varying duplex stabilities which are easily resolved by high-resolution melting. When unlabelled probes or snapback primers are used for genotyping, two melting transitions are generally observed, one for the full-length amplicon and the other for the probed region. This feature enables simultaneous variant scanning and genotyping in the same PCR reaction (Montgomery *et al.*, 2007; Zhou *et al.*, 2005). Integration of rapid-cycle PCR, qPCR, variant scanning and genotyping is illustrated in Fig. 1.8.

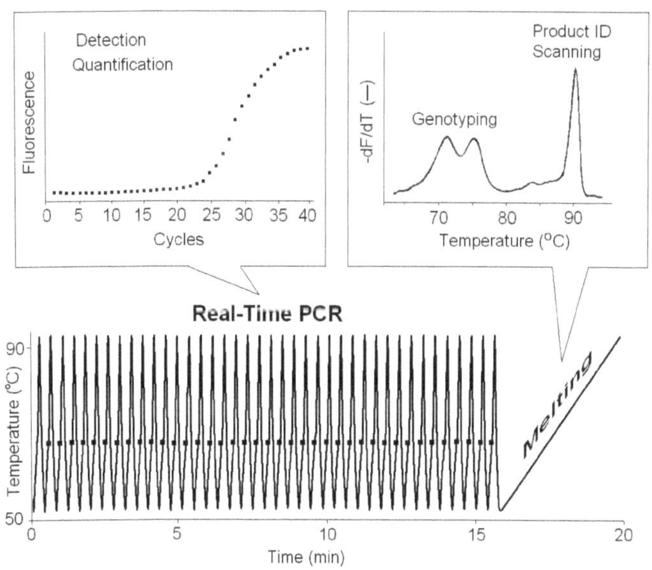

Figure 1.8 Rapid-cycle, real-time PCR and melting analysis. Momentary denaturation and annealing allows amplification in 10–20 min. Fluorescence is acquired once each cycle and is used for detection and quantification. Melting analysis is performed immediately after PCR and is used for genotyping, product identification and/or heterozygote scanning. The real-time data were obtained on a Roche LightCycler 1.5 by amplifying a 250 bp fragment of exon 2 of *PIGA* from human genomic DNA in the presence of 1X LCGreen Plus dye. After a 5 s denaturation at 95°C, 45 cycles of 95°C for 0 s, 60°C for 0 s, and 72°C for 2 s with a 2°C/s ramp between annealing and extension temperatures was performed. Fluorescence was acquired at the end of each extension step. Temperature cycling required just over 15 min, while melting analysis at 0.2°C/s required another 4 min. (Reprinted by permission of the American Society for Investigative Pathology from Lyon, E., and Wittwer, C.T. (2009). LightCycler technology in molecular diagnostics. J. Mol. Diagn. *11*, 93–101.)

Microfluidic PCR

Microfluidic PCR has the potential to amplify and quantify nucleic acids very quickly and reduce reagent demands and their associated cost. A wonderful concept, the professed 'chip PCR' has yet to gain wide application in research or diagnostics. Chip PCR was first described in 1993 (Northrup *et al.*, 1993). A microsystem for performing capillary electrophoresis was introduced around the same time (Harrison *et al.*, 1993; Manz *et al.*, 1992) and a system for incorporating both of these developments was demonstrated in 1996 (Woolley *et al.*, 1996).

However, despite wide interest in microfluidic PCR, progress has slowed from a number of setbacks. Sample and reagent adsorption onto the reaction vessel surfaces can inhibit PCR and increase the risk of carryover contamination, due to the large surface area-to-volume ratios used (Zhang and Ozdemir, 2009). Solutions for sample adsorption have been explored with differing success (Zhang *et al.*, 2006). PCR efficiency in microfluidic PCR is often compromised, and the samples used for demonstration are often less complex targets such as plasmids, bacteria or previously amplified products at high concentrations (Roper *et al.*, 2005).

There are two widely used designs for microfluidic PCR; stationary well and continuous flow. Stationary wells do not move the sample and operate in much the same way as traditional thermal cyclers. Both the sample and the device itself are heated/cooled through specific temperatures for PCR. The thermal mass is still large and thus more traditional cycling times are generally employed to perform the PCR. However, a matrix of microwells, mixing all combinations of X samples and Y targets can be obtained by microfluidics, greatly simplifying reaction preparation (Ottesen *et al.*, 2006).

Continuous-flow microfluidic PCR has some advantages over PCR performed in stationary wells. Namely, by moving the sample through fixed temperature zones, this system can achieve faster cycling times due to the fact that only the sample needs to be heated and not the entire system. A comprehensive review of the various designs and techniques involved in microfluidic PCR is available (Zhang *et al.*, 2006).

Digital PCR

An interesting extension of microfluidic PCR is digital PCR. Digital PCR combines the amplification and quantification power of PCR with limiting dilution of template targets. This allows not only for the quantification of PCR products but also for quantification of rare initial nucleic acid targets, important in such areas as cancer and prenatal diagnostics. First demonstrated by dilution PCR (Sykes *et al.*, 1992), the method was later popularized on 96-well plates (Vogelstein and Kinzler, 1999). By performing a dilution of the DNA pre-PCR, a single-template can be deposited in approximately every other well. Two probes are then used post-PCR for the determination of an allele ratio, one labelled with a green fluorescent dye and the other a red fluorescent dye. Simple comparison of well fluorescence determined the allele ratio.

The total reaction volume of the many wells required for digital PCR make 96- or even 384-well plates unwieldy for high-throughput sample analysis. However, digital PCR performed on microfluidic PCR devices has been used for single-copy DNA droplet PCR (Beer *et al.*, 2007; Kiss *et al.*, 2008), aneuploidy detection (Fan and Quake, 2007; Lo *et al.*, 2007) and absolute quantification of point variants (Oehler *et al.*, 2009).

Digital PCR improves detection specificity and sensitivity in samples with a large background of wild-type alleles compared to variant alleles and is the ultimate in allele quantification. The reduced cost associated with microfluidic devices may eventually make single-step, highly parallel individual PCR reactions for digital PCR affordable.

Conclusions

PCR has come a long way in 25 years. Based on the fundamental properties of DNA denaturation, hybridization and polymerase extension, it provides a basic tool to amplify selected DNA fragments for a myriad of applications. Real-time PCR, by combining amplification and detection, has become the gold standard of nucleic acid quantification. Melting analysis integrates an analysis component with information content that scales with the resolution of melting. Over the years PCR has become faster, better, and less expensive. The only requirements are oligonucleotide primers and the ability to temperature cycle, opening up complex genetic analysis to all scientists with a need or interest. The invention and continued innovation of PCR techniques is truly, 'magic in solution'.

References

Agarwal, K.L., Buchi, H., Caruthers, M.H., Gupta, N., Khorana, H.G., Kleppe, K., Kumar, A., Ohtsuka, E., Rajbhandary, U.L., Van de Sande, J.H., *et al.* (1970). Total synthesis of the gene for an alanine transfer ribonucleic acid from yeast. Nature *227*, 27–34.

Baltimore, D. (1970). RNA-dependent DNA polymerase in virions of RNA tumour viruses. Nature *226*, 1209–1211.

Bassam, B.J., Allen, T., Flood, S., Stevens, J., Wyatt, P., and Livak, K.J. (1996). Nucleic acid sequence detection systems: revolutionary automation for monitoring and reporting PCR products. Australasian Biotechnology *6*, 285–294.

Becker-Andre, M., and Hahlbrock, K. (1989). Absolute mRNA quantification using the polymerase chain reaction (PCR). A novel approach by a PCR aided transcript titration assay (PATTY). Nucleic Acids Res. *17*, 9437–9446.

Becker, A., Reith, A., Napiwotzki, J., and Kadenbach, B. (1996). A quantitative method of determining initial amounts of DNA by polymerase chain reaction cycle titration using digital imaging and a novel DNA stain. Anal. Biochem. *237*, 204–207.

Beer, N.R., Hindson, B.J., Wheeler, E.K., Hall, S.B., Rose, K.A., Kennedy, I.M., and Colston, B.W. (2007). On-chip, real-time, single-copy polymerase chain reaction in picoliter droplets. Anal. Chem. *79*, 8471–8475.

Bernard, P.S., Ajioka, R.S., Kushner, J.P., and Wittwer, C.T. (1998a). Homogeneous multiplex genotyping of hemochromatosis mutations with fluorescent hybridization probes. Am. J. Pathol. *153*, 1055–1061.

Bernard, P.S., Lay, M.J., and Wittwer, C.T. (1998b). Integrated amplification and detection of the C677T point mutation in the methylenetetrahydrofolate reductase gene by fluorescence resonance energy transfer and probe melting curves. Anal. Biochem. *255*, 101–107.

Bernard, P.S., and Wittwer, C.T. (2000). Homogeneous amplification and variant detection by fluorescent hybridization probes. Clin. Chem. *46*, 147–148.

Birch, D.E. (1996). Simplified hot start PCR. Nature *381*, 445–446.

Brock, T.D., and Freeze, H. (1969). *Thermus aquaticus* gen. n. and sp. n., a nonsporulating extreme thermophile. J. Bacteriol. *98*, 289–297.

Bustin, S.A., Benes, V., Garson, J.A., Hellemans, J., Huggett, J., Kubista, M., Mueller, R., Nolan, T., Pfaffl, M.W., Shipley, G.L., *et al.* (2009). The MIQE guidelines: minimum information for publication of quantitative real-time PCR experiments. Clin. Chem. *55*, 611–622.

Cardullo, R.A., Agrawal, S., Flores, C., Zamecnik, P.C., and Wolf, D.E. (1988). Detection of nucleic acid hybridization by nonradiative fluorescence resonance energy transfer. Proc. Natl. Acad. Sci. U.S.A. *85*, 8790–8794.

Chien, A., Edgar, D.B., and Trela, J.M. (1976). Deoxyribonucleic acid polymerase from the extreme thermophile *Thermus aquaticus*. J. Bacteriol. *127*, 1550–1557.

Chou, Q., Russell, M., Birch, D.E., Raymond, J., and Bloch, W. (1992). Prevention of pre-PCR mispriming and primer dimerization improves low-copy-number amplifications. Nucleic Acids Res. *20*, 1717–1723.

Collasius, M., Falk, H., Ciesler, C., and Valet, G. (1989). How to build an inexpensive cyclotherm instrument for automated polymerase chain reaction. Anal. Biochem. *181*, 163–166.

Crockett, A.O., and Wittwer, C.T. (2001). Fluorescein-labeled oligonucleotides for real-time PCR: using the inherent quenching of deoxyguanosine nucleotides. Anal. Biochem. *290*, 89–97.

D'Aquila, R.T., Bechtel, L.J., Videler, J.A., Eron, J.J., Gorczyca, P., and Kaplan, J.C. (1991). Maximizing sensitivity and specificity of PCR by pre-amplification heating. Nucleic Acids Res. *19*, 3749.

Fan, H.C., and Quake, S.R. (2007). Detection of aneuploidy with digital polymerase chain reaction. Anal. Chem. *79*, 7576–7579.

Farrar, J.S., Reed, G.H., and Wittwer, C.T. (2009). High resolution melting curve analysis for molecular diagnostics. In Molecular Diagnostics, 2nd edition, Patrinos, G., and Ansorge, W., eds. (Academic Press), pp. 229–245.

Findlay, J.B., Atwood, S.M., Bergmeyer, L., Chemelli, J., Christy, K., Cummins, T., Donish, W., Ekeze, T., Falvo, J., Patterson, D., *et al.* (1993). Automated closed-vessel system for *in vitro* diagnostics based on polymerase chain reaction. Clin. Chem. *39*, 1927–1933.

Forster, T. (1948). Zwischenmoleculare Energiewanderung und Fluoreszenz. Ann. Phys. *2*, 55–75.

Forster, T. (1965). Delocalized excitation and excitation transfer. In Modern Quantum Chemistry, Istanbul Lectures, part III, Sinanoglu, O., ed. (New York, Academic Press), pp. 93–137.

Foulkes, N.S., Pandolfi de Rinaldis, P.P., Macdonnell, J., Cross, N.C., and Luzzatto, L. (1988). Polymerase chain reaction automated at low cost. Nucleic Acids Res. *16*, 5687–5688.

Germer, S., and Higuchi, R. (1999). Single-tube genotyping without oligonucleotide probes. Genome Res. *9*, 72–78.

Green, M., Rokutanda, M., Fujinaga, K., Ray, R.K., Rokutanda, H., and Gurgo, C. (1970). Mechanism of carcinogenesis by RNA tumor viruses. I. An RNA-dependent DNA polymerase in murine sarcoma viruses. Proc. Natl. Acad. Sci. U.S.A. *67*, 385–393.

Gundry, C.N., Vandersteen, J.G., Reed, G.H., Pryor, R.J., Chen, J., and Wittwer, C.T. (2003). Amplicon melting analysis with labeled primers: a closed-tube method for differentiating homozygotes and heterozygotes. Clin. Chem. *49*, 396–406.

Harrison, D.J., Fluri, K., Seiler, K., Fan, Z., Effenhauser, C.S., and Manz, A. (1993). Micromachining a miniaturized capillary electrophoresis-based chemical analysis system on a chip. Science *261*, 895–897.

Heid, C.A., Stevens, J., Livak, K.J., and Williams, P.M. (1996). Real time quantitative PCR. Genome Res. *6*, 986–994.

Herrmann, M.G., Durtschi, J.D., Wittwer, C.T., and Voelkerding, K.V. (2007). Expanded instrument comparison of amplicon DNA melting analysis for mutation scanning and genotyping. Clin. Chem. *53*, 1544–1548.

Higuchi, R., Dollinger, G., Walsh, P.S., and Griffith, R. (1992). Simultaneous amplification and detection of specific DNA sequences. Biotechnology *10*, 413–417.

Higuchi, R., Fockler, C., Dollinger, G., and Watson, R. (1993). Kinetic PCR analysis: real-time monitoring of DNA amplification reactions. Biotechnology *11*, 1026–1030.

Holland, P.M., Abramson, R.D., Watson, R., and Gelfand, D.H. (1991). Detection of specific polymerase chain reaction product by utilizing the 5′–3′ exonuclease activity of Thermus aquaticus DNA polymerase. Proc. Natl. Acad. Sci. U.S.A. *88*, 7276–7280.

Kellogg, D.E., Rybalkin, I., Chen, S., Mukhamedova, N., Vlasik, T., Siebert, P.D., and Chenchik, A. (1994). *Taq*Start Antibody: 'hot start' PCR facilitated by a neutralizing monoclonal antibody directed against *Taq* DNA polymerase. Biotechniques *16*, 1134–1137.

Kellogg, D.E., Sninsky, J.J., and Kwok, S. (1990). Quantitation of HIV-1 proviral DNA relative to cellular DNA by the polymerase chain reaction. Anal. Biochem. *189*, 202–208.

Kiss, M.M., Ortoleva-Donnelly, L., Beer, N.R., Warner, J., Bailey, C.G., Colston, B.W., Rothberg, J.M., Link, D.R., and Leamon, J.H. (2008). High-throughput quantitative polymerase chain reaction in picoliter droplets. Anal. Chem. *80*, 8975–8981.

Klenow, H., and Henningsen, I. (1970). Selective elimination of the exonuclease activity of the deoxyribonucleic acid polymerase from Escherichia coli B by limited proteolysis. Proc. Natl. Acad. Sci. U.S.A. *65*, 168–175.

Koukhareva, I., Haoqiang, H., Yee, J., Shum, J., Paul, N., Hogrefe, R.I., and Lebedev, A.V. (2008). Heat activatable 3′-modified dNTPs: synthesis and application for hot start PCR. Nucleic Acids Symp. Ser. (Oxf.), 259–260.

Kwok, S., and Higuchi, R. (1989). Avoiding false positives with PCR. Nature *339*, 237–238.

Lay, M.J., and Wittwer, C.T. (1997). Real-time fluorescence genotyping of factor V Leiden during rapid-cycle PCR. Clin. Chem. *43*, 2262–2267.

Lebedev, A.V., Paul, N., Yee, J., Timoshchuk, V.A., Shum, J., Miyagi, K., Kellum, J., Hogrefe, R.I., and Zon, G. (2008). Hot start PCR with heat-activatable primers: a novel approach for improved PCR performance. Nucleic Acids Res. *36*, e131.

Lee, L.G., Connell, C.R., and Bloch, W. (1993). Allelic discrimination by nick-translation PCR with fluorogenic probes. Nucleic Acids Res. *21*, 3761–3766.

Liew, M., Pryor, R., Palais, R., Meadows, C., Erali, M., Lyon, E., and Wittwer, C. (2004). Genotyping of single-nucleotide polymorphisms by high-resolution melting of small amplicons. Clin. Chem. *50*, 1156–1164.

Livak, K.J., Flood, S.J., Marmaro, J., Giusti, W., and Deetz, K. (1995). Oligonucleotides with fluorescent dyes at opposite ends provide a quenched probe system useful for detecting PCR product and nucleic acid hybridization. PCR Methods Appl. *4*, 357–362.

Lo, Y.M., Lun, F.M., Chan, K.C., Tsui, N.B., Chong, K.C., Lau, T.K., Leung, T.Y., Zee, B.C., Cantor, C.R., and Chiu, R.W. (2007). Digital PCR for the molecular detection of fetal chromosomal aneuploidy. Proc. Natl. Acad. Sci. U.S.A. *104*, 13116–13121.

Lyon, E., and Wittwer, C.T. (2009). LightCycler technology in molecular diagnostics. J. Mol. Diagn. *11*, 93–101.

Manz, A., Harrison, D.J., Verpoorte, E.M.J., Fettinger, J.C., Paulus, A., Ludi, H., and Widmer, H.M. (1992). Planar chips technology for miniaturization and integration of separation techniques into monitoring systems: capillary electrophoresis on a chip. J. Chromatogr. *593*, 253–258.

Montgomery, J., Wittwer, C.T., Palais, R., and Zhou, L. (2007). Simultaneous mutation scanning and genotyping by high-resolution DNA melting analysis. Nat. Protoc. *2*, 59–66.

Morrison, L.E., Halder, T.C., and Stols, L.M. (1989). Solution-phase detection of polynucleotides using interacting fluorescent labels and competitive hybridization. Anal. Biochem. *183*, 231–244.

Morrison, T.B., Weis, J.J., and Wittwer, C.T. (1998). Quantification of low-copy transcripts by continuous SYBR Green I monitoring during amplification. Biotechniques *24*, 954–958, 960, 962.

Mullis, K., Faloona, F., Scharf, S., Saiki, R., Horn, G., and Erlich, H. (1986). Specific enzymatic amplification of DNA *in vitro*: the polymerase chain reaction. Cold Spring Harb. Symp. Quant. Biol. *51*, 263–273.

Mullis, K.B. (1991). The polymerase chain reaction in an anemic mode: how to avoid cold oligodeoxyribonuclear fusion. PCR Methods Appl. *1*, 1–4.

Mullis, K.B., and Faloona, F.A. (1987). Specific synthesis of DNA *in vitro* via a polymerase-catalyzed chain reaction. Methods Enzymol. *155*, 335–350.

Nazarenko, I.A., Bhatnagar, S.K., and Hohman, R.J. (1997). A closed tube format for amplification and detection of DNA based on energy transfer. Nucleic. Acids Res. *25*, 2516–2521.

Newton, C.R., Holland, D., Heptinstall, L.E., Hodgson, I., Edge, M.D., Markham, A.F., and McLean, M.J. (1993). The production of PCR products with 5′ single-stranded tails using primers that incorporate novel phosphoramidite intermediates. Nucleic Acids Res. *21*, 1155–1162.

Northrup, M.A., Ching, M.T., White, R.M., and Watson, R.T. (1993). DNA amplification in a microfabricated reaction chamber. In Proceedings of Transducers 1993 (Chicago, USA), pp. 924–926.

Oehler, V.G., Qin, J., Ramakrishnan, R., Facer, G., Ananthnarayan, S., Cummings, C., Deininger, M., Shah, N., McCormick, F., Willis, S., *et al.* (2009). Absolute quantitative detection of ABL tyrosine kinase domain point mutations in chronic myeloid leukemia using a novel nanofluidic platform and mutation-specific PCR. Leukemia *23*, 396–399.

Ottesen, E.A., Hong, J.W., Quake, S.R., and Leadbetter, J.R. (2006). Microfluidic digital PCR enables multigene analysis of individual environmental bacteria. Science *314*, 1464–1467.

Ozawa, T., Tanaka, M., Ikebe, S., Ohno, K., Kondo, T., and Mizuno, Y. (1990). Quantitative determination of deleted mitochondrial DNA relative to normal DNA in parkinsonian striatum by a kinetic PCR analysis. Biochem. Biophys. Res. Commun. *172*, 483–489.

Pang, S., Koyanagi, Y., Miles, S., Wiley, C., Vinters, H.V., and Chen, I.S. (1990). High levels of unintegrated HIV-1 DNA in brain tissue of AIDS dementia patients. Nature *343*, 85–89.

Piatak, M., Jr., Luk, K.C., Williams, B., and Lifson, J.D. (1993). Quantitative competitive polymerase chain reaction for accurate quantitation of HIV DNA and RNA species. Biotechniques *14*, 70–81.
Ririe, K.M., Rasmussen, R.P., and Wittwer, C.T. (1997). Product differentiation by analysis of DNA melting curves during the polymerase chain reaction. Anal. Biochem. *245*, 154–160.
Roper, M.G., Easley, C.J., and Landers, J.P. (2005). Advances in polymerase chain reaction on microfluidic chips. Anal. Chem. *77*, 3887–3893.
Saiki, R.K., Gelfand, D.H., Stoffel, S., Scharf, S.J., Higuchi, R., Horn, G.T., Mullis, K.B., and Erlich, H.A. (1988). Primer-directed enzymatic amplification of DNA with a thermostable DNA polymerase. Science *239*, 487–491.
Saiki, R.K., Scharf, S., Faloona, F., Mullis, K.B., Horn, G.T., Erlich, H.A., and Arnheim, N. (1985). Enzymatic amplification of beta-globin genomic sequences and restriction site analysis for diagnosis of sickle cell anemia. Science *230*, 1350–1354.
Sanger, F., Nicklen, S., and Coulson, A.R. (1977). DNA sequencing with chain-terminating inhibitors. Proc. Natl. Acad. Sci. U.S.A. *74*, 5463–5467.
Solinas, A., Brown, L.J., McKeen, C., Mellor, J.M., Nicol, J., Thelwell, N., and Brown, T. (2001). Duplex Scorpion primers in SNP analysis and FRET applications. Nucleic Acids Res. *29*, E96.
Spiegelman, S., Burny, A., Das, M.R., Keydar, J., Schlom, J., Travnicek, M., and Watson, K. (1970a). Characterization of the products of DNA-directed DNA polymerases in oncogenic RNA viruses. Nature *227*, 563–567.
Spiegelman, S., Burny, A., Das, M.R., Keydar, J., Schlom, J., Travnicek, M., and Watson, K. (1970b). DNA-directed DNA polymerase activity in oncogenic RNA viruses. Nature *227*, 1029–1031.
Spiegelman, S., Burny, A., Das, M.R., Keydar, J., Schlom, J., Travnicek, M., and Watson, K. (1970c). Synthetic DNA–RNA hybrids and RNA–RNA duplexes as templates for the polymerases of the oncogenic RNA viruses. Nature *228*, 430–432.
Sykes, P.J., Neoh, S.H., Brisco, M.J., Hughes, E., Condon, J., and Morley, A.A. (1992). Quantitation of targets for PCR by use of limiting dilution. Biotechniques *13*, 444–449.
Temin, H.M., and Mizutani, S. (1970). RNA-dependent DNA polymerase in virions of Rous sarcoma virus. Nature *226*, 1211–1213.
Thelwell, N., Millington, S., Solinas, A., Booth, J., and Brown, T. (2000). Mode of action and application of scorpion primers to mutation detection. Nucleic Acids Res. *28*, 3752–3761.
Tyagi, S., Bratu, D.P., and Kramer, F.R. (1998). Multicolor molecular beacons for allele discrimination. Nat. Biotechnol. *16*, 49–53.
Tyagi, S., and Kramer, F.R. (1996). Molecular beacons: probes that fluoresce upon hybridization. Nat. Biotechnol. *14*, 303–308.
Verlaan-de Vries, M., Bogaard, M.E., van den Elst, H., van Boom, J.H., van der Eb, A.J., and Bos, J.L. (1986). A dot-blot screening procedure for mutated ras oncogenes using synthetic oligodeoxynucleotides. Gene *50*, 313–320.
Vogelstein, B., and Kinzler, K.W. (1999). Digital PCR. Proc. Natl. Acad. Sci. U.S.A. *96*, 9236–9241.
Vossen, R.H., Aten, E., Roos, A., and den Dunnen, J.T. (2009). High-resolution melting analysis (HRMA): more than just sequence variant screening. Hum. Mutat. *30*, 860–866.
Wang, A.M., Doyle, M.V., and Mark, D.F. (1989). Quantitation of mRNA by the polymerase chain reaction. Proc. Natl. Acad. Sci. U.S.A. *86*, 9717–9721.
Weier, H.U., and Gray, J.W. (1988). A programmable system to perform the polymerase chain reaction. DNA *7*, 441–447.
Whitcombe, D., Theaker, J., Guy, S.P., Brown, T., and Little, S. (1999). Detection of PCR products using self-probing amplicons and fluorescence. Nat. Biotechnol. *17*, 804–807.
Wittbrodt, J., and Erhardt, W. (1989). An inexpensive and versatile computer-controlled PCR machine using a Peltier Element as a thermoelectric heat pump. Trends Genet. *5*, 202–203.
Wittwer, C.T., Fillmore, G.C., and Garling, D.J. (1990). Minimizing the time required for DNA amplification by efficient heat transfer to small samples. Anal. Biochem. *186*, 328–331.
Wittwer, C.T., Fillmore, G.C., and Hillyard, D.R. (1989). Automated polymerase chain reaction in capillary tubes with hot air. Nucleic Acids Res. *17*, 4353–4357.
Wittwer, C.T., and Garling, D.J. (1991). Rapid cycle DNA amplification: time and temperature optimization. Biotechniques *10*, 76–83.
Wittwer, C.T., Herrmann, M.G., Moss, A.A., and Rasmussen, R.P. (1997a). Continuous fluorescence monitoring of rapid cycle DNA amplification. Biotechniques *22*, 130–131, 134–138.

Wittwer, C.T., Rasmussen, R.P., and Ririe, K.M. (2009). Rapid PCR and melting analysis. In The PCR Revolution: Basic Technologies and Applications, Bustin, S.A., ed. (Cambridge, Cambridge University Press).
Wittwer, C.T., Reed, G.B., and Ririe, K.M. (1994). Rapid cycle DNA amplification. In The Polymerase Chain Reaction, Mullis, K., Ferre, F., Gibbs, R., ed. (Deerfield Beach, FL), pp. 174–181.
Wittwer, C.T., Reed, G.H., Gundry, C.N., Vandersteen, J.G., and Pryor, R.J. (2003). High-resolution genotyping by amplicon melting analysis using LCGreen. Clin. Chem. *49*, 853–860.
Wittwer, C.T., Ririe, K.M., Andrew, R.V., David, D.A., Gundry, R.A., and Balis, U.J. (1997b). The LightCycler: a microvolume multisample fluorimeter with rapid temperature control. Biotechniques *22*, 176–181.
Woolley, A.T., Hadley, D., Landre, P., deMello, A.J., Mathies, R.A., and Northrup, M.A. (1996). Functional integration of PCR amplification and capillary electrophoresis in a microfabricated DNA analysis device. Anal. Chem. *68*, 4081–4086.
Zhang, C., and Xing, D. (2007). Miniaturized PCR chips for nucleic acid amplification and analysis: latest advances and future trends. Nucleic Acids Res. *35*, 4223–4237.
Zhang, C., Xu, J., Ma, W., and Zheng, W. (2006). PCR microfluidic devices for DNA amplification. Biotechnol. Adv. *24*, 243–284.
Zhang, Y., and Ozdemir, P. (2009). Microfluidic DNA amplification--a review. Anal. Chim. Acta *638*, 115–125.
Zhou, L., Errigo, R.J., Lu, H., Poritz, M.A., Seipp, M.T., and Wittwer, C.T. (2008). Snapback primer genotyping with saturating DNA dye and melting analysis. Clin. Chem. *54*, 1648–1656.
Zhou, L., Myers, A.N., Vandersteen, J.G., Wang, L., and Wittwer, C.T. (2004). Closed-tube genotyping with unlabeled oligonucleotide probes and a saturating DNA dye. Clin. Chem. *50*, 1328–1335.
Zhou, L., Wang, L., Palais, R., Pryor, R., and Wittwer, C.T. (2005). High-resolution DNA melting analysis for simultaneous mutation scanning and genotyping in solution. Clin. Chem. *51*, 1770–1777.

qPCR Inhibition and Amplification of Difficult Templates

2

Jack M. Gallup

Abstract

One of the least acknowledged problems with PCR, RT-PCR and qPCR is reaction inhibition.[1] Addressing or eliminating inhibition is central to allowing qPCR to be modelled by the least complex mathematics, and enables more effective troubleshooting of amplifications from difficult templates such as AT- or GC-rich sequences, repetitive sequences, and templates with prohibitive secondary structures. In the absence of inhibition, additives aimed at improving PCR, RT-PCR and qPCR performance can be assessed more directly, allowing investigators to identify and utilize better primer/probe designs, enzymes and mastermixes, and formulate better reverse transcription reactions. In addition to inhibition, RNA integrity is a major concern which can be addressed by using appropriate optical assessments and the 3′:5′ assay. To address inhibition, commercial kits for removing inhibitory substances have been developed in addition to the SPUD assay and the P-Q assay development/project management software tool. Although reagent choice alone plays a large part in determining the success or failure of reverse transcription, PCR, RT-PCR or qPCR, this chapter briefly explores some of the current strategies for detecting, avoiding and/or eliminating inhibition during reverse transcription, PCR, RT-PCR and qPCR. It also discusses strategies to amplify difficult templates and optimize reverse transcription reactions.

Introduction

Given the classically interrelated/interdependent nature of reverse transcription, RT-PCR, PCR and qPCR, anything that affects the performance of one of these procedures, generally affects them all. Keeping this in mind, perhaps one of the single most revelatory statements

[1]Abbreviations and symbols: RT, reverse transcription and/or reverse transcriptase enzyme; PCR, conventional polymerase chain reaction; RT-PCR, conventional reverse transcriptase polymerase chain reaction; qPCR, quantitative real-time reverse transcription polymerase chain reaction (also abbreviated qRT-PCR and RT-qPCR elsewhere in the literature); C_q quantification cycle (formerly known as threshold cycle (C_t), crossing point (C_p) or take-off point (TOP); Cp, second derivative-determined maximal increase of fluorescence acceleration; PREXCEL-Q (software program), is abbreviated P-Q; The symbol, Æ, is used to indicate that empirical testing is expected for procedures on the part of investigators; The symbol, ЖК, indicates the required use of *in silico* tools to design or evaluate nucleic acid sequences.

regarding qPCR was made by Dr Kenneth Livak when he said (on ScienceWatch.com in July 2008):

> ...The reason that the method – (this $2^{-\Delta\Delta CT}$ method) – works is actually pretty interesting. Whenever people do the mathematics for these sorts of things, you have to make all these assumptions so that you can get a relatively straightforward mathematical expression. It turns out that the actual process of PCR is much more complex than these assumptions would indicate, but the simple approximation that we made in the paper (Applied Biosystems User Bulletin #2. ABI PRISM 7700 Sequence Detection System, 2001) still manages to give you very good data. That's fascinating to me. How is it that these mathematical expressions can sometimes be very useful, even though we've come to understand the underlying processes are far more complicated? It turns out that our original assumptions may not be precisely accurate, but there's still something about the overall process that enables the equation to provide good quantitative results.
>
> Kenneth Livak

Livak's comments, when taken in conjunction with recent publications on qPCR amplification log-logistic sigmoid modelling, shed new light on certain key aspects of fluorogenic real-time qPCR (e.g. efficiency estimations, and the true nature of the *log-linear region* of qPCR amplification curves) (Tichopad *et al.*, 2003; Rutledge, 2004; Guescini *et al.*, 2008; Ritz *et al.*, 2008; Rutledge *et al.*, 2008; Spiess *et al.*, 2008). Although we may have captured and harnessed for use, lightning in a tube, we are still hypothesizing about which sigmoidal function most faithfully models the contours of that lightning. Three-, four- and five-parameter sigmoidal curve-fitting mathematical models have been proposed (Richards, 1959; Birch, 1999; Liu *et al.*, 2002; Tichopad *et al.*, 2003; Rutledge *et al.*, 2004,2008; Peters *et al.*, 2006; Kebreab *et al.*, 2007, Guescini *et al.*, 2008; Spiess *et al.*, 2008). Ultimately, these and other recent insights into the character and behaviour of the fluorogenic qPCR process all point to the high importance of (1) meticulous nucleic acid isolation, (2) RNA integrity and quantity assessment and (3) sample handling and storage procedures. According to Dr Russell Higuchi (one of the original developers of qPCR) in a recent keynote address he gave at the qPCR Symposium USA in 2008:

> The most important lesson learned, is that PCR is capable of providing reproducible quantification – putting to rest any question of whether DNA and RNA can be quantified using an amplification strategy. The most difficult aspect of PCR, however, remains sample preparation.
>
> Russell Higuchi

On this topic, Stephen A. Bustin, Rebecca Hands, Tania Nolan and others have remained highly vocal for several years now – urging qPCR users to view nucleic acid sample preparation as the most critical step in the entire procedure. When thoroughly addressed and conscientiously performed, good sample preparation is essential to unlocking qPCR's true potential. By preparing all nucleic acid samples in a highly defined and careful manner, we allow the enzymatic reactions on which reverse transcription, RT-PCR, PCR and qPCR depend to occur as intended (Bustin, 2003, 2005, 2008; Fleige and Pfaffl, 2006; Nolan *et al.*,

2006; Doma and Parker, 2007; Swango *et al.*, 2007; Bustin *et al.*, 2009; Nolan and Bustin, 2009; Murphy and Bustin, 2009).

Numerous factors collude differentially at the molecular/thermodynamic level to afford RT-PCR, PCR and qPCR varying degrees of success at achieving its characteristic *2-to-the-n-process* behaviour $[X_n = X_0(2)^n]$. Such factors include *in silico* primer/probe design, sequence considerations and their temperature-dependent properties, dNTP usage, choice of RNA-dependent reverse transcriptase and thermostable DNA-dependent DNA polymerase enzymes, use or non-use of uracyl-DNA glycosylase enzyme, Mg^{2+} concentration, choice of machine platform for fluorescent signal detection, and the use or non-use of ROX or fluorescein as the endogenous reference dye. To add to this, the recent identification of additives (formulations and mixtures of which remain proprietary) that enhance the PCR amplification of difficult templates, the development of processes that aid in removing or lessening the effect(s) of inhibitors co-purified with nucleic acids during their isolation, and the availability of more thermostable RT enzymes (active even at 70°C, e.g. Thermo-X™ Reverse Transcriptase, Life Technologies) have dramatically improved reverse transcription, RT-PCR, PCR and qPCR performance.

Additionally, the Excel 2003-based program, P-Q, can be used to reliably identify and avoid inhibitory phenomena for qPCR assays and automate the many basic, yet time-consuming calculations involved in each part of the experimental workflow. P-Q assumes that good primer/probe designs are already in place for all targets and that the most efficient reverse transcription reaction formulations have already been established (for two-step qPCR). P-Q identifies the inhibition-free dynamic dilution ranges for each standard and sample for each target of interest and provides easy to-follow printouts for every set-up aspect of the qPCR process. P-Q is not a final qPCR data processing or reference gene validation program; it does not duplicate the function or purpose of geNorm, BestKeeper, GPR, LightCycler® Relative Quantification Software by Roche, REST, Q-Gene, qCalculator, SuperArray Biosciences (SAB) Excel-based data processing files, Dart-PCR, Gene Expression Macro by Bio-Rad, qPCR DAMS, Biogazelle's qBase and qBasePlus programs (Hellemans *et al.*, 2007; Pfaffl *et al.*, 2009; Vandesompele *et al.*, 2009) or the new web-based application called 'QPCR' (Pabinger, *et al.*, 2009). P-Q, rather, is used to develop and manage any kind of qPCR assay and is able to consistently identify the non-inhibitory ranges within which to run each standard and sample, on a per-target basis. Further, concerns regarding the inability to efficiently amplify difficult templates (due to high GC or AT content, problematic template secondary structure or repetitive sequences), choice of priming strategy, RT enzyme and mastermix are all dealt with most effectively only after sample-related reverse transcription, RT-PCR, PCR and qPCR inhibition has been addressed and/or eliminated at the very outset. Since C_q values can be influenced to appear anywhere from 8 to 50 cycles solely on account of varying degrees of reaction inhibition, acknowledging, addressing and combating inhibition must remain foremost in our minds preceding any qPCR endeavour.

The mathematical importance of eliminating inhibition from qPCR assays

Although thermodynamic algorithms for predicting various thermodynamic properties of nucleic acids such as melting temperature T_m (i.e. as applied to primers, probes and associated amplicons) have become more sophisticated and better estimators of reality, there is still significant room for improvement. Such are the current difficulties associated with PCR and qPCR.

Amplification of a specific target nucleic acid template within perfectly rendered serial dilutions of a particular qPCR sample is often observed to exhibit different partial amplification efficiencies. The average becomes the efficiency estimate for the overall target amplification reaction; a value which agrees almost exactly with the (Rutledge *et al.*, 2008) 'E_{max}' initial amplification reaction efficiency or exponential amplification value determination. Nevertheless, this is still inherently problematic in that investigators must ultimately choose a particular sample and standard (dilution-range-related) window within which to harvest C_q values at a chosen threshold which is used directly in quantification calculations to obtain final results (Ritz *et al.*, 2008). When standards and samples are still inhibitory to the assay, the chosen window(s) of measurement become less valid. Overall, stressing the use of extreme caution when interpreting data generated by any incompletely understood process is not out of place, and cannot be overemphasized. From a safely distant, general point of view, given that it does not exist elsewhere in nature per se, qPCR can be empirically described as:

> A series of enzyme-dependent processes that takes place in a closed system as a multiphasic, multitemperature-dependent forward progression that plays out against a back-drop of ever-changing reaction kinetics as it proceeds through time in the face of stochastically diminishing reactants and components resulting in the exponential accrual of a specifically intended biopolymer end-product (amplicon).

Accepting qPCR as a process that is still undergoing scrutiny, one which is currently hailed as both 'the most powerful tool in all of molecular biology' and 'kinetically misunderstood' (e.g. 'not exponential in character') ultimately urges us to pay much more attention to sample preparation and sample-related assay inhibition. Recent articles (Fleige and Pfaffl, 2006; Bustin *et al.*, 2009; Nolan and Bustin, 2009; Murphy and Bustin, 2009) by noted qPCR experts speak extensively on the importance of standardizing initial sample handling, RNA and DNA isolation, RNA and DNA quantity and quality assessments (e.g. also using the 3′:5′ assay), nuclease treatments and incorporating the use of the SPUD assay to detect qPCR inhibition in every qPCR reaction (Nolan *et al.*, 2006; Glaser, 2007). Until such things become common practice, data generated by qPCR assays must be continued to be viewed with a degree of doubt (Bustin, 2005).

C_q-based standard curve method vs. sigmoidal curve-fitting and Cy_0 methods

On the notion that qPCR is still a process that has not yet met its quintessential kinetic description, Rutledge *et al.* (2008) state:

> ...The most prominent approach [to analysing qPCR data] has been based upon analysis of the 'log-linear region,' founded upon the presumption that amplification efficiency is constant within this region. A recently developed sigmoidal model has provided new insights that challenge such historically held views, dictating that amplification efficiency is not only dynamic, but is linearly coupled to amplicon DNA

quantity. Called 'linear regression of efficiency' or LRE, this kinetic-based approach redefines amplification efficiency as the maximal efficiency (E_{max}) generated at the onset of thermocycling.

The LRE method promises to be the wave of the future for qPCR data processing. It examines the fluorescence events directly to determine key parameters (F_o, F_{max}, E_{max}) using a four-parameter sigmoidal mathematical model to decipher qPCR fluorescence data. This idea builds on what others have proposed earlier (Liu *et al.*, 2002; Tichopad *et al.*, 2003; Peirson *et al.*, 2003; Swillens *et al.*, 2004). However, a newer sigmoidal model (posited by Guescini *et al.*, 2008) is a modified standard curve-based method (named Cy_0) that does not require an assumption of uniform reaction efficiency between standards and unknowns and does not require a singular chosen threshold (Fig. 2.1). The Cy_0 method uses five-parameter sigmoidal Richards curve fitting model that resolves the tangent line for each sigmoidal amplification curve (Richards, 1959; Guescini *et al.*, 2008). These tangents pass through the calculated inflection point of each amplification curve (calculated from each curve's PCR efficiency) and intersect at a point in common at (or near) the x-axis. Small variations in the x-axis cycle (Cy_0) values for samples harbouring different degrees of inhibition yet containing the same amount of target template theoretically circumvent the problems associated with interpreting C_q values from differentially inhibited biological samples (at least for the SYBR Green I-based qPCR examples used to illustrate their model). The standard curve approach is used for this method since no universally accepted mathematical model for determining PCR efficiency currently exists. The Cy_0 method is based on the premise that amplification efficiency is not constant throughout the PCR; it is high in the exponential phase and gradually declines toward the plateau phase, whereas most other current methods of PCR efficiency assessment report 'overall' efficiency as a single value.

Additional recent publications on qPCR efficiency evaluation have concentrated on the analysis of individual shapes of fluorescence plots to estimate a dynamic efficiency value. At present, there is no genuine method for evaluating qPCR efficiency from a single amplification. Each of the many proposed methods is differentially prone to either overestimating or underestimating the 'true' PCR efficiency. By contrast, the standard curve method is based on a simple approximation of data obtained by comparing standard dilutions to unknown samples wherein qPCR efficiency estimations are based on the slope of standard (dilution) curves. However, Guescini *et al.* (2008) maintain that this approach does not account for PCR efficiencies in individual target samples, and that their proposed Cy_0 method overcomes this limitation by evaluating single amplification variations using Richards curve-fitting to determine values which minimally depend on the PCR kinetic. But, in the final evaluation of their own Cy_0 method, Guescini *et al.* (2008) demonstrate that, when comparing the C_q, second derivative (Cp)[2] and sigmoidal curve-fitting (SCF) methods

---

[2]Definition of second derivative Cp method: When crossing point (Cp) values were acquired by using the second derivative maximum method as calculated by the LightCycler® Software 3.5 (Roche Diagnostics), the Cp is defined as the number of PCR cycles elapsed when maximal acceleration of signal fluorescence increase is reached. For example the earlier the fluorescence increase, the higher is the concentration of the measured mRNA

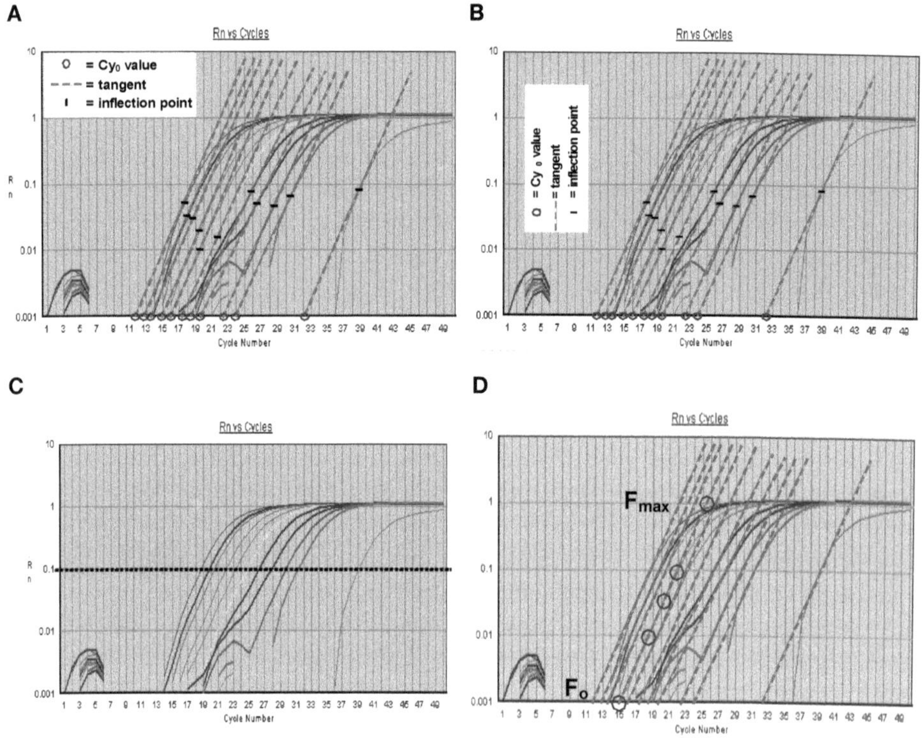

Figure 2.1 (A) Cy_0 method for determining quantitative differences between amplification curves (the point of inflection on each singular amplification plot is not fixed). This method does not assume equal efficiency for all reactions. (B) Graph A rotated 90 degrees showing the relationship between C_q vs. LOG of dR_n (ROX-normalized reporter fluorescence in this case). (C) General approach to fluorescence-based estimation of efficiency of single amplification curves. Evaluation of sample amplification efficiencies are determined by examining a window of amplification F_{Cq} and relative fluorescence units (RFU) values between F_o and F_{max}. Sigmoidal curve-fitting methods are based on this approach. This method does not assume equal efficiency for all reactions when curves are evaluated on a singular basis. (D) Threshold-based C_q method of amplification analysis wherein efficiencies are determined from the slope of target standard curves (the most prevalent method). The quantification cycle (C_q) for each reaction is determined by the point of intersection of each amplification plot with the same threshold line drawn through the most LOG-linear region of all same-target amplifications. Theoretically, at C_q, the same number of amplicons (Xn) has been generated by every sample. This method assumes equal efficiency for all reactions.

with the Cy_0 method using the criteria of precision and accuracy, under optimal amplification conditions, the four methods were equally precise and accurate. Further, there is no good way to calculate standard error or establish confidence intervals for singular Cy_0 method amplification evaluations (S. Gardner, Iowa State University, 2009, unpublished). Their conclusion thus reaffirms the importance of analysing all samples and standards only under non-inhibitory, high-efficiency-of-amplification qPCR conditions; both premises on which the P-Q program is based and which is particularly adept at achieving for every target in each appropriately prepared, diluted and cared-for nucleic acid sample set. Use of the SPUD assay (on each single sample) in conjunction with commercial nucleic acid sample isolation kits (designed to remove and/or minimize inhibitory contaminants in final nucleic acid isolates) and the P-Q assay-development project-management tool are some direct ways to address sample-related qPCR inhibition (Nolan *et al.*, 2006; Gallup and Ackermann, 2006; Lazic *et al.*, 2007; Gallup and Ackermann, 2008; Sow *et al.*, 2009; Sponseller *et al.*, 2009; Olivier *et al.*, 2009).

The quasi-exponential nature of kinetic fluorogenic qPCR efficiency estimates

Acknowledging qPCR as a process that is still evolving in its description, and accepting the idea of 'continuously efficient exponential accrual of product' by qPCR to be neither spectre nor absolute truth, we nonetheless strive to interpret qPCR fluorescence data in the most responsible, precise and mathematically meaningful ways possible. The observed 'quasi-exponential' accrual of fluorescent signal in concurrence with the 'quasi-exponential' accumulation of amplicon product over increasing qPCR cycles constitute solidly manifest, non-imaginary phenomena, even though the process is not continuously efficient.

Perhaps owing to the fact that commercial mastermixes have been painstakingly optimized over the years by talented others to the point that any 'Rutledge/Guescini-like' dissimilarity with a continuously efficient 2-to-the-*n* process (among appropriately prepared and diluted standards and samples) has been, for the most part, masked on account of highly refined product formulations (e.g. mastermixes containing the most qPCR-conducive dNTP and Mg^{2+} concentrations, fluorescence stabilizers, ROX, BSA, betaine and other additives, *Taq* and/or RT enzyme amounts, optimal pH and ionic strength, etc.). We are still allowed, mathematically, to treat well-rendered, non-inhibited, highly efficient fluorogenic qPCR results as having emanated from a 2-to-the-*n*-like process (or 'E_{AMP}-to-the *n*' process) even though it may occur, kinetically, in a manner more mathematically akin to the sigmoid amplification models that Rutledge, Guescini and others have proposed (Liu *et al.*, 2002; Tichopad *et al.*, 2003; Rutledge *et al.*, 2004, 2008; Peters *et al.*, 2006; Kebreab *et al.*, 2007; Spiess *et al.*, 2008; Guescini *et al.*, 2008). Therefore, trusting in the overall impression that (given good primer/probe designs, proper sample preparation, reverse transcription and PCR sample and reaction formulations), we can still use a relatively simplistic lens through which to view results of the qPCR process. Much of our attention should focus on things that can help or hurt the activity of the enzymes involved; RNA-dependent DNA-polymerase reverse transcriptase (RT) enzyme (e.g. for first-strand, cDNA synthesis) and thermostable DNA-dependent DNA-polymerase enzyme (e.g. *Taq* for PCR)

Causes of inhibition

When reverse transcription, RT-PCR, PCR and qPCR reactions occur with maximal efficacy and specificity, results worthy of terms such as 'accuracy,' 'precision' and 'robustness' can be reasonably anticipated. But, since qPCR is man-made and relies largely on a combination of enzymes and biomolecules that have been disengaged from their native sources, we must be willing to accept that, without the advent of finely tuned, commercially available mastermixes to carry these reactions out, qPCR would most likely amount to nothing more than an inconsequential molecular skirmish. To get these processes to work, we have taken (and altered) RT enzymes from avian and murine viruses, *Taq* from *Thermus aquaticus* (and other bacterially derived thermostable DNA-dependent DNA polymerases from hot-spring-dwelling bacteria such as *Tfl* from *Thermus flavus*, *Ultma* from *Thermotoga maritima*, *Vent* polymerase from *Thermococcus litoralis* and others such as *Pwo* and *rTth*), and the primers and probes we use (some with added labels and other chemical modifications, etc.) are now largely machine synthesized, while the very nucleic acids we study have been kidnapped and purified away from their native source material.

The 'sample effect'

Even as a highly tweaked application, qPCR, surprisingly, is most dramatically knocked out of equilibrium by the very biological samples it purports to assess. Each experimentally introduced sample introduces its own degree and assortment of qPCR-inhibitory and qPCR-friendly carryover contaminants, affecting, skewing, influencing and/or enhancing the entire qPCR process in often inexplicable ways; all of which can be termed the '*sample effect*.' And each inhibitor can affect each different DNA polymerase and RT enzyme differently (Nolan *et al.*, 2006; Bustin *et al.*, 2009; Nolan and Bustin, 2009; Murphy and Bustin, 2009) so the need to empirically test each different reagent combination could appear daunting at first. But, rather than an apparently formidable road-block, this complication also exposes some of the greatest points of control we, as investigators, have on the entire process. How we treat and what we do with the sample going into the system to begin with is central to how well we allow that system to operate in the end. Method of RNA or DNA isolation, quantity and quality assessments, nuclease treatments, methods of storage preceding qPCR, reverse transcription conditions and priming strategies are all pivotal considerations – each one of them directly affecting final results. On these grounds, certainly, most of us who perform qPCR understand that its usefulness extends beyond a mere 'yes' or 'no' genetic event indicator and further see the wisdom in establishing universal guidelines for virtually every facet of the technique. To that end, the MIQE guidelines have been created – which, in addition to specifying the details that must be included about each assay in order for qPCR results to be appropriately published, also represent a collection of best standard practices suggested by the experts in the field (see also Chapter 8, on the MIQE Guidelines) (Bustin *et al.*, 2009). But, even after samples have been prepared according to the best suggested practices, it is the ever-present notion of *sample effect* which always brings us face-to-face with the assay's most highly ignored problem of *inhibition*. Therefore, after sample isolation, nuclease treatment and template quality assessments, addressing and/or eliminating inhibition needs to be considered first, prior to performing reverse transcription, PCR, RT-PCR and qPCR. Most investigators using qPCR ignore inhibition entirely (Bustin, 2005, 2009; Nolan *et al.*, 2006). Although a number of substances have been identified as inhibitory to some or all of the RT and PCR processes, many such inhibitory contaminants and/or conditions remain as yet, unidentified (Tichopad *et al.*, 2004).

Inhibitors of RT, PCR and qPCR inhibitors in soil and plant material

In plant and soil samples, inhibitors include humic acids and related compounds such as polysaccharides, polyphenols (including tannin) and haem, which tend to co-purify along with nucleic acids upon isolation. However, while a recent study showed that many well-known plant polysaccharides (arabinogalactan, carrageenan, dextran, gum guar, gum karaya, gum locust bean, inulin, mannan, pectin, starch and xylan) were not inhibitory to PCR, two tested acidic polysaccharides (dextran sulfate and gum ghatti) were (Demeke and Adams,

1992). But, the inhibitory effect of gum ghatti can be reversed with the addition of 0.5% Tween 20 (when the polysaccharide–DNA ratio was 500:1) while the inhibitory effect of dextran sulfate (at 50:1) was reversed by addition of 0.25–0.5% Tween 20, 5% DMSO or 5% polyethylene glycol 400, although none of the additives helped when dextran sulfate was present at a ratio of 100:1 (dextran sulfate–DNA) (Demeke and Adams, 1992). Phenol–chloroform extraction is largely ineffective at eliminating unwanted contaminants from plant- and soil-derived nucleic acids, while other methods are laborious and time-consuming. To address such contaminants, commercial products (e.g. by MO BIO Laboratories, Inc.) offer various options for removing inhibitors directly from DNA samples. Currently, there are few effective solutions for high-throughput extractions of plant leaf DNA, and many methods require multiple steps and additional expensive materials. Older methods are laborious, and kits based on spin columns are expensive and are often not designed with high-throughput potential in mind. Column-based methods also often yield DNA samples that still contain polyphenolics and polysaccharides – making the DNA unsuitable for PCR amplification. The QuickExtract™ Plant DNA Extraction Solution (from EPICENTRE) provides a simple, rapid, and scalable plant-leaf DNA extraction method for PCR-based applications, including Random Amplification of Polymorphic DNA (RAPD) and Short Tandem Repeat (STR) analyses end-point PCR and qPCR. The kit enables the low-cost processing of hundreds of samples simultaneously without centrifugation, spin columns or use of caustic organic solvents. It is also amenable to high-throughput applications and use with liquid handling robotic systems. However, when such kits are not available, or do not work for a particular plant sample type, addressing sample-related inhibition of RT, RT-PCR, PCR or qPCR processes must be dealt with by using strategically precise sample dilutions, such as those suggested by the use of P-Q (Gallup and Ackermann, 2006; Kawashima *et al.*, 2006; Lazic *et al.*, 2007; Gallup and Ackermann, 2008; Olivier *et al.*, 2009; Sow *et al.*, 2009; Sponseller *et al.*, 2009). Æ

RT, PCR and qPCR inhibitors in animal material

The list of inhibitory contaminants in animal nucleic acid sample isolates currently include IgG, haemoglobin, haem, porphyrin, other unknown blood components, fat, heparin (from peritoneal mast cells), glycogen, polysaccharides, various proteins, Ca^{2+}, DNA or RNA concentration and DNA (and possibly RNA) binding proteins (Rossen *et al.*, 1992; Rijpens *et al.*, 1996; Wilson *et al.*, 1997; Thellin *et al.*, 1999; Schmittgen *et al.*, 2000; Suzuki *et al.*, 2000; Pfaffl *et al.*, 2001; Tichopad *et al.*, 2004; Gallup and Ackermann, 2006). MicroRNA (miRNA) is not thought to be a contributing factor to qPCR inhibition since high thermocycling temperatures (94–95°C) most likely prevent the formation of stable RNA-binding (RISC) complexes which might otherwise associate with template RNA. Identifying all possible inhibitors of RT, RT-PCR, PCR and qPCR becomes unnecessary as it is realized that the primary goal is to eliminate whatever it is that is impeding these processes. Standard TRIzol isolation of many nucleic acid species often results in carryover products, some of which are extremely inhibitory to RT enzymes, RT-PCR, PCR and qPCR. Column-based purification methods (e.g. such as those offered by Qiagen, Marligen, Invitrogen, etc.), are often variable in their ability to achieve the desired result (of no inhibitory potential or behaviour) in the final isolates. Empirical testing of different isolation methods per chosen sample type is the best advice that can be given on this topic. Once a successful method has been identified, use it universally for all such similar sample types (Bustin *et al.*, 2008; Bustin *et al.*, 2009). It is also highly important to note here that different DNA polymerases exhibit different tolerances for different inhibitors. For example, *Taq* polymerases (including Ampli*Taq* Gold) are completely inhibited in the presence of 0.004% (v/v) blood, whereas other polymerases maintain almost complete activity in the presence of 20% (v/v) blood (e.g. *Tfl*, *Pwo* and *rTth* DNA polymerases). *Tfl* and *Tth* DNA polymerases are also the most resistant to the inhibitory effects of K^+ and Na^+ and biological samples (Al-Soud and Rådström, 1998; Poddar *et al.*, 1998; Bustin *et al.*, 2008, 2009). Æ

In addition to its known causes, qPCR inhibition is even further complicated by the idea that it can be introduced to different degrees by different samples, in turn affecting the performance of each different enzyme or reagent differently, leaving many of us to wonder how to actually use qPCR with enough confidence and ergonomic ease to perform it correctly. The answer here, as suggested to some extent already, co-mingles with the need for us to engender the belief in a standardized initial approach that emphasizes eradication of RT and PCR inhibition as a top priority. To do this, we need to commence each qPCR endeavour by first envisaging each qPCR experiment and its associated experimental sample set as a unique system in and of itself each time a new group of samples is preened for such use. In other words, each new group of samples from which nucleic acids are to be isolated, prepared and deployed into qPCR, must always first be examined for integrity, then tested for inhibitory behaviour (as a group over an appropriately informative dilution series) for each target of interest by qPCR before any dynamically sound, final reactions are assembled.

Sample isolation and RT and/or qPCR-inhibitory material

There is a mistaken notion that column-based purification always yields nucleic acid isolates harbouring the least amount of RT and/or PCR-inhibitory material and/or potential. This is simply not the case. Column-based, TRIzol-based or other guanidinium isothiocyanate (GIT)-based nucleic acid isolation methods vary with sample type (and are differentially complicated by the extent of expertise of the human operator) in their ability to eliminate inhibitory components from the final nucleic acid sample isolates used in qPCR. But, the method of isolation and the quality of operator technique are only two of several factors involved in successfully eradicating assay inhibition in the final reaction formulations. Let alone the fact that mere choice of thermostable DNA-dependent DNA polymerase used for the PCR phase can solve some major inhibitory aspects of the assay alone (Bustin *et al.*, 2009), using all samples and standards at and/or within non-inhibitory, dynamically relevant, LOG-linear-amplification-capable dilution ranges is the most crucial step, and something that can be attained in every qPCR endeavour by using P-Q. All of the considerations mentioned above, especially the preliminary qPCR testing of representative mixtures of small portions of each experimental sample in each particular qPCR study, for each target of interest – all of these things are addressed and calculated by the program for every kind of qPCR (singular probe-based, TaqMan hydrolysis probe-based and SYBR Green-based one- and two-step qPCR). The program calculates all sample and standard dilutions, reverse transcription reaction formulations (for two-step qPCR) and all mastermix set-ups for each part of the qPCR process including Test Plate(s), Final Sample Plate(s) and NRC Plate(s), saving precious time.

Operator-introduced variation

In addition to the many quality control factors already discussed, no two human operators will perform any one of the myriad steps involved in exactly the same way, so a perfunctory preliminary ritual (postlude to template integrity and quantity assessments) of testing experimental sample mixtures, I suggest, is a practice that is uniquely poised to be able to shed the most light on each new qPCR sample system as it arises. I believe this should become the unspoken requisite standard activity, prelude to all qPCR endeavours; the eventual, universal introductory philosophy of operation and a tradition of the trade. Such an initially required, responsible practice would seem to demand status as an absolute necessity in all RT-PCR, PCR and qPCR practices (see also the MIQE Guidelines). Moreover, variables introduced to each PCR by the human operator during the procedural workflow highlight the need for universal guidelines. Those guidelines would help to ensure some level of normalcy for a technique that will continue to be riddled with the potential for generating disastrous, misleading, even nonsensical outcomes for several more years, should a standardized initial approach and good sample preparation practices not become the immediate, universal *modus operandi*. Most of the caveats encountered with qPCR are thus currently addressed by the MIQE guidelines, the SPUD inhibition assay, inhibitory-substance removal products offered by companies such as MO BIO Laboratories Inc. and others, by proper choice of assay reagents and enzymes, and, I also contend, by using the P-Q program. Æ

The 'inhibitory characteristic' of a sample

Before discussing P-Q, it is important to first describe what I call the 'inhibitory characteristic' of a sample or a group of samples (Fig. 2.2). Owing to the fact that nucleic acids isolated from different sources (plant material, animal material, soil, contaminated water, blood, faeces, etc.) by different methods introduce into the samples different amounts of contaminating substances, some of which are inhibitory to the RT enzyme or the qPCR phase to different degrees, there is a clear need to identify the dilution threshold at which such inhibition lets up for each representative sample or group of samples. A main feature of the P-Q program is that this blanket dilution for all samples is the very first thing to be identified before the assay proceeds any further. The 'inhibitory characteristic' of each similarly isolated sample set (from the same biological source) is determined by using a mixture of all similar sample types (e.g. a small portion of all similar-tissue RNA extracts mixed together) as the serially diluted sample that is examined on a Test Plate. The underlying idea here is: 'what better average representative is there of a sample or sample set other than the very samples themselves?' Such sample mixtures are called 'Stock I' in the P-Q vernacular, and when such mixtures are examined over an 11-point dilution series spaced evenly on a LOG scale (from no sample dilution to 1:50,000 or more) for each target of interest, the amplification plot of C_q vs. log of dilution for each target implicitly represents the expected average qPCR behaviour of each individual sample – given identical isolation methodologies for all such similar samples. Establishing the minimum dilution all samples must undergo to simply avoid the 'group inhibitory characteristic' is something all qPCR assays need to incorporate – no matter how *pure* the nucleic acid material is thought to be going into the qPCR. Since any sample, no matter how painstakingly and meticulously prepared, may still harbour inhibitory potential, observing the extent of inhibitory characteristic is an absolute

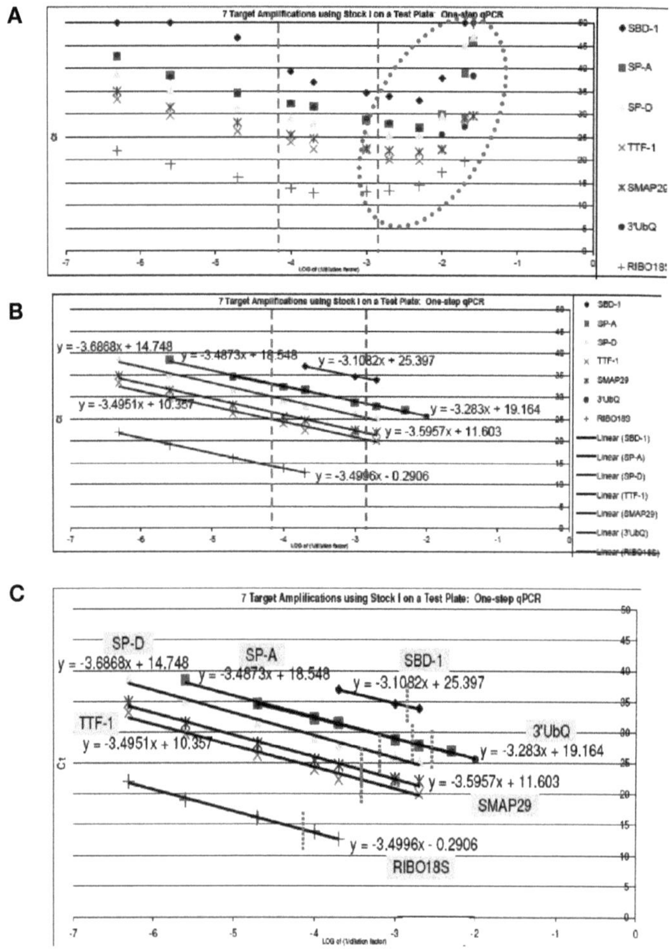

Figure 2.2 A Stock I dilution series exposing the 'inhibitory characteristic' for a group of sheep total lung RNA samples isolated by TRIzol. A portion of 56 experimental samples was mixed to create the Stock I sample, and the inhibitory characteristic of the sample group for each of seven targets was examined on a Test Plate. (A) The inhibitory regions are indicated within the circular dotted line. (B) The P-Q-resolved valid, log-linear dilution ranges for each of the seven targets. (C) The smaller vertical dotted lines represent where P-Q calculated each of the 56 samples to be used within each target's respective valid standard curve range. These lines represent a particular ng/µl in-well concentration of RNA sample in the final RT-qPCR reactions in each case. From Gallup *et al.*, 2008, *International Journal of Biomedical Sciences.*

necessity for all qPCR endeavours; one that should never be left out no matter how well the operator feels she or he is performing the technique, since quality of technique, samples, reagents and other materials can change from day to day.

Once the inhibitory-characteristic-avoidance dilution has been determined and applied to all samples, regardless of their resulting concentrations, it is then that they are further diluted to the target-specific ng/µl concentrations at which they can be appropriately used within each of their target standard curve ranges. The standards (and standard curve ranges) themselves are also each diluted to encompass ranges already demonstrated to be safely outside of the sample group (average) inhibitory characteristic for each different target. Proof of the validity of this approach has not failed yet in our hands (Grubor *et al.*, 2004; Kawashima *et al.*, 2006; Gallup and Ackermann, 2006, 2008; Lazic *et al.*, 2007; Gallup *et al.*, 2009; Olivier *et al.*, 2009; Sow *et al.*, 2009; Sponseller *et al.*, 2009). Further, it is a very straightforward, time-saving and cost-efficient approach to qPCR when 3′:5′ or SPUD assays are not yet being run concurrently. The idea of exposing the inhibitory characteristic of a sample or group of samples (whether it be RNA, DNA or cDNA) has even worked for a mixture of different isolates from different tissue types, e.g. thirteen different tissue total RNA isolates from two different chickens exposed the inhibitory characteristic for all sample types at once – and the final qPCR was built around that information yielding beautiful results for each of three targets of interest (*Gallus gallus* Gallinacin 1, Gallinacin 2 and 18S ribosomal RNA) within each of the thirteen different tissue types (C.B. Brockus, J.M. Gallup, unpublished). Too often, investigators will guess at the first dilution all of their samples will undergo before deploying them into qPCR. Going blindly into the assay from the outset, I feel is a mistake that is still being made far and wide in the qPCR world. After the inhibitory characteristic-eliminating dilution has been applied to each sample, the dilution that each sample undergoes from there, per each target of interest, and the dilutions of the Stock I material to create the final standard curves for each target are all spelled out by the P-Q program; taking the tedium and error out of the dilution calculation process. More succinctly, P-Q eliminates the common, often unspoken practice of blindly choosing some random blanket pre-qPCR dilution for all samples regardless of their genesis and, instead, replaces it with a tangible, logical process; one with reason at its core. Grabbing sample pre-qPCR (pre-RT and pre-PCR) dilution values out of thin air is a highly untenable, leap-of-faith approach.

Inhibition issues specific to RNA and RT reactions

The efficacy of reverse transcription reactions has been a controversial topic for quite some time, so it is important to address some of the issues involved. Being that the topics of RNA sample integrity, quality and quantity assessment is already dealt with exhaustively and extensively elsewhere (Nolan *et al.*, 2008; Bustin *et al.*, 2009), this section on RNA and RT will briefly deal with issues related to the formulation of good reverse transcription reactions. Since the issue of secondary template structure has, by analogy, already been dealt with in the previous sections, this section is written bearing in mind that concerns over GC-rich and AT-rich template sequences apply to RNA templates in much the same way as they apply to DNA templates. For qPCR, these concerns are lessened to a great extent due to the generally shorter length of the amplicons and target sequences involved. **Æ Ж**

The fidelity of reverse transcription (RT) reactions

More thermostable RT enzymes have become available over the years, and some are even reported to maintain good activity up to 70°C (e.g. *rTth* reverse transcriptase/DNA polymerase, ABI and Thermo-X™ Reverse Transcriptase, Invitrogen). Using a higher temperature during reverse transcription most likely translates into less secondary structure in the template, which therefore increases the chance that the RT enzyme will be able to stay on task, yielding full-length cDNAs. So, again, it is largely the choice of RT enzyme here that ultimately decides the success-rate of any reverse transcription step. Careful considerations regarding T_m of gene-specific or other priming used for reverse transcription should, however, always strive to be compatible with the subsequent PCR phase (where cycling temperature often include steps ranging from 95°C for denaturation to between 58°C and 72°C for annealing and extension). Most reverse transcription reaction temperatures have typically been 42°C, 48°C, 50°C, and up to 60°C recently – with the advent of enzymes offered by Quanta BioSciences, Invitrogen and others. The ability to use shorter times for reverse transcription (e.g. 3 min) is also an option with this new generation of RT enzymes. As well, predicting the secondary structure of RNA templates beforehand in order to improve priming strategy at lower (normally) RT-prohibitive temperatures has become more feasible with *in silico* tools such as SHAPE (Legiewicz *et al.*, 2009). **Æ Ж**

What does a well-rendered reverse transcription reaction require, besides a good RT enzyme (e.g. SuperScript III, iScript, qScript, etc. **Æ**)? Much recent attention (Cambridge Healthtech Institute's 3rd, 4th and 5th Annual qPCR Conferences) has been paid to different priming strategies during RT. The combination of poly[dT]$n_{(12-18)}$ with random hexamers, octamers, nonamers or pentadecamers has surfaced as the most effective approach to reverse transcription reaction priming which also avoids 3′ bias (Bustin, 2003; Bustin *et al.*, 2009). But it must be remembered that most poly-A tails are missing from mRNAs isolated from FFPE sections as a result of the aldehyde cross-linking/fixation derivatizations of adenine, so priming using oligo-[dT] is not advised with such samples. The effect of formalin fixation on RNA involves the addition of mono-methylol [$-CH_2OH$] groups to all four bases at various rates but with the additional dimerization of adenine groups by methylene bridging. However, most of the methylol groups can be removed by simply elevating the temperature in a buffer solution, and successful isolation of total RNA after extensive proteinase K digestion (up to 5 days) may be due to the continuing removal of these groups over time, restoring the template activity of the RNA (Masuda *et al.*, 1999; Lewis *et al.*, 2001; Bustin *et al.*, 2009). Although, gene-specific priming (GSP) is still the most efficient priming of all (as used in one-step qPCR applications) if one is only investigating a small number of targets (since a single RT reaction has to be run for each GSP when performing two-step qPCR), many investigators, however, still choose to first convert RNA isolates (sometimes after linear amplification) to cDNA due to nucleic acid stability and long-term storage considerations. Conversion of RNA to cDNA by RT constitutes the last 'snapshot' of the transcriptome to be taken. And taking a fair, unbiased [representative] version of such a picture is crucial since all final results truly depend upon the clarity, resolution and precision of the original impression. Poorly rendered reagent or sample-inhibited RT reactions can result in skewed, warped and/or distorted snapshots of the original transcriptomic profile, predisposing to misleading final qPCR results. Only well-rendered RT reactions have the best chance at reflecting the relative abundance of each particular transcriptomic target of interest.

So, what are the factors to consider when formulating good RT reactions? In past research, I have experienced success with reverse transcription reactions using the following line of questioning:

1 How and when do I measure my RNA samples on a spectrophotometer or NanoDrop? Should I use RiboGreen? PicoGreen?

Answer. According to the MIQE guidelines quantification of RNA in the extracted samples is important because it is advisable that approximately the same amounts of RNA be used when comparing different samples. However, there are several quantification procedures in common use, including spectrophotometry (NanoDrop; Thermo Scientific), microfluidic analysis (Agilent Technologies' Bioanalyzer, Bio-Rad Laboratories' Experion), capillary gel electrophoresis (Qiagen's QIAxcel), or fluorescent dye detection (Ambion/Applied Biosystems' RiboGreen). Each method produces different results, making it unwise to try to compare data obtained with the different methods. The preferred method for quantifying RNA uses fluorescent RNA-binding dyes (e.g. RiboGreen), which are best for detecting low target concentrations. In any case, it is advisable to measure all samples with a single method only and to report this information (Bustin *et al.*, 2009). In addition, always measure the RNA samples post-DNase treatment (to avoid including gDNA in the readings) using a facsimile of *the buffer the RNA is currently in* (post-DNase) as the blanking (zeroing) buffer for the spectrophotometer and/or NanoDrop (see example in Fig. 2.3). The other considerations here involve linearity and limit of detection. The trustworthy limit of detection is pretty much the same for both spectrophotometer and NanoDrop $OD_{260\,nm}$ readings; between 0.05 and 0.0625. This corresponds to a 2 to 2.5 ng/µl RNA sample (when measured at full strength). However, a spectrophotometer's linear range begins to dramatically deteriorate with absorbance readings of 1.2 and higher, while the upper limit for NanoDrop readings would be equivalent to spectrophotometer readings of greater than 40. On a spectrophotometer, readings of RNA isolated from samples rich in RNA (e.g. liver and spleen) can be diluted 1:100 before absorbance readings are taken, while samples with medium amounts of RNA (such as lung and kidney) can be diluted 1:50 for absorbance readings. Samples with lower amounts of RNA present (dendritic cell preparations,

112 ul	Nuclease-free water
16 ul	10X TURBO DNase Buffer
32 ul	2U/ml TURBO DNase Enzyme
Put at 37°C for 30 minutes	
16 ul	TURBO Inactivation Reagent
vortex every 10 seconds for 2 minutes	
microfuge for 3 minutes @ 10,000 x g	
128 ul	Recovered off of top after spin
20 ul	40 U/µl RNase inhibitor
1132 ul	Nuclease-free water
1280.0 ul	**Custom Zeroing Solution**

Figure 2.3 The 'blanking buffer' for nucleic acid quantification. To 'zero' or 'blank' spectrophotometers or NanoDrop instruments appropriately, the buffer used should be identical to what the nucleic acid samples are in at the time of their $A260_{nm}/A280_{nm}$ assessments. (This particular custom zeroing solution is then diluted with water to the same extent that samples are diluted with water upon their final spectrophotometer or NanoDrop absorbance measurements if final sample dilutions are needed at this point, e.g. for spectrophotometer readings. NanoDrop measurements would generally not require additional dilution at this point.)

brainstem and/or other neural tissues) should be diluted less to obtain readings above 0.05 and, ideally, near the most accurate point of a spectrophotometer's ability to interpret sample transmittance (e.g. optimal is a reading of ~0.301). On a NanoDrop, as long as one is careful to 'zero' the instrument correctly and to properly wash the sample probe and lower pedestal between sample readings, any full-strength sample can be read without dilution to a good degree of precision. It is important that these measurements are taken only after DNase treatments have been performed and only after the instruments have been zeroed (blanked) with the exact buffer, water or complex mixture that the RNA samples actually currently reside in. Water will work in situations where RNA has been eluted from a column post-DNase treatment (e.g. as in SuperArray Biosciences applications and methodologies). However, our typical spectrophotometer and/or NanoDrop zeroing buffer is a complex mixture as shown in Fig. 2.3.

2 *How much RT (reverse transcriptase) enzyme should be used per ml in the reverse transcription reaction?*
Answer. Between 3 and 8 units/µl, as empirically discovered in my own work. Never use more than 8 units/µl since RT enzymes, even denatured, may still harbour the potential to bind cDNA molecules and cause sporadic results during qPCR. I use 5 units SSIII RT/µl.

3 *How should RNA be DNase-treated?*
Answer. Using your favourite DNase of choice, either on-column or off-column treatments seem to be OK. However, I have found the most effective DNase treatment to be an off-column method using an excess of TURBO-DNase[3] (Life Technologies) in the following sample and reagent proportions (per 50 µl reaction):

- 5 µl 10× TURBO DNase buffer
- 10 µl TURBO DNase (2 units/µl) enzyme
- 35 µl RNA sample

Incubate for 30 min at 37°C, then add:

- 5 µl TURBO inactivation reagent

Vortex every 10 s for 2 min, then spin for 3 min at 10,000 × *g*. Remove 40 µl of sample while totally avoiding the bottom 15 µl (discard).

[3]This approach is not described in the product literature for this DNase, but I have found it to always result in either undetectable gDNA-contamination signal or generation of gDNA-contamination C_q values at least 13 cycles or more away from the genuine mRNA target C_q values. Therefore, our gDNA signals have a negligible impact on our final quantification calculations. The key here is the 3-min spin, as opposed to the 1.5-min spin suggested in the TURBO DNase product literature. This helps to pack down the Inactivation Reagent in the bottom of the tube much better so that, when we take the 40 µl of sample away from the 55 µl reaction (in a 200 µl DNase-treatment tube), we can better avoid the Inactivation Reagent. And, avoiding this reagent is important since it is a proprietary polymer which binds protein and divalent metal ions in solution – which would adversely affect downstream RT and PCR reactions.

4 How do I use these DNase-treated RNA samples in one-step qPCR?
Answer. First dilute the samples 1:10 (with inclusion of an RNase inhibitor as part of the 1:10 dilution) as follows:

- 40 µl DNase-treated RNA sample
- 353.75 µl nuclease-free water w/0.1 mM EDTA, pH 6.75 (or just nuclease-free water)
- 6.25 µl RNase inhibitor (40 units/µl)

Resulting in 400 µl of '1:10-diluted RNase-inhibitor-protected RNA samples' each of which contains 0.4 to 0.625 units/µl final RNase inhibitor concentration. At this point, use P-Q to determine the *inhibitory characteristic* of the sample set and proceed from there.

5 How do I use these DNase-treated RNA samples in two-step qPCR, and how much RNA should I add per ml in each of the reverse transcription reactions?
Answer. A different post-DNase dilution scheme is used for RNA samples that are to be converted to cDNA prior to use in qPCR. For this approach, using nuclease-free water, first dilute all DNase-treated column-purified RNA isolates 1:20, and all DNase-treated TRIzol-purified RNA isolates 1:50. Then normalize all samples from there to the same ng/µl concentration using either nuclease-free water or a buffer containing RNase inhibitor (similar to what is used for dilution in point 4 above). For reverse transcription, all reactions must receive the exact same volume and amount of the ng/µl-normalized RNA samples. For effective, uninhibited RT reactions, this final RNA/µl amount can range from 0.5 to 80 ng RNA/µl. In my own empirical studies, I have found 30 ng total RNA/µl to be optimal and I would suggest never using more than 70 ng total RNA/µl RT reaction since, after that point, one begins to invite increasingly unacceptable amounts of undesirable inhibitory materials and/or inhibitory behaviour into the RT party mix, and no one likes inhibitory grumps at an RT party. Due to the differential dilutions that each of the RNA samples must undergo to arrive at the same ng/µl concentration preceding reverse transcription, in samples that have undergone the least amount of dilution (to attain the desired ng/µl concentration preceding reverse transcription), inhibitory sample contaminants or behaviour(s) begin to rear their head more and more, and differentially so, when total RNA ng/µl reverse transcription reaction is ≥50 ng total RNA/µl, e.g. for TRIzol-isolated RNA from ovine whole lung tissue (J.M. Gallup, unpublished). The extent of these inhibitory phenomena has been demonstrated to be highly correlated with the type of sample and the type of RNA isolation method used (Rossen *et al.*, 1992; Rijpens *et al.*, 1996; Al-Soud and Rådström, 1998; Thellin *et al.*, 1999; Schmittgen *et al.*, 2000; Suzuki *et al.*, 2000; Pfaffl *et al.*, 2001; Tichopad *et al.*, 2003, 2004, Gallup and Ackermann, 2006, 2008; Bustin *et al.*, 2009). In many cases, RNA samples are not originally plentiful or concentrated enough to allow investigators to add this much RNA per µl reverse transcription reaction in the first place, however. But, the specific pre-dilutions post-DNase treatment mentioned here (one dilution for column isolates and another more extensive dilution for TRIzol isolates) are absolutely required in order to help minimize and/or eliminate RT enzyme inhibition, and to help attain more equivalent reverse transcription reaction efficiencies across the board for all samples being converted to cDNA. The goal here is to first dilute back any inhibitory materials in the DNase-treated RNA samples, then adjust each RNA sample, from

that point, to the same ng/µl concentration for the purpose of being able to add the same exact volume (and therefore the same ng) of DNase-treated RNA sample per each reverse transcription reaction. This allows each sample to undergo conversion to cDNA under as identical conditions as possible in order that all reactions can be reasonably thought to have occurred at very similar efficiencies. Given that no reverse transcription priming strategy is 'best' or 'optimal' and that none have been demonstrated to achieve unbiased linear replication of the transcriptome, maximizing reverse transcription efficiency in any way possible is the least we must do.

6 *How do we know reverse transcription reactions have occurred at acceptable efficiencies?*
Answer. Most people don't, but here again note the MIQE guidelines suggestion for RNA first: 'The preferred method for quantifying RNA uses fluorescent RNA-binding dyes (e.g. RiboGreen), which is best for detecting low target concentrations. It is advisable to measure all samples using a single method only and to report this information (Bustin *et al.*, 2009). To quantify cDNA post-RT, alkaline hydrolysis of RNA using 0.1 M NaOH for 20 min at 70°C can be performed, and the remaining cDNA assessed by RiboGreen. Using RNAse I is ineffective in removing all RNA and/or double-stranded regions thereof, so alkaline hydrolysis of the RNA is used instead. Afterwards, the cDNA is still unaffected and PCR ready (Libus and Štorchová, 2006).

7 *How should cDNA samples be diluted post-reverse transcription?*
Answer. First, dilute all samples at least 1:2 (if rare targets are expected) or 1:5 (if normal expression of targets is generally expected) or 1:10 (if robust expression of all targets is expected) across the board using nuclease-free water. Then use P-Q to determine the *inhibitory characteristic* of the samples as a group, and proceed from there.

8 *What thermocycling conditions should be used for reverse transcription?*
Answer. Most modern variants of the MMLV reverse transcriptase enzyme engineered and/or manipulated by several different companies to exhibit more robust catalytic properties function optimally at ~50°C. For example, Invitrogen SuperScript™ III RT enzyme and Quanta BioSciences qScript™ RT enzyme (which, until recently, was also sold by Bio-Rad under the name 'iScript™' RT enzyme). 60°C is allowed when using SSIII, and up to 70°C can be used with Invitrogen's Thermo-X™ Reverse Transcriptase. Æ

Difficult templates

The difficulties encountered in amplifying various templates by conventional PCR, RT-PCR and by qPCR have centred on concerns regarding excessive target sequence GC and/or AT content, repetitive target sequences, extent of DNA superstructures and RNA template secondary structures (as a result of high GC-content), and priming strategies for reverse transcription reactions. These challenges are common to all PCR-based methods, and addressing them for one, to a large extent, addresses them for all, with the exception that such issues are generally magnified by conventional PCR, PCR used for sequencing, and RT-PCR approaches on account of their generation of typically longer amplicons (>250 bp) as opposed those generated by qPCR (typically 60 to 150 bp). Naturally, the longer the sequences sought to be amplified by any of these techniques, the more one needs to be concerned with the sequence-related issues at hand. Because there is no universal set of optimal conditions for all PCR methods, each PCR is likely to require specific optimization of reaction conditions per each different template, primer and/or probes chosen. Lack of optimization can result in non-detectable PCR products or low amplification efficiencies for a chosen template, presence of non-specific bands or products, formation of primer dimers that compete with a chosen template or primer set for amplification, and errors in nucleotide incorporation. To be sure, preliminary *in silico* mining (use of computer programs, web-based tools and databases) of any and all sequence information has become a major aspect of modern RT-PCR, PCR and qPCR assay designs. ЖK

GC-rich templates

In general, the problem with GC-rich sequences is that they can form robust secondary structures (stem loops and/or hairpin formations) on account of the triple hydrogen bonding capability across a nucleic acid helix. GC-rich regions resist denaturation and prevent annealing of PCR primers, resulting in the production of shorter readable sequences than other templates as DNA polymerases are limited or entirely prevented from copying through the complex secondary template structures during RT, RT-PCR, PCR and qPCR. It is generally advised to use higher temperatures during reverse transcription to improve specificity and reduce GC-related secondary structure during first-strand synthesis and to employ higher denaturation temperatures (e.g. 98°C as opposed to 94 or 95°C) during the PCR phase to allow for more complete denaturation and more efficient and specific priming and subsequent amplification. However, since temperature adjustment alone is not sufficient in many situations, different additional strategies can be employed to amplify difficult templates. One strategy is the substitution of inosine (dITP) for dGTP in primers and/or probes to improve their performance on GC-rich templates (by limiting their own ability to form hairpin or stem–loops) (Bustin *et al.*, 2009) while another strategy involves the use of a number of additives which aid in amplifying GC-rich sequences, most of which cause helix destabilization (Altshuler, 2006). Indeed, it is most likely a particular combination of two or more such additives that comprise many of the commercially available qPCR 'enhancing solutions.' One research team recently indicated that using a combination of three additives – betaine, dimethyl sulfoxide, and 7-deaza-2′-deoxyguanosine (7-deaza-2′-dGTP) – dramatically and universally improved the amplification of DNA sequences of three disease genes with GC contents ranging from 67% to 79%. The key concentrations of each of the additives in this particular study were: 1.3 M betaine (mono)hydrate (Sigma), 5% DMSO (Sigma), and 50 μM 7-deaza-2′-dGTP (Roche Diagnostics).

The benefit of using all three additives together is thought to be that they prevent intramolecular stable stem loops in GC-rich templates (due to strong G–C pairing), with each individual chemical acting in a slightly different way. DMSO is thought to disrupt basepairing, destabilizing helical formation and thereby lowering effective T_m, while betaine, called an 'isostabilizing' agent, equalizes the contribution of G-C and A-T basepairing to the stability of the DNA duplex (e.g. causing A-T hydrogen bonding to have similar strength as G-C hydrogen bonding). 7-deaza-2′-dGTP, substituting dGTP, reduces the strength or number of hydrogen bonds with complementary dCTP and prevents formation of other undesirable Watson–Crick and non-Watson–Crick hydrogen bonded basepairings. Ideally, each additive (as a mixture) can be employed at concentrations that do not inhibit the particular reverse transcriptase or DNA polymerase enzyme used for RT or PCR, allowing any synergistic effect of such reagent combinations to collude in resolving complicated amplifications of GC-rich DNA sequences (Jung *et al.*, 2002; Altshuler, 2006; Musso *et al.*, 2006). It has also been noted (Musso *et al.*, 2006) that PCR products obtained with the addition of betaine, DMSO, and 7-deaza-2′-dGTP do not need further purification prior to TA cloning or automated DNA sequencing since the additives do not affect topoisomerase I ligation activity – and appear to assist polymerization in subsequent sequencing reactions even using different *Taq* polymerases. The carefully calibrated use of such mixtures provides an affordable and reliable means to improve the analysis of difficult DNA sequences such as those found in inherited diseases caused by genetic defects of such GC-rich regions in genomic DNA. The addition of 2–5% glycerol, 0.05% to 0.1% w/v nuclease-free BSA and 10–16 mM tetrapropylammonium chloride to this mixture appears to be helpful as well (Gelfand, 1989; Frackman *et al.*, 1998; Choi *et al.*, 1999; Grunenwald, 2003; Musso *et al.*, 2006; Shaik *et al.*, 2008).

Widely varying opinions abound as to what enhancers or combination of additives are the best for each different situation. Some studies have demonstrated 2-pyrrolidone to be an especially potent enhancer across a wide variety of amplicons, demonstrating particularly good performance with GC-rich targets, followed in efficacy by acetamide (for medium GC-content targets), and lesser effectiveness found with other low-molecular-weight amides such as *N*-methylpyrrolidone, β-hydroxyethylpyrrolidone, propionamide, N,*N*-dimethylacetamide, *N*-methylformamide and *N,N*-dimethylformamide (Chakrabarti and Schutt, 2001). Another recent study discusses an apparently universally applicable PCR-enhancing mixture whose 5× formulation is 2.7 M betaine, 6.7 mM DTT, 6.7% DMSO, and 55 μg/ml BSA, which is stable at –20°C for at least 3 months and works optimally for a variety of thermostable DNA polymerases (in addition to *Taq*) when the final PCR reaction buffer also contains 65 mM TRIS-Cl, 16.6 mM $(NH_4)_2SO_4$, 3.1 mM $MgCl_2$, and 0.01% (v/v) Tween 20 at a pH of 8.8. Such a mixture of low-cost components employed as a universally helpful PCR enhancer solution is surely economically helpful for large-scale projects and routine applications requiring continuously reliable PCR results (Ralser *et al.*, 2006). Æ

AT-rich templates

Templates which have long homopolymeric regions, very AT-rich or GC-rich base compositions, or which contain long short tandem repeats (STR) regions, are all difficult to sequence and/or work with during PCR and qPCR. AT-rich templates are problematic in that they require lower RT, PCR or qPCR cycling temperatures (during first strand synthesis by RT or during the annealing/extension phase of PCR/qPCR) often enticing non-specific products to co-amplify with intended target sequences (Jex *et al.*, 2008). Such priming for AT-rich template varieties has even been carried out at 37°C (Altshuler, 2006). An advantage of AT-rich regions is that one can use a lower temperature during RT with lower risk of RNA secondary structure formations. Presently, successful qPCR for AT-rich targets is a function of choosing the right product from the right company. For example for AT-rich regions and single-nucleotide polymorphisms (SNPs), Biosearch Technologies has created the BHQplus™ probe, which is a novel, short-length probe for qPCR that has been engineered with a duplex stabilizing technology terminated with the Biosearch Black

Hole Quencher. The BHQplus probe permits the design of shorter oligonucleotides that can detect more difficult targets such as AT-rich regions. Although reportedly cheaper than MGB™ probes, both probe varieties form highly stable duplexes with DNA targets allowing for shorter probe designs making them highly useful in detecting single nucleotide polymorphisms (SNP) and allelic discrimination assays. Similarly, incorporating locked nucleic acids (LNA; e.g. designed by Exiqon) into primer/probe designs helps to maintain primer/probe T_m while allowing for shorter primer/probe designs in cases where one wishes to avoid AT-rich template regions. The *in silico* resource for designing such probes for SNP genotyping assays is the free, on-line, web-based software program called RealTimeDesign™ (Daniels *et al.*, 2008). ЖÆ

Repetitive DNA

Strategic use of *in silico* modelling for qPCR primer design helps to avoid regions of repetitive DNA and provides a level of genomic resolution greater than that expected from standard cytogenetic assays. The most critical factor for successful probing of repetitive DNA elements is primer design. To guarantee optimal primer design, high-resolution gene maps can be constructed using information available in publications and online databases and repositories such as UCSC Genome Bioinformatics (http://genome.ucsc.edu/), NCBI (http://www.ncbi.nlm.nih.gov/), Roe Lab Human Genomic DNA Sequencing Progress (http://www.genome.ou.edu/hum_totals.html), Egenome [http://egenome.chop.edu/], Human BAC Resource [http://www.ncbi.nlm.nih.gov/genome/cyto/hbrc.shtml], Human Organized Whole Genome Database (http://wwwalis.tokyo.jst.go.jp/HOWDY/), YACs on chromosome 22 (http://www.biologia.uniba.it/rmc/2-YAC-BAC/YAC/YAC-22.html) and Genatlas (http://www.dsi.univ-paris5.fr/genatlas/). The use of 1 M betaine in reaction mixes has been shown to be highly advantageous to PCR when sequencing through repetitive template regions (such as AACCCC-type repeats found in telomeres). 1 M betaine in reaction mixtures has also demonstrated reduction of band compressions in other repeat elements resulting in higher accuracies of sequencing. Due to the shorter amplicon sizes generated by qPCR overall, repetitive sequence issues can most often be dealt with by merely designing primers and/or primer and probes to adjacent, less-problematic sequence regions and/or by employing any one of the different probing strategies currently available (e.g. BHQplus™, MGB™, LNA) (Weksberg *et al.*, 2005; Yadav *et al.*, 2002, Daniels *et al.*, 2008). ЖÆ

In silico sequence information data mining (using state-of-the-art web tools and computer programs) is absolutely required to identify unique sequences within regions of interest and to avoid complex repetitive regions. Using such information in conjunction with known (previously published) locations of previously used primers, deletions, repeat sequences, known genes and pseudogenes, final sequences (e.g. for qPCR) can then be selected to lie mostly within exonic and/or intervening regions from known or putative gene sequences (e.g. using the UniSTS comprehensive database of sequence tagged sites to identify UniSTS markers) (Johnson *et al.*, 1999). Selected sequences can then be aligned against (for example) the human genome using the BLAT program (Kent, 2002) to ensure that only contiguous sequences with 100% homology to one unique location are chosen. A single primer set can then be designed for each of the unique sequences, e.g. using Primer Express by Applied Biosystems or Beacon Designer by Stratagene, etc. (Weksberg *et al.*, 2005). A principal challenge is targeting primer sets with the proper specificity; to avoid amplifying repetitive sequences and gene homologues. If left unaccounted for, non-specific amplifications are the top reason an assay fails since primers and probes can be inadvertently bound by repetitive elements within target samples and taken out of play entirely. To address this, NCBI database queries using electronic PCR to define unique amplicons through fuzzy matching strategies have been employed. In addition, the web-based Biosearch Technologies RealTimeDesign™ software program was created by Celadon Laboratories to handle high-volume sequence screening and design and can process up to 10 different targets at once. After primer/probe design, all sequence information is stored in the software's archives, which is a centralized database that tracks the progress of each project and allows review of each assay later on. The program operates entirely through a web browser so there is nothing to install or maintain, and it was recently reported that it keeps the inherent assay failure rate down to 8% – when 'success' is defined as a positive amplification within two cycles of when it is expected (Weksberg *et al.*, 2005; Daniels *et al.*, 2008). Ж

Additives and enhancers for PCR, RT-PCR and qPCR

Many formulations of PCR enhancing solutions and their constituent additive proportions are proprietary, but, for the most part, their chemical identities are known. PCR additives and enhancing agents have been used to increase the yield, specificity and consistency of PCR reactions. Although such additives may improve amplification, it is impossible to tell which ones will be useful for which part of an application without testing each of them singularly and in combination, empirically, in each situation (Frackman *et al.*, 1998). It is additionally important to note that some commercial DNA polymerase formulations are actually a mixture of polymerase enzymes; one with a good proofreading function and another possessing aggressive transcriptional activity (in order to maximize fidelity and product yield). The effect of various additives and/or combinations of additives on such enzyme mixtures must therefore be tested empirically with these additional complex features in mind (Bustin *et al.*, 2009). Æ A relatively current list of the most-used additives and their benefits to various (similar) aspects of RT, RT-PCR, PCR and qPCR is as follows:

DMSO (dimethyl sulfoxide)

Potential benefits of use for similar facets of RT, RT-PCR, PCR and/or qPCR: DMSO at 2–5% (typical use suggested is 5%) may be necessary for amplification of some templates, however 10% DMSO can reduce *Taq* or other DNA polymerase activities by up to 50%, so it should not be used routinely. DMSO is thought to reduce secondary structure and is particularly useful for GC-rich templates. Its benefits must be tested empirically in each situation (Gelfand, 1989; Smith *et al.*, 1990; Bookstein *et al.*, 1990; Varadaraj *et al.*, 1994; Schmerer, 2003). Æ

Glycerol (propane-1,2,3-triol)

Potential benefits of use for similar facets of RT, RT-PCR, PCR and/or qPCR: in some cases, 5% glycerol exhibits the same benefits as DMSO, while in other cases it appears to have no beneficial or detrimental effects at all (Choi *et al.*, 1999; Schmerer, 2003). Æ

Betaine (mono)hydrate (*N,N,N*-trimethylglycine = [carboxymethyl] trimethylammonium) (mono)hydrate

Potential benefits of use for similar facets of RT, RT-PCR, PCR and/or qPCR: typical suggested range of use is 1.0–1.7 M. It was demonstrated (Frackman *et al.*, 1998) by NMR analysis that the PCR additive provided by many well-known companies is betaine or betaine mono(hydrate), of which a 5 M solution can be purchased from Sigma-Aldrich (cat. no. B-0300). Other products thought to consist largely of (or most likely containing) betaine include the 'GC-RICH solution enhancer' from Roche, 'TaqMaster enhancer' from Eppendorf, 'GC-melt' from Clontech, EPICENTRE's '*TAQ*Xpedite GREEN Real-Time PCR MasterMix' and 'FailSafe enhancer' (formerly 'MasterAmp PCR Enhancement Technology') from EPICENTRE. Betaine is generally used at a final concentration of 1.0–1.7 M. It has also been observed that betaine serves to strengthen primer–template complexes in AT-rich regions (Melchoir *et al.*, 1973; Gefland, 1989; Smith *et al.*, 1990; Rees *et al.*, 1993; Varadaraj *et al.*, 1994; Weissensteiner *et al.*, 1996; Baskaran *et al.*, 1996; Hengen, 1997; Henke *et al.*, 1997; Wilson, 1997; Frackman *et al.*, 1998; Choi *et al.*, 1999; Al-Soud and Rådström, 2000; Shammas, 2001; Bartlett and Stirling, 2003; Grunenwald, 2003; Schmerer, 2003; Weksberg *et al.*, 2005; Musso *et al.*, 2006). It should also be mentioned that it has also been demonstrated that although product yield can be reduced by use of betaine, product specificity is often greatly improved (Frackman *et al.*, 1998). Æ

Researchers evaluating betaine have found that ~1.2 M betaine and 4 mM $MgCl_2$ worked the best. Note that the use of magnesium is always a parameter open to adjustment in every PCR situation. In general, more magnesium increases (*Taq*) DNA polymerase activity at the expense of specificity, while lower magnesium generally increases (*Taq*) DNA polymerase specificity but lessens enzyme activity. (Such helpful posts are common on the web these days, and one such highly informative interactive website is the BioTechniques Molecular Biology Forum, which is a science-based bulletin board for techniques, tips and questions concerning molecular biology, cell biology, microscopy and bioinformatics that is excellent for RT, RT-PCR, PCR and qPCR troubleshooting. It can be found at: http://molecularbiology.forums.biotechniques.com/.)

Trehalose (2-(hydroxymethyl)-6-[3,4,5-trihydroxy-6-(hydroxymethyl)tetrahydropyran-2-yl]oxy-tetrahydropyran-3,4,5-triol $C_{12}H_{22}O_{11}\cdot 2H_2O$) or simply: two D-glucose molecules linked by an α,α-1,1–glycosidic bond

Potential benefits of use for similar facets of RT, RT-PCR, PCR and/or qPCR: trehalose has also been reported to be extremely helpful with GC-rich templates when used at a concentration of 0.2 M. Trehalose belongs to the same class of compounds to which betaine belongs, known as 'compatible solutes' which stabilize cells, cellular components and enzymes when exposed to extreme conditions. Trehalose is synthesized by a variety of eukaryotes, conferring tolerance against desiccation, dehydration, temperature extremes and oxidation. Addition of trehalose increases the enzymatic activity of several euthermal enzymes used for cDNA synthesis or restriction digestion of DNA. Trehalose also enhances the priming specificity in differential-display RT-PCR through high-temperature priming and use of a thermally activated reverse transcriptase. In addition to the other known low-molecular-weight compounds that enhance the PCR of difficult templates such as DMSO, other sulfoxides (such as DMS), formamide, non-ionic detergents and betaine, trehalose is also a potent PCR enhancer for GC-rich templates. Its use has helped investigators avoid false negatives during PCR genotyping. The PCR-enhancing properties of trehalose are ascribed to its ability to lower the melting temperature of DNA and to its ability (as a compatible solute) to thermo-stabilize *Taq* DNA polymerase (Spiess *et al.*, 2004). Æ

7-Deaza-2′-deoxyguanosine triphosphate (dC7GTP or 7-deaza-2′-dGTP)

Potential benefits of use for similar facets of RT, RT-PCR, PCR and/or qPCR: The base analogue 7-deaza-2′-deoxyguanosine triphosphate (7-deaza-2′-dGTP) has been demonstrated to facilitate amplification of templates with stable secondary structures when used in place of dGTP at a ratio of 3:1. Or, when used in conjunction with deoxyinosine triphosphate (dITP), it has been demonstrated to be used most effectively at a ratio of 4:1 (7-deaza-2′-dGTP:dITP) (Motz *et al.*, 2000). Also, beneficial for sequencing GC-rich RNA, is the substitution of ITP with GTP to limit band compression on RNA sequencing gels. 7-deaza-2′-dGTP has also been used in DNA sequencing for the resolution of band compressions due to GC-rich template regions. Amplification of some DNA regions is often hindered due to high GC content, and replacement of dGTP with 7-deaza-2′-dGTP in the mastermix has helped in many reported cases (Mills *et al.*, 1979; Barr *et al.*, 1986; Mizusawa *et al.*, 1986; McConlogue *et al.*, 1988; Motz *et al.*, 2000; Jun *et al.*, 2002; Musso *et al.*, 2006). Æ

A common problem with the Sanger chain termination method for automated DNA sequencing is ambiguous base-calling on account of band compressions caused by anomalies in the migration behaviour of certain DNA fragments in polyacrylamide sequencing gels. The compressions are thought to be due to intramolecular (non-Watson–Crick, Hoogsteen and/or reverse-Hoogsteen) basepairing between guanine and cytosine residues. To reduce the resulting undesired secondary structures, mixtures of the two nucleotide analogues 7-deaza-2′-dGTP and dITP (deoxyinosine triphosphate) used to replace dGTP in cycle sequencing reactions in combination with varying buffer conditions have been demonstrated to resolve band compressions and dramatically improve reading length. An optimal ratio of 7-deaza-2′-dGTP:dITP of 4:1 has been demonstrated with AmpliTaq FS DNA polymerase (ABI). Using this 4:1 nucleotide analogue mixture in place of dGTP (sometimes with addition of up to 15 mM $(NH_4)_2SO_4$) is highly effective in generating heterogeneous DNA chains by reducing and/or eliminating problematic intramolecular basepairing during PCR, thus allowing for more accurate base-calling. At least 77% of all such problematic compressions have been resolved at this ratio – whereas other sequence-related compression phenomena must be resolved empirically. Similarly assisted by the use of 7-deaza-2′-dGTP and dITP is the use of PCR to resolve CpG island sequences. CpG islands are GC-rich sequences found in >60% of the promoters of genes in higher eukaryotes. Such CG dinucleotides are substrates for DNA methylases which methylate the 5-carbon of the Cytosine pyrimidine ring resulting in transcriptional silencing of promoters. Because of high GC content, CpG islands demonstrate strong base–base interactions leading to superstructures with higher melting temperatures. As a result, *Taq* polymerases (e.g. ‘sequenases’) fall off the template DNA, causing early termination of the PCR. In addition, particularly when dealing with DNA extracted from microdissected tissue, the addition of 7-deaza-2′-dGTP is strongly recommended for analysis of CpG islands given the limited amount of starting material from such samples. In connection with ITP and dITP, respectively, their replacement of GTP or dGTP results in inosine-cytosine basepairing characterized by the formation of only two Watson–Crick hydrogen bondings, which are less stable than the triple hydrogen bondings between G and C. The result is a substantial reduction in the formation of intramolecular 'hairpin' type structures normally stabilized by Watson–Crick basepairing, and also produces a more accurate size distribution of DNA in electrophoresis gels (Barr *et al.*, 1986; Motz *et al.*, 2000; Jung *et al.*, 2002). This particular explanation has been refuted by others, however. There have also been reports of several additives, such as DMSO, formamide, and betaine not improving PCR performed on GC-rich templates and in lieu of DMSO and betaine, both, just 7-deaza-2′-dGTP is used as the sole enhancer based on its ability to weaken base–base interactions, thus solving superstructures and improving PCR performance in general (Mizusawa *et al.*, 1986; Jung *et al.*, 2002). **Æ**

BSA (bovine serum albumin)

Potential benefits of use for similar facets of RT, RT-PCR, PCR and/or qPCR: the suggested range of use is generally 10 to 50 μg/ml, and in other cases, 0.05–0.1% w/v (0.5–1 mg/ml). BSA (at 0.75 mg/ml) has proven particularly useful when attempting to amplify ancient DNA or templates which contain PCR inhibitors such as melanin (Pääbo *et al.*, 1988; Kreader *et al.*, 1996; Henegariu *et al.*, 1997; Giambernardi *et al.*, 1998; Frackman *et al.*, 1998; Schmerer, 2003; Stone *et al.*, 2006; King *et al.*, 2009). BSA, though having no direct effect on the enzymatic reaction per se, is nonetheless thought to be able to stabilize enzymes and neutralize inhibitory contaminants that may be present in DNA or RNA template preparations or in reaction buffers (Ralser *et al.*, 2006). **Æ**

Formamide (methanamide)

Potential benefits of use for similar facets of RT, RT-PCR, PCR and/or qPCR: formamide is generally used at 1–5% and 10% formamide is reported (Gelfand, 1989) to have no effect on the activity of *Taq* polymerase, however, Sarkar *et al.* (1990) found that 1.25% formamide worked as well as 2.5% and 5%, and no amplification was seen at all with 10% formamide, so it is wise to avoid using concentrations of formamide greater than is absolutely necessary for optimal amplification (Gefland, 1989; Sarkar *et al.*, 1990; Altshuler, 2006). **Æ**

Tetramethylammonium chloride (TMAC)

Potential benefits of use for similar facets of RT, RT-PCR, PCR and/or qPCR: TMAC is generally used at a final concentration of 15–100 mM to eliminate non-specific priming. TMAC has is also used to reduce potential DNA-RNA mismatch (Hung *et al.*, 1990; Chevet *et al.*, 1995). TMAC has is also used to reduce potential DNA-RNA mismatch and improve the stringency of hybridization reactions (Wood *et al.*, 1985; Jacobs *et al.*, 1988, Kovárová *et al.*, 2000). **Æ**

Other tetraalkylammonium derivatives, e.g. tetraethylammonium chloride (TEA-Cl) and tetrapropylammonium chloride (TPrA-Cl)

Potential benefits of use for similar facets of RT, RT-PCR, PCR and/or qPCR: other tetratalkylammonium chloride derivatives (TAACls) used at 10–16 mM have also demonstrated great promise as a new class of additives which contribute to the robustness of real-time PCR monitored by SYBR Green I dye and doubly serve as stabilizers of the SYBR Green I-dye itself in PCR mixtures (and in other applications). In addition to enhancing PCR performance, TAACls (e.g. tertapropylammonium chloride) have been demonstrated to allow amplification of longer genomic fragments in the presence of SYBR Green I and to help decrease amplification of non-specific products often observed even under optimal conditions. Further, since it has been shown that SYBR Green I degradation products are potent inhibitors of PCR (since SYBR Green I is generally not chemically stable after about a month), use of TAACls to stabilize/preserve SYBR Green I dye seems to be well warranted (Demeke and Adams, 1992; Chevet *et al.*, 1995; Kovárová *et al.*, 2000; Shaik *et al.*, 2008). Æ

Non-ionic detergents, e.g. Triton X-100, Tween 20 or Nonidet P-40 (NP-40)

Potential benefits of use for similar facets of RT, RT-PCR, PCR and/or qPCR: non-ionic detergents stabilize *Taq* polymerase and may also suppress the formation of secondary template structure. 0.1–1% Triton X-100, Tween 20 or NP-40 may increase yield but may also increase non-specific amplification. As little as 0.01% SDS contamination of template DNA (left over from extraction procedure) can inhibit PCR by reducing *Taq* polymerase activity to as low as 10%. However, inclusion of 0.5% Tween-20 or -40 will effectively neutralize this effect (Gelfand, 1989). Also, in plants, the inhibitory effects of gum ghatti, and dextran sulfate inhibition can be reversed by the addition of 0.25–0.5% Tween 20, 5% DMSO or 5% polyethylene glycol 400, but none of the additives help when dextran sulfate is present at a ratio of 100:1 (dextran sulfate–DNA) (Bachmann *et al.*, 1990; Demeke and Adams, 1992). Æ

PREXCEL-Q

My own experience with qPCR always brings to mind a particular series of events that solidified my resolve to create a unique computer program for qPCR assay development and project management.

In late 2001, I was performing one-step qPCR on a particular total RNA sample TRIzol-isolated from sheep lung (experimentally infected with bovine respiratory syncytial virus, bRSV) and the resulting qPCR yielded no viral signal (no C_q value generated). The lung sample involved was by all other accounts (by necropsy, immunohistochemistry, histopathology, gross pathology, lesion score, clinical signs, days post-inoculation etc.) a clearly infected lung sample and thus should be harbouring easily detectable amounts of the ssRNA bRSV viral nucleic acid of interest. In my pilot qPCR for this bRSV target, the primers and hydrolysis probe (designed to a portion of the bRSV nucleocapsid sequence) had previously demonstrated greater than 92% amplification efficiency with bRSV RNA TRIzol-extracted from purified bRSV. As well, the qPCR reagents and RNA preparative reagents were all new so there was a small period of confusion as to why the assay would all-of-a-sudden, fail. My conclusion was to run a dilution series on the sample and look for the target again by qPCR. As a result, we saw that with increasing pre-qPCR sample RNA dilution (after the most concentrated samples showed no amplification, like the first time), the bRSV signal actually got progressively stronger until it reached a point where it began to respond correctly to dilution (i.e. higher C_q values with more sample dilution). It was this inverse/counter-intuitive manner in which the third, fourth and fifth C_q values appeared that provided the clue as to the susceptibility of qPCR to inhibition (simply) on account of how much sample is added to the reaction. In the dilution study, the first two most concentrated samples (the full-strength RNA sample, and the 1:10-diluted RNA sample), showed no amplification (no C_q values), but then, at a pre-qPCR sample dilution of 1:20, a C_q of 43.55 appeared, and at a dilution of 1:30, a C_q of 34.2, at 1:40 a C_q of 27.45, and the final tested sample dilutions of 1:80 and 1:160 gave respective C_q values of 28.53 and 29.56. These observations led me to conclude that any nucleic acid sample subjected to reverse transcription, RT-PCR, PCR or qPCR of any variety can harbour varying degrees of inhibitory material or potential (or perhaps stimulatory material or potential); therefore the practice of using one randomly chosen blanket nucleic acid sample dilution preceding reverse transcription, RT-PCR, PCR or qPCR (e.g. 1:5 or 1:10) for all targets, is not advisable. Furthermore, some samples can be so inhibitory that RT and PCR-based assays can generate entirely believable false-negative results – even when millions of copies of target template are present.

In addition to the things I learned from this basic sample dilution study, I also recalled that I had not observed such inhibitory phenomena during qPCR performed on purified bRSV inocula a week or so earlier. After all, the inocula was a differently 'purified' viral nucleic acid sample, presumably containing only minimal left-over material from the ovine turbinate cell culture in which the virus was allowed to replicate. These events immediately explained the many other similar observations I had made over the previous 2 years performing qPCR on sheep lung total RNA, concurrent with work performed by a colleague on equine dendritic and macrophage cell culture column-purified RNA isolates. It was the cumulative teaching of these events that justified the investment of numerous hours (between 1997 and 2008) toward building an automatic Excel 2003-based computer program that could be used to quickly identify the dilution ranges within which nucleic acid samples and standards could most likely be expected to behave optimally during reverse transcription, RT-PCR, PCR and qPCR. Such a tool would ultimately function to help reveal the true relative initial abundance of any genomic or transcriptomic target of interest; given any tissue type, any nucleic acid extraction/isolation procedure *and any human operator*. This is the way that the concept of 'addressing sample-related RT, RT-PCR, PCR and qPCR inhibition' first presented itself to me and became such a serious endeavour – one which I considered to be worthy of in-depth attention.

As a consequence of the time-intensive genesis of P-Q, it has become extremely alien for me to contemplate qPCR of any kind without first viewing it entirely through the lens of the program. Creation of P-Q has allowed 'inhibition' to remain, in my mind, the first thing to look for, confront and do away with post-haste before attempting to improve assay performance in any other way. It makes no sense to fine-tune reverse transcription, RT-PCR, PCR or qPCR (in terms of addressing choice of RT, first-strand priming, template secondary structure, AT- or GC-rich or repetitive sequence issues) when inhibition is still influencing part or all of the technique to begin with. I contend that, once we are in the clear, working safely outside and away from inhibition and its posse of accomplices, only then can we logically and gainfully endeavour to tweak and optimize the more subtle and powerful aspects of this (and any other similarly prone) technique. Further, with the recent publication of the MIQE guidelines and recent formation of the RDML Consortium, for the first time since Mullis's discovery of PCR in 1983, we have the means by which to identify the best standard practices for qPCR; from sample preparation to data acquisition and processing. In that same spirit, as a supportive member of the RDML Consortium and as an advocate of the MIQE guidelines, from *the inhibition-sensitive, premeditatedly precise nucleic acid sample and standard dilution-dependent* vantage of P-Q have I written this chapter.

Empirically, the P-Q program is a new software program intended for use as a universally applicable qPCR assay-development and project-management tool (Gallup and Ackermann, 2006, 2008; Bustin, 2008; Sow *et al.*, 2009). P-Q uses C_q values generated from a Test Plate on which all targets are examined within a serially diluted experimental sample mixture to determine final qPCR reaction set-up parameters. It is unique in that it is the only program that starts from the very beginning of a qPCR assay – allowing the entire qPCR process to be included – from A_{260nm} sample readings to final reaction set-ups for all qPCR approaches – spelling out each reaction and reaction component for the user in the form of easy-to-read printouts that are used as set-up recipes in the lab. There is a great need for proper experiment design and set-up, prior to running qPCR experiments. This part of the workflow is often neglected and no similar tools are currently available. With the dramatically increasing numbers of laboratories using qPCR, a basic program such as P-Q is highly useful in helping to reduce perfunctory qPCR set-up-related errors. In addition, many who perform reverse transcription, PCR and/or qPCR do not account for inhibitory phenomena and therefore, do not achieve reactions that are maximally efficient. The P-Q program was also designed keeping best-practices in mind, such as those suggested specifically by Bustin, Nolan, Hands and colleagues over the past several years (Bustin, 2003, 2005, 2008; Fleige and Pfaffl, 2006; Nolan *et al.*, 2006; Swango *et al.*, 2007; Doma and Parker, 2007; Bustin *et al.*, 2009; Nolan and Bustin, 2009; Murphy and Bustin, 2009). P-Q's most recent incarnation includes buttons instead of keystroke commands to increase ease of use.

Even in the absence of an absolute understanding of every facet of fluorogenic qPCR, in addition to the recent cited works which rather astutely refute the formerly accepted notion of its continuously efficient exponential phase, the technique has nevertheless attained nearly 'number one' status as the world's premier quantitative gene expression investigative tool. Amid the various debates going on as to whether or not one qPCR approach is better than another, or whether template integrity itself is the underlying cause of many published results, I believe the underlying problems facing every study using reverse transcription, RT-PCR, PCR or qPCR to be fairly universal. It is because of this that I offer the idea that any type of *goo* can be analysed correctly (i.e. nucleic acids by qPCR) only when we identify the dilution range within which that goo is allowed the best chance to tell us the truth in the first place. This idea, taken in conjunction with Dr Stephen A. Bustin's recent observations (during a webinar in March, 2006) that 'no one looks for inhibition of the assay' and 'a recent survey of practices revealed that only 6% of researchers test their nucleic acid samples for the presence of inhibitors' (Bustin, 2005) suggested to me several years ago that the program would be rather important to finish. And so it was finished in the form of 35 interwoven Microsoft Excel 2003 files, of which a Java/GUI-based version is currently under way. The program extols the virtues of something called the *CRUD rule*: '**C**$_q$ values are only **R**eliable in **U**ninhibited qPCR reactions wherein high-quality samples and standards have undergone premeditatedly precise, dynamically sound **D**ilutions.' It stresses that all measurements be taken for each target (in samples and corresponding standards) only within preliminarily demonstrated, log-linear-amplification-capable, high-efficiency-of-amplification, valid dynamic sample/target dilution ranges. And, as previously stated, these valid dynamic sample and standard dilution ranges for each target are identified beforehand by running a qPCR Test Plate using a progressively diluted experimental sample mixture as the representative sample for all targets. The governing idea here is that, once 'CRUD' is not allowed to interfere with the 'goo' in the reverse transcription and PCR reactions on which man-made qPCR relies, only then can we accept the entire process to be potentially capable of yielding quantitatively informative or biologically relevant information. When used in concert with the P-Q program, the SPUD assay can be used to demonstrate proof of principle; that P-Q indeed attains a 'no-inhibition' status in each final standard and sample reaction. Combined use of both SPUD and CRUD seems to be a good idea for qPCR (Nolan *et al.*, 2006; Gallup and Ackermann, 2006, 2008; Sow *et al.*, 2009). And indeed, the two methods cross-corroborate one another (Gallup *et al.*, 2009).

What the P-Q program does can also be stated as follows: It allows any qualified operator to first nudge (by Test Plate analysis) any similarly isolated set of nucleic acid samples (reverse-transcribed or not) into its unique zone of maximal qPCR believability, and is then recursively used to design all final reactions (for all samples and targets) within that zone, on a per-target basis. It then spells out all final sample and standard dilutions, reverse transcription reaction formulations (for two-step qPCR), mastermix assemblies and provides concise printouts for immediate in-lab use for each step. The performance of the final P-Q-rendered qPCR is thus determined primarily by quality of sample preparation, precision of pre-qPCR sample quantity assessment and fidelity of primer/probe design.

Mathematically, when inhibition is absent, the Gompertz four-parameter (fixed inflection point) sigmoidal model (Kebreab *et al.*, 2007) converges with the Richards-like five-parameter (variable inflection point) sigmoidal model (Spiess *et al.*, 2008) and the Cy_o method (Guescini *et al.*, 2008) since inhibition has been effectively eliminated by precise differential sample dilutions predicated on two P-Q-required criteria: (1) First diluting all samples (the same amount) to just beyond the 'inhibition threshold' revealed as a 'sample group inhibitory characteristic' by an initial/preliminary Test Plate analysis (for each target) performed on a mixture of all of the experimental samples, and (2) A precisely calculated dilution of all samples differentially, from there, to a common ng/µl concentration on a per-sample, per-target basis preceding final qPCR. With these conditions met, standard curve estimates of efficiency (E) and/or exponential amplification (E_{AMP}) agree closely with fluorescence-based estimates of initial reaction E_{max} to the degree that C_q values become acceptably precise representatives of relative target abundance within each such appropriately diluted sample. In a general sense, then, P-Q allows each similarly derived sample set to be used as the material by which to fine-tune an entire system by using the experimental sample mixture itself as the averaging feature or gauge by which the system is self-calibrated and self-measured. In so doing, each such system is allowed to achieve its own state of identity or dynamic equilibrium (wherein the most valid homo-dynamic or hetero-dynamic dilution ranges for all samples and targets of interest within each of the samples have been carefully and empirically established preceding final qPCR) (Fig. 2.4). Under such conditions, the C_q values generated by such a system would seem to be worthy of scientific consideration. In short, P-Q seems to sufficiently address all of the dilution-related ills of qPCR (Fig. 2.5).

When the entire ritual of qPCR is viewed, internalized and accepted as a naturally semi-stochastic and highly operator-dependent activity, I have thus found that it is most accurately, precisely and robustly performed as a highly defined process in two parts: (1) preliminary Test Plate(s) and (2) Final Plate(s) wherein each part uses an experimental

A

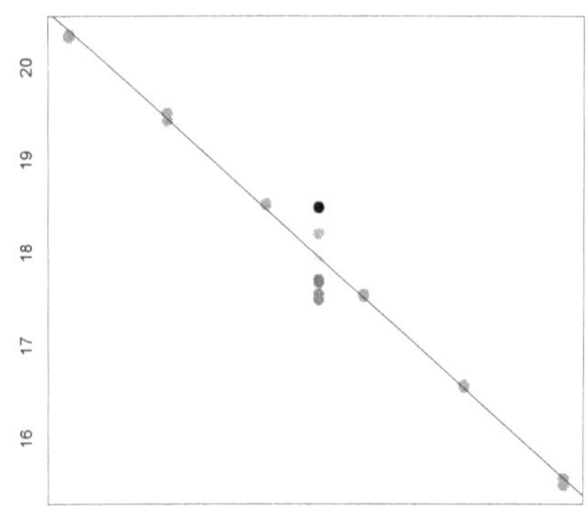

B

SampleType	Threshold; dRn.	Relative dilution	Avg, C_q; dRn.
Unknown	0.0385	0.06666667	18.075
Unknown	0.0385	0.06666667	18.614
Unknown	0.0385	0.06666667	18.510
Unknown	0.0385	0.06666667	18.665
Unknown	0.0385	0.06666667	17.965
Unknown	0.0385	0.06666667	17.795
Stock I	0.0385	0.33333333	15.795
Stock I	0.0385	0.17325800	16.835
Stock I	0.0385	0.09005501	17.815
Stock I	0.0385	0.04680825	18.795
Stock I	0.0385	0.02432971	19.735
Stock I	0.0385	0.01264595	20.615

C

D

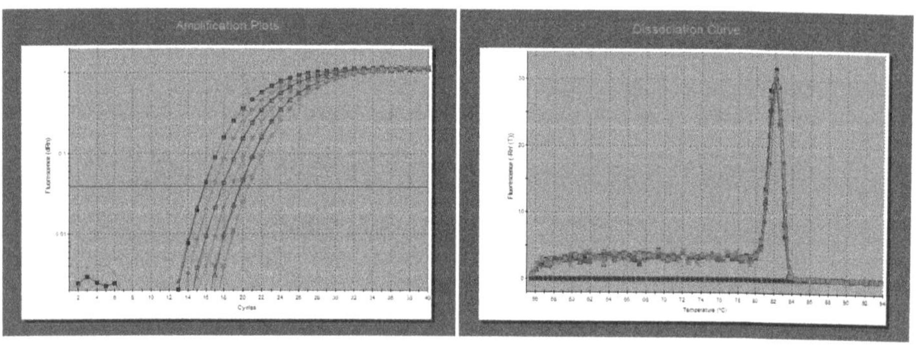

Figure 2.4 Example of a SYBR-Green based assay using P-Q to establish the good, valid dynamic dilution range for samples and standards. Shown is the standard curve for a single target which was generated using an experimental sample mixture; a small portion of each sample was mixed to form the assay's 'Stock I' (the sample amplified with 96.9% efficiency according to the slope of this standard curve). (A) Graph of the standard curve for PS7r1 using the Stock I sample mixture, and C_q values of six unknowns (duplicate C_q values plotted). (B) P-Q-identified valid Stock I dilution range (in black rectangle) and associated real-time qPCR (averaged duplicate) values generated for PS7r1 using a one-step SYBR Green-based qPCR method (averaged C_q values shown). (C) The associated Stratagene Mx3005P amplification plots for PS7r1 at the six chosen Stock I dilutions (each performed in duplicate). (D) The Mx3005P dissociation curve analysis for this run. Provided courtesy of Dr Stuart Gardner (Iowa State University).

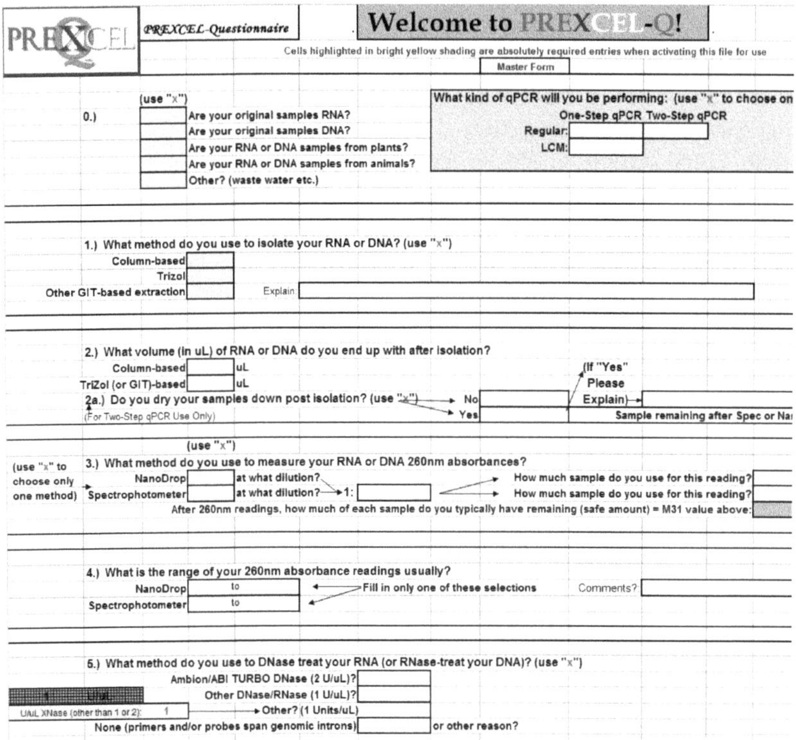

Figure 2.5 A partial glimpse of the Questionnaire interface for the PRExcel-Q program. The screen-capture image here represents only of a small portion of the entire program. The entire P-Q program consists of 35 Excel 2003 files which interdependently calculate all set-up parameters for all qPCR approaches. The printouts generated by the program are fail-safe recipes for setting up every part of the qPCR workflow. P-Q is not a qPCR data analysis software.

sample mixture (termed 'Stock I') as the 'language', gauge or measure by which all sample and standard dilutions and standard curve ranges are established for each target of interest within each entire, particular sample set. It turns out – as revealed by using and tweaking the program for several years – that the point or range at (or within) which all samples and standards have been diluted to exist within their valid dynamic ranges, most (if not all) of the qPCR problems associated with 'inhibition' are, as if by implicit definition, *gone*. In other words, dynamically appropriate dilution ranges, it seems, can only be identified when the qPCR is behaving 'correctly' in the first place. The qPCR simply would not be able to exhibit any of the healthy manifestations of a good *2-to-the-n*-like process (high-amplification efficiency and dilution-responsive LOG-linear-amplification-like-capability) if inhibitory materials or processes were still waging an influence. The program tells each investigator whether the samples (and targets within the samples) are even worth investigating in the first place on account of pilot results obtained up-front from the initial Test Plate. Given good primer and probe designs to begin with, P-Q can receive, process and find the useful range for every qPCR sample and target in any qPCR sample set, as long as care has been taken by the operator(s) beforehand to treat all samples identically during each pre-qPCR preparatory manoeuvre and/or sample manipulation while also observing all suggested practices for maintaining high sample integrity according to the MIQE guidelines and any of the other sample-handling suggestions described in great detail in several recent publications (Bustin, 2003, 2005, 2008; Fleige and Pfaffl, 2006; Nolan *et al.*, 2006; Bustin *et al.*, 2009; Nolan and Bustin, 2009; Murphy and Bustin, 2009).

It is additionally important to acknowledge that not all qPCR reactions are meant to be – in which cases, testing alternative primer/probe designs or choosing (or formulating) another mastermix become the most immediately reasonable recourses. In fact, reagent choice alone is the factor which decides the fate of all qPCR reactions (see the effect of the choice of mastermix as a very typical example of this in Fig. 2.6). An extremely dramatic example of this occurred to me in 2006, after I had been stubbornly using company 'C's' one-step mastermix kit for the previous 5 years, and then finally agreed to try company 'D's' formulation (at the urging of a very-convincing company representative). Upon comparing the two mastermixes under optimal conditions for each, I found that company D's mix improved our qPCR sensitivity 2000-fold for certain targets, and over 1,000,000-fold for other targets (Fig. 2.7). Since reagent choice alone can make or break reverse transcription, RT-PCR, PCR and qPCR, never be afraid to shop around; the fate of each such endeavour is reagent-choice dependent first and foremost, always. ÆЖ

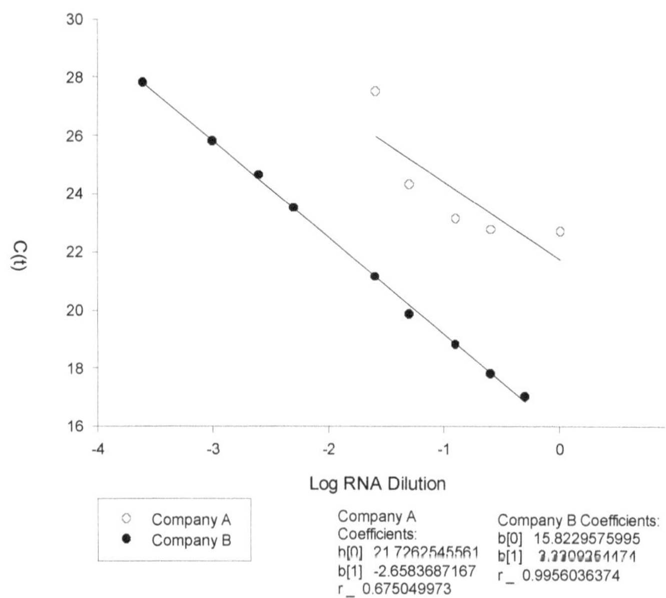

Figure 2.6 Comparison of one-step SYBR Green-based qPCR mastermix from company A and company B on RimM expression profiles from 48-h-old bacterial biofilm cells. This represents a typical example of the effect of reagent choice on qPCR. By reagent choice alone, qPCR can behave very poorly (even not at all) to extremely well. Such a dramatic effect on assay performance due solely to reagent choice(s) applies to reverse transcription, amplifying problematic AT- and/or GC-rich sequences and repetitive DNA templates, PCR, RT-PCR and qPCR reactions. The extent of success or failure of any qPCR assay is always, undeniably, a function of reagent choice(s). Provided courtesy of Dr. Lindsay Nielsen (Iowa State University).

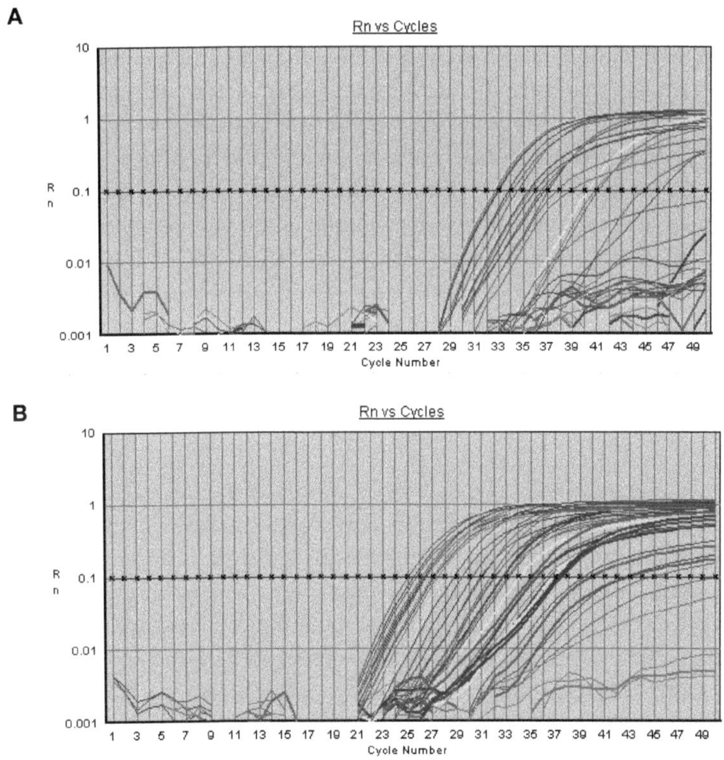

Figure 2.7 Example of the dramatic effect of one-step mastermix choice on qPCR performance. (A) Company C. (B) Company D. Using the same samples in a one-step qPCR application, when company 'C's' one-step mastermix formulation was compared with company 'D's' one-step mastermix formulation, the mastermix from company D proved to be 2000 to 1,000,000 times more sensitive than the correlate mastermix from company C.

P-Q proof of principle in minimizing the 'operator effect'

Perhaps equal to the *sample effect* in its ability to destroy a perfectly good universe is something I will call the '*operator effect*.' No writing about reverse transcription, RT-PCR, PCR and qPCR is ever complete without acknowledging the mammoth influence that different human operators bring to the table in affecting the success of these techniques. Toward this end, I relate to you my description of another real-life scenario in which two human operators, using identical methodologies, performing identical nucleic acid isolations from identical biological samples, did not obtain the same valid qPCR sample/standard dynamic dilution ranges when using P-Q and its suggested regimen. In this particular case, one investigator's standards and samples used in qPCR (for the identical targets as the second investigator's) exhibited a valid dynamic dilution range for all targets starting at 1:150 out to 1:10,000 (final sample dilution in-well) while the other investigator, performing the identical series of techniques on the *same* animal and tissue type, using identical reagents and methods for column-based RNA isolation, DNase treatment, RT reaction formulation and qPCR primer sequences ended up with final samples and standards that, in qPCR, exhibited a valid dynamic dilution range that only existed between 1:3,390 and 1:10,000 (in-well). Without P-Q, these two folks would have thought that one qPCR experiment worked while the other did not. The alarming disparity in the overall global dynamics between these two very-similarly performed assays was ultimately attributed partially to different human operator and largely to the use of a different SYBR Green-based mastermix for the qPCR. Both researchers completed their assays with good final results (which cross-corroborated each other's findings) and learned of the superiority of one company's mastermix over another, and additionally exposed 'RNase ZAP' as a potential problem when it was not rinsed from the homogenizing probe in-between samples. This particular occurrence is but one of many real-life situations in which the broad-range usefulness of the P-Q computer program has been demonstrated. Given the many qPCR studies that have been successfully designed and assisted by the program, it is clear at this point that it can be confidently used to identify the valid dilution range(s) for seemingly any set of samples and standards prepared by any operator or group of operators. The program virtually tailors itself each time to each new situation; each different sample set, per each different chosen method of nucleic acid isolation, per each different particular degree of coordination and/or technical prowess of the human operator(s) involved, per each chosen qPCR approach. It is on account of this that I believe the same *goo* can speak the same language on all planets, in all labs, if treated, handled, prepared, pre-tested, diluted and analysed correctly by any capable operator using P-Q for qPCR assay-development and project- management. Its use in conjunction with the Nolan–Bustin SPUD assay and other process which can remove inhibitory materials from nucleic acid isolates used in reverse transcription, PCR, RT-PCR or qPCR, appears worthy of consideration (Nolan *et al.*, 2006; Gallup *et al.*, 2009). Æ

The 3′:5′ and SPUD assays

Dr Tania Nolan emphasizes the importance of quality control and optimization experiments, especially for real-time qPCR. Noticing that RNA isolated from FFPE-archived material is highly degraded, Dr Nolan created a high-throughput SYBR Green I-based quantification method for evaluating the integrity of such RNA templates known as the 3′:5′ assay. The original aim of her method was to ensure that qPCR would be performed only on samples of similar integrity. Using the 3′:5′ assay, she demonstrated that quality of RNA template can significantly affect the outcome of qPCR used for gene expression studies. The 3′:5′ assay begins reverse transcription of an RNA target using oligo-(dT) priming. This reverse transcription starts from the poly-A tail and moves from 3′ to 5′. Then, two sets of primers are used; one which targets the 3′ region and the other targeting the 5′ region of the same specific cDNA template, which are amplified either in separate reactions or in a duplex assay using amplicon-specific probes. If the RNA is intact, then quantifying the 5′ and 3′ ends of the cDNA generated from the same target cDNA should yield the same quantity of target. But, if the RNA is degraded somewhere between the 5′ and 3′ assay sites, then the reverse transcriptase will stall at the break point, and the assay will yield more 3′ cDNA. Dr Nolan observed that not all transcripts in a sample degrade at the same rate and that transcripts even appear to degrade at different rates in different tissues. Her studies into these matters continue.

In addition to the 3′:5′ assay, regarding contaminants in nucleic acid samples that can inhibit the enzymatic reactions required for qPCR, Dr Nolan and colleagues developed the

SPUD assay. The SPUD assay involves running a control qPCR using a synthesized 101 bp amplicon (from the potato genome) in parallel with an identical qPCR reaction that contains both the spiked potato amplicon (typically 22,000 copies of it per reaction) and the biological sample or standard. If the sample contains an inhibitor, it will take more cycles for the potato amplicon fluorescence signal to reach C_q. In the presence of inhibition, the C_q will appear later. But, additionally confounding such analyses is the possibility that inhibitory agents may affect the 3′ and the 5′ qPCR assay to different degrees, about which Dr Nolan has pointed out that it is never safe to assume that certain inhibitors will always inhibit all qPCR assays to the same degree. Therefore, correcting for inhibitors by quantifying a reference gene is really not a solution. While the SPUD assay can identify the presence of an inhibitor, the results of the 3′:5′ assay may also reflect the variable effects of an inhibitor, resulting in either more 5′ amplicons or more 3′ amplicons. As a suggested quality control practice, Dr Nolan has thus recommended using both 3′:5′ and SPUD assays on every RNA sample (Nolan *et al.*, 2006; Glaser, 2007).

qPCR and music

As a side-note to some of the seriousness that abounds within the world of troubleshooting and properly formulating good RT, RT-PCR, PCR and qPCR reactions, there arises the entertaining relationship between the classical PCR equation '$X_n = X_o(2)^n$' and the formula for frequency (in Hertz) of musical notes (tones on a standard size, $A_{440\ \mathrm{Hz}}$ piano), $Hz_n = Hz_o(2)^{n/12}$. What could possibly be learned from such a relationship? Plenty. In particular, when qPCR reactions are less than 100% efficient (e.g. when exponential amplification or amplicon *doubling* falls short of a factor of 2 per cycle), its governing formula becomes $X_{Cq} = X_o(E_{AMP})^{Cq}$. C_q values generated at less than 100% amplification efficiency, when compared with the same target concentration differences represented by C_q values generated at 100% efficiency (or $E_{AMP} = 2$), appear further and further apart. The rate of C_q appearance (tempo) slows down – yet, the successive C_q values are actually signifying or playing the exact same melody, only on higher keys than the corresponding 100% efficient reaction. The other difference here, to fully accommodate the analogy, is that the piano has been stretched, and the same '*Taq*-hand' is merely playing the same tune only over a longer time period using larger physical key distances, due to the stretching of the keyboard (the C_q values and/or piano keys are physically farther apart). The higher the frequency of the note on a piano, the less discerning the human ear is at telling which note is which. The same thing applies to high C_q values generated by qPCR (e.g. C_q values above 38) when <5 initial target template copies generate increasingly sporadic C_q values. Music-wise and qPCR-wise – the same story is being told: ~1-copy amplifications have been experimentally shown to exhibit C_q value spreads of nearly six cycles, and, if we 'hear' them correctly, we know what melody they are singing. That is to say, if a soprano's vibrato is executed around the intended target tone well enough, a well-tempered audience knows she is actually singing a high 'C' – but she is using a range of surrounding tones to physically create the *vibrato* effect, and the use of such a spread of tones does not offend the ear, since the ear is less discerning in that frequency range. Similarly, the signature of single-copy templates is experimentally indicated once a certain degree of C_q variance (or *vibrato*) has been obtained (e.g. when observing 10 replicates of single-copy qPCR reactions, Cq values anywhere from 43 to 48 can be expected at 70% efficiency) (Lockey *et al.*, 1998). If *Poisson probability distribution* bears me out on this, then the amplification of 1 copy of template on a qPCR machine capable of detecting fluorogenic real-time qPCR signals above background at a threshold cycle precisely indicating 10 billion amplicon copies made, this single copy C_q will always appear at 34.37 ± (1.15 *vibrato*) cycles. And, without vibrato, this C_q value would be exactly $10/\log_{10}(2)$ or 33.219. The repeating interval for qPCR template doubling happens with every successive C_q, whereas frequency doubling (in Hz) on the piano, happens every 12 notes (e.g. 1 cycle on a piano), thus the implied superscript '1/12' on the number '2' in the frequency equation '$Hz_n = Hz_o(2^{1/12})^n$' is the only difference from the doubling relationship describing PCR (the Hz equation uses '1/12' instead of '1' as the superscript of '2'). Finally, when inhibition is allowed to invade either RT or subsequent PCR and/or qPCR reactions, the piano has effectively been knocked out of tune, and atonal mathematic discord abounds. We must therefore strive in earnest to perform only *in-tune qPCR*. Describing qPCR in musical terms is sometimes useful in bringing tangibility to some of the abstract contours of the technique; it all cries out for a qPCR symphony!

Appendix

A brief review of some of the maths referred to in this chapter:

For PCR, the simple expression for a *2-to-the-n* process involved

$$X_n = X_o(2)^n$$

For qPCR, efficiency of the reaction must be considered

$$X_n = X_o(E + 1)^n \text{ or } X_o = X_n(E + 1)^{-n}$$

Efficiency of amplification: $E = 10^{(-1/m)} - 1$

Exponential amplification ('amplicon doubling per cycle') value: $E_{AMP} = (E + 1) = 10^{(-1/m)}$
Thus, in log base 10 vernacular, when $X_o = 1$, since it can be shown by simple rearrangment that $X_{Cq} = E_{AMP}{}^{b}$ (in which case Cq = b):

$$X_o = X_{Cq}E_{AMP}{}^{-Cq} = E_{AMP}{}^{b} \times E_{AMP}{}^{-Cq} = E_{AMP}{}^{b-Cq}$$

(When b = the intercept of an 'absolute copy' standard curve, and if b equals the C_q value that truly represents 'one copy' of target, the expression $X_o = E_{AMP}{}^{b-Cq}$ can be used to calculate copies instead of relative quantity values). See also Livak *et al.*, 2001. Also for qPCR, since the linear relationship can be described as:

$$[y = mx + b] = [C_q = m^*\log(X_o) + b], \text{ and this rearranges to } (C_q - b)/m = \log(X_o).$$

Thus, after taking the inverse log of both sides

$$X_o = 10^{((Cq-b)/m)} \text{ which again} = E_{AMP}{}^{b-Cq}$$

For correlating target copy number with relative dilution standard curves
(A) Absolute standard curve (for target x): $X_a = (E_{AMPa})^{ba-Cqa}$
(B) Relative dilution sample standard curve (for target x): $X_s = (E_{AMPs})^{bs-Cqs}$
where a = absolute template (either a linearized plasmid containing target x sequence or a synthetic amplicon of the target sequence) used for copy number-based standard curves, and s = target x amplified from within a biological/experimental sample.
Then, to interpret biological/experimental sample target C_q values in terms of copies, b_s is replaced with b_n, which is calculated as follows:

$$(C)\ b_n = b_a{}^*\log(E_{AMPa})/\log(E_{AMPs})$$

b_n is then put into above equation (B) to get $(E_{AMPs})^{bn-Cqs}$, which can then be used to estimate initial copies of target x in qPCR performed on biological/experimental samples – all based on the information gleaned from 'absolute' (linearized plasmid or PCR amplicon product) material assessed for the same target. However, it is extremely important to note that *absolute copy number* estimates are always only as good as the care initially taken with the original absolute template (linearized plasmid, PCR amplicon or organismal extract) to which the biological/experimental samples are being compared. Any 'absolute' measurement of copy number by qPCR *is always relative* to the integrity, quality and accuracy of the absolute sample's own quantitative assessment and physical manipulation along the way to its becoming the material used as template for an 'absolute' standard curve in the first place. Moreover, it has recently been demonstrated, rather eloquently, that linearized plasmid (containing target sequence) and synthetic target amplicons, when used as templates for absolute standard curves, give much more accurate results in estimating copy numbers than do supercoiled plasmids containing target sequence (Hou *et al.*, 2010). The entire, consolidated equation relating absolute template copy number and biological sample copy number (both being assessed for target x using the same primers and or primers/probe) thus becomes:

$$X_{o\ sample} = E_{AMPs}{}^{[b_a \cdot \log_{E_{AMPs}}(E_{AMP_a}) - C_{q_s}]}$$

qPCR reaction fluorescence in terms of a 4-parameter sigmoidal equation where:

$$Fc = \frac{F_{max}}{1+\left(\frac{F_{max}}{F_0}-1\right)(E_{max}+1)^{-C}}$$

F_c = fluorescence of the reaction at a chosen threshold cycle, F_o = initial reaction baseline fluorescence, F_{max} = maximal reaction fluorescence, E_{max} = maximal initial reaction efficiency (Rutledge *et al.*, 2008).

qPCR reaction fluorescence in terms of a Richards 5-parameter sigmoidal equation

$$F_x = F_b + (F_{max}/(1+\exp(-(1/b)*(x-c)))^d)$$

where Fx = the reaction fluorescence at cycle x, $Fmax$ = the maximal reaction fluorescence value, x = the fractional cycle of the turning (inflection) point of the curve, d = the Richards coefficient and Fb = the background reaction fluorescence, while b and c help determine the shape of each curve (Guescini *et al.*, 2008; Ritz *et al.*, 2008).

Musical note frequency-qPCR equation relationship
For amplicon accrual:

$$X_{Cq} = X_o(2)^{Cq} \text{ or } X_{Cq} = X_o(2^1)^{Cq}$$

For musical frequencies:

$$Hz_n = Hz_o(2)^{n/12} \text{ or } Hz_n = Hz_o(2^{1/12})^n$$

To calculate the frequency (in Hz) of the lowest 'A' (on a piano) if starting at A-440 Hz (the 'A' above 'middle C'), we first see that the lowest 'A' note is 48 piano keys below A 440 Hz. So, if our starting tone (Hz_o) is 440 Hz, and n = –48, the equation for the lowest piano 'A' would be $Hz = 440(2^{1/12})^{-48} = 27.5$ Hz.

In other words, from the lowest piano 'A' (27.5 Hz) to the 'A' above 'middle C' (the A at 440 Hz), four cycles of 12 tones has elapsed or intervened; thus, 27.5 Hz doubled four times equals 440 Hz (e.g. $27.5 \times 2^4 = 440$ Hz). Each doubling of tonal frequency (in Hz) is characterized by the passing of 12 notes. Conversely, if we wanted to calculate Hz for A-440 when starting at the lowest 'A' on the piano (27.5 Hz), the equation would take on the form: $27.5(2^{1/12})^{48} = 440$ Hz.

The appearance of C_q values during qPCR differs from this tonal frequency relationship only in a few semi-esoteric regards: (1) choice of instrument and materials; (2) tonal frequency doubling (measured in Hz) is instead called 'amplicon doubling' in qPCR, and this 'doubling'; and (3) occurs every one C_q in qPCR instead of every twelve piano keys in tonal frequencies.

Acknowledgements

The author would like to acknowledge Jan and Lyn, brothers Jeff and Jim, wife Christina and daughter Kalyn for their support and understanding during the writing of this chapter. Special thanks go to Dr Mark Ackermann for his friendship and professional support of this work.

References

Applied Biosystems User Bulletin #2. (2001). ABI PRISM 7700 Sequence Detection System, SUBJECT: Relative Quantitation of Gene Expression: Calculating the Input Amount. (Carlsbad, CA: Applied Biosystems), p. 8.

Al-Soud, W.A., and Rådström, P. (1998). Capacity of nine thermostable dna polymerases to mediate dna amplification in the presence of PCR-inhibiting samples. Appl. Environ. Microbiol. *64*, 3748–3753.

Al-Soud, W.A., and Rådström, P. (2000). Effects of amplification facilitators on diagnostic PCR in the Presence of blood, feces, and meat. J. Clin. Microbiol. *38*, 4463–4470.
Altshuler, M.L. (2006). PCR, Troubleshooting, M.L.: The Essential Guide (Wymondham, UK: Caister Academic Press).
Bachmann, B., Lüke, W., and Hunsmann, G. (1990). Improvement of PCR amplified DNA sequencing with the aid of detergents. Nucleic Acids Res. *18*, 1309.
Barr, P.J., Thayer, R.M., Laybourn, P., Najarian, Seela, R.C., F., and Tolan, D.R. (1986). 7-deaza-2′-deoxyguanosine-5′-triphosphate: Enhanced Resolution in M13 Dideoxy Sequencing. BioTechniques *4*, 428–432.
Bartlett, J.M.S., and Stirling, D. (2003). Methods in Molecular Biology: PCR Protocols (UK: Humana Press).
Baskaran, N., Kandpal, R.P., Bhargava, A.K., Glynn, M.W., Bale, A., and Weissman, S.M. (1996). Uniform amplification of a mixture of deoxyribonucleic acids with varying GC content. Gen. Res. *6*, 633–638.
Birch, C.P.D. (1999). A new generalized logistic sigmoid growth equation compared with the richards growth equation. Ann. Bot. *83*, 713–723.
Bookstein, R., Lai, C.-C., Hoang, T., and Lee, W.-H. (1990). PCR-based detection of a polymorphic BamHI site in intron 1 of the human retinoblastoma (RB) gene. Nucleic Acids Res. *18*, 1666.
Bustin, S.A. (2003). A–Z of Quantitative PCR (La Jolla, CA: International University Line, Biotechnology Series).
Bustin, S.A. (2005). Real-time, Fluorescence-based quantitative PCR: a snapshot of current procedures and preferences. Expert Rev. Mol. Diagn. *5*, 493–498.
Bustin, S.A. (2008). Real-time polymerase chain reaction – towards a more reliable, accurate and relevant assay. Eur. Pharm. Rev. *6*, 19–27.
Bustin, S.A. (2008). Real-time quantitative PCR – opportunities and pitfalls. Eur. Pharm. Rev. *4*, 18–23.
Bustin, S.A., and Nolan, T. (2009). Analysis of mRNA expression by real-time PCR. In Real-time PCR: Current Technology and Applications, Logan, J., Edwards, K., and Saunders, N., ed. (Norfolk: Caister Academic Press), pp. 111–135.
Bustin, S.A., Benes, V., Garson, J.A., Hellemans, J., Huggett, J., Kubista, M., Mueller, R., Nolan, T., Pfaffl, M.W., Shipley, G.L., Vandesompele, J., and Wittwer, C.T. (2009). The MIQE Guidelines: Minimum Information for Publication of Quantitative Real-Time PCR Experiments. Clin. Chem. *55*, 611–622.
Chakrabarti, R., and Schutt, C.E. (2001). The enhancement of PCR amplification by low molecular weight amides. Nucleic Acids Res. *29*, 2377–2381.
Chevet, E., Lemaftre, G., and Katinka, M.D. (1995). Low concentrations of tetramethyl-ammonium chloride increase yield and specificity of PCR. Nucleic Acids Res. *23*, 3343–3344.
Choi, J.-S., Kim, J.-S., Joe, C.-O., Kim, S., Ha, K.-S., and Park, Y.-M. (1999). Improved cycle sequencing of GC-rich DNA template. Exp. Mol. Med. *31*, 20–24.
Daniels, D. (2008). qPCR Methods Expand rapidly: high-throughput platforms and real-time profiling advance discipline. Gen. Engin. Biotech. News *28*, 19.
Demeke, T., and Adams, R.P. (1992). The effects of plant polysaccharides and buffer additives on PCR. BioTechniques *12*, 332–334.
Doma, M.K., and Parker, R. (2007). RNA quality control in eukaryotes. Cell *131*, 660–668.
Fleige, S., and Pfaffl, M.W. (2006). Review: RNA integrity and the effect on the real-time qRT-PCR performance. Mol. Aspects Med. *27*, 126–139.
Frackman, S., Kobs, G., Simpson, D., and Storts, D. (1998). Betaine and DMSO: enhancing agents for PCR. Promega Notes *65*, 27–30.
Gallup, J.M., Kawashima, K., Lucero, G., and Ackermann, M.R. (2005). New quick method for isolating RNA from laser captured cells stained by immunofluorescent immunohistochemistry; RNA suitable for direct use in fluorogenic TaqMan one-step real-time RT-PCR. Biol. Proced. Online *7*, 70–92.
Gallup, J.M., and Ackermann, M.R. (2006). Addressing fluorogenic real-time qPCR inhibition using the novel custom EXCEL file system 'FocusField2–6GallupqPCRSet-upTool-001' to attain consistently high fidelity qPCR reactions. Biol. Proced. Online *8*, 87–155.
Gallup, J.M., and Ackermann, M.R. (2008). The 'PRExcel-Q Method' for qPCR. Int. J. Biomed. Sci. *4*, 273–293.
Gallup, J.M., Sow, F.B., Van Geelen, A., and Ackermann, M.R. (2009). SPUD qPCR Assay Confirms PRExcel-Q Software's Ability to Avoid qPCR Inhibition. Curr. Issues Mol. Biol. *12*, 129–134.
Gelfand, D.H. (1989). *Taq* DNA polymerase. In PCR Technology: Principles and Applications for DNA Amplification, Erlich, H.A., ed. (New York: Stockton Press), pp. 17–22.

Giambernardi, T.A., Rodeck, U., and Klebe, R.J. (1998). Bovine serum albumin reverses inhibition of RT-PCR by melanin. BioTechniques *25*, 564–566.

Glaser, V. (2007). Application of real-time qPCR broadens: technology's ability to quantify RNA Levels makes it chief among the many tools used to understand gene silencing. Gen. Engin. Biotech. News *27*, 1–5.

Grubor, B.M., Gallup, J.M., Meyerholz, D.K., Crouch, E.C., Evans, R.B., Brogden, K.A., Lehmkuhl, H.D., and Ackermann, M.R. (2004). Enhanced surfactant protein and defensin mRNA levels and reduced viral replication during parainfluenza virus type 3 pneumonia in neonatal lambs. Clin. Diag. Lab. Immunol. *11*, 599–607.

Grunenwald, H. (2003). Optimization of polymerase chain reactions. In Methods in Molecular Biology: PCR Protocols, Bartlett, J.M.S., and Stirling, D., ed. (UK: Humana Press), pp. 89–99. doi:10.1385/1-59259-384-4:89.

Guescini, M., Sisti, D., Rocchi, M.B.L., Stocchi, L., and Stocchi, V. (2008). A new real-time PCR method to overcome significant quantitative inaccuracy due to slight amplification inhibition. BMC Bioinform. *9*, 326–355. doi:10.1186/1471–2105–9–326.

Hellemans, J., Geert, M., De Paepe, A., Frank, S., and Vandesompele, J. (2007). qBase relative quantification framework and software for management and automated analysis of real-time quantitative PCR data. Genome Biol. *8*, R19.

Henegariu, O., Heerema, N.A., Dlouhy, S.R., Vance, G.H., and Vogt, P.H. (1997). Multiplex PCR: critical parameters and step-by-step protocol. BioTechniques *23*, 504–511.

Hengen, P.N. (1997). Methods and reagents – optimizing multiplex and LA-PCR with betaine. Trends in Biochem. Sci. *22*, 225–226.

Henke, W., Herdel, K., Jung, K., Schnorr, D., and Loening, S.A. (1997). Betaine improves the PCR amplification of GC-rich DNA sequences. Nucleic Acids Res. *25*, 3957–3958.

Hou, Y., Zhang, H., Miranda, L., and Lin, S. (2010). Serious overestimation in quantitative PCR by circular (supercoiled) plasmid standard: microalgal pcna as the model gene. PLoS ONE *5*, e9545. doi:10.1371/journal.pone.0009545.

Huggett, J.F., Novak, T., Garson, J.A., Green, C., Morris-Jones, S.D., Miller, R.F., and Zumla, A. (2008). Differential susceptibility of PCR reactions to inhibitors: an important and unrecognised phenomenon. BMC Res. Notes *1*, 70.

Hung, T., Mak, K., and Fong, K.A. (1990). A specificity enhancer for polymerase chain reaction. Nucleic Acids Res. *18*, 4953.

Jacobs, K.A., Rudersdorf, R., Neill, S.D., Dougherty, J.P., Brown, E.L., and Fritsch, E.F. (1988). The thermal stability of oligonucleotide duplexes is sequence independent in tetraalkylammonium salt solutions: application to identifying recombinant DNA clones. Nucleic Acids Res. *16*, 4637–4650.

Jex, A.R., Hu, M., Littlewood, D.T.J., Waeschenbach, A., and Gasser, R.B. (2008). Using 454 technology for long-PCR based sequencing of the complete mitochondrial genome from single *Haemonchus contortus* (Nematoda). BMC Genomics 9, 11–24. doi:10.1186/1471-2164-9-11.

Johnson, A., Minoshima, S., Asakawa, S., Shimizu, N., Shizuya, H., Roe, B.A., and McDermid, H.E. (1999). A 1.5-Mb contig within the cat eye syndrome critical region at human chromosome 22q11.2. Genomics 57, 306–309.

Jossinet, F., Ludwig, T.E., and Westhof, E. (2007). RNA structure: bioinformatic analysis. Cur. Opin. Microbiol. *10*, 279–285.

Jung, A., Ruckert, S., Frank, P., Brabletz, T., and Kirchner, T. (2002). 7-Deaza-2′-deoxyguanosine allows PCR and sequencing reactions from CpG islands. Mol. Pathol. 55, 55–57.

Kawashima, K., Meyerholz, D.K., Gallup, J.M., Grubor, B., Lazic, T., Lehmkuhl, H.D., and Ackermann, M.R. (2006). Differential expression of ovine innate immune genes by preterm and neonatal lung epithelia infected with respiratory syncytial virus. Viral Immunol. *19*, 316–323.

Kebreab, E., Schulin-Zeuthen, M., Lopez, S., Soler, J., Dias, R.S., de Lange, C.F.M., and France, J. (2007). Comparative evaluation of mathematical functions to describe growth and efficiency of phosphorus utilization in growing pigs. J. Anim. Sci. *85*, 2498–2507. doi:10.2527/jas.2006–738.

Kent, W.J. (2002). BLAT– The BLAST-Like alignment tool. Genome Res. *12*, 656–664.

King, C.E., Debruyne, R., Kuch, M., Schwarz, C., and Poinar, H.N. (2009). A quantitative approach to detect and overcome PCR inhibition in ancient DNA extracts. BioTechniques 47, 941–949. doi: 10.2144/000113244.

Kovárová, M., and Dráber, P. (2000). New specificity and yield enhancer of polymerase chain reactions. Nucleic Acids Res. *28*, e70.

Kreader, C.A. (1996). Relief of amplification inhibition of PCR with bovine serum albumin or T4 gene 32 protein. Appl. Environ. Microbiol. *62*, 1102–1106.

Lazic, T., Wyatt, T.A., Matic, M., Meyerholz, D.K., Grubor, B., Gallup, J.M., Kersting, K.W., Imerman, P.M., Almeida-De-Macedo, M., and Ackermann, M.R. (2007). Maternal alcohol ingestion reduces surfactant protein A expression by preterm fetal lung epithelia. Alcohol *41*, 347–355.

Legiewicz M., and Rausch, J. (2009). SHAPE technology: a modern framework for RNA structure derivation. Eur. Pharm. Rev. *3*, 13–18.

Lewis, F. (2003). Extraction of total RNA from formalin-fixed paraffin-embedded tissue. In A–Z of Quantitative PCR, Bustin, S.A., ed. (La Jolla, CA: International University Line, Biotechnology Series), pp. 591–601.

Lewis, F., Maughan, N.J., Smith, V., Hillan, K., and Quirke, P. (2001). Review Article: Unlocking the archive – gene expression in paraffin-embedded tissue. J. Pathol. *195*, 66–71. doi:10.1002/path.921.

Libus, J., and Štorchová, H. (2006). Quantification of cDNA generated by reverse transcription of total RNA provides a simple alternative tool for quantitative RT-PCR normalization. BioTechniques *41*, 156–164. doi:10.2144/000112232.

Liu, W., and Saint, D.A. (2002). Validation of a quantitative method for real time PCR kinetics. Biochem. Biophys. Res. Commun. *294*, 347–353.

Livak, K.J., and Schmittgen, T.D. (2001). Analysis of relative gene expression data using real-time quantitative PCR and the $2^{-\Delta\Delta CT}$ method. Methods *25*, 402–408. doi:10.1006/meth.2001.1262.

Lockey, C., Otto, E., and Long, Z. (1998). Real-time fluorescence detection of a single DNA molecule. BioTechniques *24*, 744–746.

Logan, J., Edwards, K., and Saunders, N. (2009). Real-time PCR: Current Technology and Applications (Norfolk: Caister Academic Press).

Luu-The, V., Paquet, N., Calvo, E., and Cumps, J. (2005). Improved real-time RT-PCR method for high-throughput measurements using second derivative calculation and double correction. BioTechniques *38*, 287–293.

Masuda, N., Ohnishi, T., Kawamoto, S., Monden, M., and Okubo, K. (1999). Analysis of chemical modification of RNA from formalin-fixed samples and optimization of molecular biology applications for such samples. Nucleic Acids Res. *27*, 4436–4443.

McConlogue, L., Brow, M.A., and Innis, M.A. (1988). Structure-independent DNA amplification by PCR using 7-deaza-2′-deoxyguanosine. Nucleic Acids Res. *16*, 9869.

McDowell, D.G., Burns, N.A., and Parkes, H.C. (1998). Localised sequence regions possessing high melting temperatures prevent the amplification of a DNA mimic in competitive PCR. Nucleic Acids Res. *26*, 3340–3347.

Melchior, W.B., Jr., and Von Hippel, P.H. (1973). Alteration of the relative stability of dA·dT and dG·dC basepairs in DNA. Proc. Natl. Acad. Sci. U.S.A. *70*, 298–302.

Mills D.R., and Kramer, F.R. (1979). Structure-independent nucleotide sequence analysis. Proc. Natl. Acad. Sci. U.S.A. *76*, 2232–2235.

Mizusawa, S., Nishinmuar, S., and Seela, F. (1986). Improvement of the dideoxy chain termination method of DNA sequencing by use of deoxy-7-deazaguanosine triphosphate in place of dGTP. Nucleic Acids Res. *14*, 1319–1324.

Morrogh, M., Olvera, N., Bogomolniy, F., Borgen, P.I., and King, T.A. (2007). Tissue preparation for laser capture microdissection and RNA extraction from fresh frozen breast tissue. BioTechniques *43*, 2–6.

Motz, M., Pääbo, S., and Kilger, C. (2000). Improved cycle sequencing of GC-rich templates by a combination of nucleotide analogs. BioTechniques *29*, 268–270.

Murphy, J., and Bustin, S.A. (2009). Reliability of real-time reverse transcription PCR in clinical diagnostics: gold standard or substandard? Expert Rev. Mol. Diagn. *9*, 187–197.

Musso, M., Bocciardi, R., Parodi, S., Ravazzolo, R., and Ceccherini, I. (2006). Betaine, dimethyl sulfoxide, and 7-deaza-dGTP, a powerful mixture for amplification of GC-rich DNA sequences. J. Mol. Diag. *8*, 544–550.

Mutter, G.L., Zahrieh, D., Liu, C., Neuberg, D., Finkelstein, D., Baker, H.E., and Warrington, J.A. (2004). Comparison of frozen and RNA later solid tissue storage methods for use in RNA expression microarrays. BMC Genomics 5, 88–95.

Nolan, T., and Bustin, S.A. (2009). The importance of sample quality for qPCR. Eur. Pharm. Rev. *1*, Article 1.

Nolan, T., Hands, R.E., and Bustin, S.A. (2006). Quantification of mRNA using real-time RT-PCR. Nat. Protoc. *1*, 1559–1582.

Nolan, T., Hands, R.E., Ogunkolade, B.W., and Bustin, S.A. (2006). SPUD: a qPCR assay for the detection of inhibitors in nucleic acid preparations. Anal. Biochem. *351*, 308–310.

Olivier, A., Gallup, J., de Macedo, M.M.M.A., Varga, S.M., and Ackermann, M. (2009). Human respiratory syncytial virus A2 strain replicates and induces innate immune responses by respiratory epithelia of neonatal lambs. Int. J. Exp. Path. *90*, 431–438.

Pääbo, S., Gifford, J.A., and Wilson, A.C. (1988). Mitochondrial DNA sequences from a 7000-year old brain. Nucleic Acids Res. *16*, 9775–9787.

Pabinger, S., Thallinger, G.G., Snajder, R., Eichhorn, H., Rader, R., and Trajanoski, Z. (2009). QPCR: application for real-time PCR data management and analysis. BMC Bioinform. *10*, 268–277. doi:10.1186/1471–2105–10–268.

Peirson, S.N., Butler, J.N., and Foster, R.G. (2003). Experimental validation of novel and conventional approaches to quantitative real-time PCR data analysis. Nucleic Acids Res. *31*, e73.

Peters, W.S., and Baskin, T.I. (2006). Tailor-made composite functions as tools in model choice: the case of sigmoidal vs bi-linear growth profiles. Plant Methods *2*, 11–24. doi:10.1186/1746–4811–2–11.

Pfaffl, M.W., Lange, I.G., Daxenberger, A., and Meyer, H.H. (2001). Tissue-specific expression pattern of estrogen receptors (ER): quantification of ER alpha and ER beta mRNA with real-time RT–PCR. Acta. Pathol. Microbiol. Immunol. Scand. *109*, 345–355.

Pfaffl, M.W., Vandesompele, J., and Kubista, M. (2009). Data analysis software. In Real-time PCR: Current Technology and Applications, Logan, J., Edwards, K., and Saunders, N., ed. (Norfolk: Caister Academic Press), pp. 65–83.

Poddar, S.K., Sawyer, M.H., and Connor, J.D. (1998). Technical Note: Effect of inhibitors in clinical specimens on *Taq* and Tth DNA polymerase-based PCR amplification of influenza A virus. J. Med. Microbiol. *47*, 1131–1135.

Ralser, M., Querfurth, R., Warnatz, H.-J., Lehrach, H., Yaspo, M.-L., and Krobitsch, S. (2006). An efficient and economic enhancer mix for PCR. Biochem. Biophys. Res. Commun. *347*, 747–751.

Rees, W.A., Yager, T.D., Korte, J., and Von Hippel, P.H. (1993). Betaine can eliminate the basepair composition dependence of DNA melting. Biochemistry *32*, 137–144.

Richards, F.J. (1959). A flexible growth function for empirical use. J. Exp. Bot. *10*, 290–300.

Rijpens, N.P., Jannes, G., Van Asbroeck, M., Rossau, R., and Herman, L.M.F. (1996). Direct detection of *Brucella* spp. in raw milk by PCR and reverse hybridization with 16S–23S rRNA spacer probes. Appl. Environ. Microbiol. *62*, 1683–1688.

Ritz, C., and Spiess, A.-N. (2008). qpcR: an R package for sigmoidal model selection in quantitative real-time polymerase chain reaction analysis. Bioinformatics *24*, 1549–1551. doi:10.1093/bioinformatics/btn227.

Rossen, L., Norskov, P., Holmstrom, K., and Rasmussen, F.O. (1992). Inhibition of PCR by components of food sample, microbial diagnostic assay and DNA-extraction solutions. Int. J. Food Microbiol. *17*, 37–45.

Rutledge, R.G. (2004). Sigmoidal curve-fitting redefines quantitative real-time PCR with the prospective of developing automated high-throughput applications. Nucleic Acids Res. *32*, e178. doi:10.1093/nar/gnh177.

Rutledge, R.G., and Stewart, D. (2008). A kinetic-based sigmoidal model for the polymerase chain reaction and its application to high-capacity absolute quantitative real-time PCR. BMC Biotech. *8*, 47–74. doi:10.1186/1472–6750–8–47.

Rutledge, R.G., and Stewart, D. (2008). Critical evaluation of methods used to determine amplification efficiency refutes the exponential character of real-time PCR. BMC Mol. Biol. *9*, 96–107. doi:10.1186/1471-2199-9-96.

Sarkar, G., Kapelner, S., and Sommer, S.S. (1990). Formamide can dramatically improve the specificity of PCR. Nucleic Acids Res. *18*, 7465.

Schmerer, W.M. (2003). Reduction of shadow band synthesis during PCR amplification of repetitive sequences from modern and ancient DNA. In Methods in Molecular Biology: PCR Protocols, Bartlett, J.M.S. and Stirling, D., ed. (UK: Humana Press), pp. 309–313. doi:10.1385/1–59259–384–4:309.

Schmittgen, T.D., and Zakrajsek, B.A. (2000). Effect of experimental treatment on housekeeping gene expression: validation by real-time, quantitative RT-–PCR. J. Biochem. Biophys. Meth. *20*, 69–81.

Shaik, G.M., Dra'berova', L., Dra'ber, P., Boubelı'k, M., and Dra'ber, P. (2008). Tetraalkylammonium derivatives as real-time PCR enhancers and stabilizers of the qPCR mixtures containing SYBR Green I. Nucleic Acids Res. *36*, e93. doi:10.1093/nar/gkn421.

Shain, E.B., and Clemens, J.M. (2008). A new method for robust quantitative and qualitative analysis of real-time PCR. Nucleic Acids Res. *36*, e91. doi:10.1093/nar/gkn408.

Shammas, F.V. (2001). Improvement of quantitative PCR reproducibility by betaine as determined by fluorescence-based method. BioTechniques *30*, 950–954.

Smith, K.T., Long, C.M., Bowman, B., and Manos, M.M. (1990). Using cosolvents to enhance PCR amplification. Amplifications 5, 16–17.

Sow, F.B., Gallup, J.M., Meyerholz, D.K., and Ackermann, M.R. (2009). Gene profiling studies in the neonatal ovine lung show enhancing effects of VEGF on the immune response. Dev. Comp. Immunol. *33*, 761–771.

Sow, F.B., Gallup, J.M., Sacco, R.E., and Ackermann, M.R. (2009). Laser capture microdissection revisited as a tool for transcriptomic analysis: application of an Excel-based qPCR preparation software (PRExcel-Q). Int. J. Biomed. Sci. *5*, 105–124.

Spiess, A.-N., Mueller, N., and Ivell, R. (2004). Trehalose is a potent PCR enhancer: lowering of dna melting temperature and thermal stabilization of *Taq* polymerase by the disaccharide trehalose. Clin. Chem. *50*, 1256–1259.

Spiess, A.-N., Feig, C., and Ritz, C. (2008). Highly accurate sigmoidal fitting of real-time PCR data by introducing a parameter for asymmetry. BMC Bioinform. *9*, 221–232. doi:10.1186/1471-2105-9-221.

Sponseller, B.A., de Macedo, M.M., Clark, S.K., Gallup, J.M., and Jones, D.E. (2009). Activation of peripheral blood monocytes results in more robust production of IL-10 in neonatal foals compared to adult horses. Vet. Immunol. Immunopathol. *127*, 167–173.

Stone, A.C. (2004). Extraction and Amplification of Ancient DNA. In PCR technology: Current Innovations, Weissensteiner, T., Griffin, H.G., and Griffin, A.M., ed. (Boca Raton, FL: CRC Press), pp. 5–9.

Suzuki, T., Higgins, P.J., and Crawford, D.R. (2000). Control selection for RNA quantitation. BioTechniques *29*, 332–337.

Swango, K.L., Hudlow, W.R., Timken, M.D., and Buoncristiani, M.R. (2007). Developmental validation of a multiplex qPCR assay for assessing the quantity and quality of nuclear DNA in forensic samples. Forensic Sci. Int. *170*, 35–45.

Swillens, S., Goffard, J.-C., Maréchal, Y., de Kerchove d'Exaerde, A., and Housni, H.E. (2004). Instant evaluation of the absolute initial number of cDNA copies from a single real-time PCR curve. Nucleic Acids Res. *32*, e53.

Thellin, O., Zorzi, W., Lakaye, B., De Borman, B., Coumans, B., Hennen, G., Grisar, T., Igout, A., and Heinen, E. (1999). Housekeeping genes as internal standards: use and limits. J. Biotechnol. *75*, 291–295.

Tichopad, A., Dilger, M., Schwarz, G., and Pfaffl, M.W. (2003). Standardized determination of real-time PCR efficiency from a single reaction set-up. Nucleic Acids Res. *31*, e122. doi:10.1093/nar/gng122.

Tichopad, A., Pfaffl, M.W., and Didier, A. (2003). Tissue-specific expression pattern of bovine prion: quantification using real time RT–PCR. Mol. Cell. Probes *17*, 5–10.

Tichopad, A., Didier, A., and Pfaffl, M.W. (2004). Inhibition of real-time RT–PCR quantification due to tissue specific contaminants. Mol. Cell. Probes *18*, 45–50.

Untergasser, A., Lefever, S., Pattyn, F., Hellemans, J., and Vandesompele, J. (2008). A new standard for qPCR data. RDML. Euro. Biotech. News *7*, 40–41.

Vandesompele, J., Kubista, M., and Pfaffl, M.W. (2009). Reference gene validation software for improved normalization. In Real-time PCR: Current Technology and Appl. ications, Logan, J., Edwards, K., and Saunders, N., ed. (Norfolk: Caister Academic Press), pp. 47–63.

Varadaraj, K., and Skinner, D.M. (1994). Denaturants or cosolvents improve the specificity of PCR amplification of a G + C-rich DNA using genetically engineered DNA polymerases. Gene *140*, 1–5.

Weissensteiner, T., and Lanchbury, J.S. (1996). Strategy for controlling preferential amplification and avoiding false negatives in PCR typing. BioTechniques *21*, 1102–1108.

Weksberg, R., Hughes, S., Moldovan, L., Bassett, A.S., Chow, E.W.C., and Squire, J.A. (2005). A method for accurate detection of genomic microdeletions using real-time quantitative PCR. BMC Genomics *6*, 180–189. doi:10.1186/1471-2164-6-180.

Wilson, I.G. (1997). Minireview: inhibition and facilitation of nucleic acid amplification. Appl. Env. Microb. *63*, 3741–3751.

Wood, W.I., Gitschier, J., Lasky, L.A., and Lawn, R.M. (1985). Base composition–independent hybridization in tetramethylammonium chloride: a method for oligonucleotide screening of highly complex gene libraries. Proc. Natl. Acad. Sci. U.S.A. *82*, 1585–1588.

Yadav, S.P., Haqqi, T., Zhao, X., and Panciu, A. (2002). Sequencing in the presence of Betaine: Improvement in sequencing of localized repeat sequence regions. ABRF abstract at http://www.abrf.org/Other/ABRFMeetings/ABRF2002/2002Abstracts.html. The Lerner Res. Institute, Cleveland, OH.

Significance of Controls and Standard Curves in PCR

3

Ian Kavanagh, Gerwyn Jones and Saima N. Nayab

Abstract

Whilst qPCR is a powerful technique, the results achieved using this method are valid only if the appropriate controls have been included in the experiment. Careful selection of controls and proper optimization of qPCR conditions promise generation of highly specific, repeatable, reproducible and sensitive data. This chapter discusses the strategies for preparing both negative and positive controls for PCR, when they should be employed and how to interpret the information they provide. It also highlights the significance of standard curves for determining the initial starting amount of the target template and for assessing assay efficiency, precision, sensitivity, and dynamic range. It also provides guidance on how to prepare standards, interpret standard curves and troubleshoot inefficient qPCR reactions.

Controls for conventional and quantitative PCR and RT-PCR

In any experiment, unintended influences such as researcher error and environmental changes can creep into a workflow, altering the outcome of the resulting data. Therefore, to thoroughly test any hypothesis, an experiment must be planned and carried out with carefully selected controls in an attempt to eliminate any alternative explanations of the results. The two simplest forms of controls use either a set of samples that are expected to demonstrate no effect (negative control) or a set of samples that are expected to demonstrate a definite effect (positive controls).

Negative controls

A negative control as the name suggests, is a sample that is known to give a negative result. This control can provide the baseline result that is obtained when a test does not produce a measurable result that can be claimed as positive. In PCR, the negative control is typically used to identify whether there is either contamination of the reaction, or any non-specific amplification that would ultimately affect the quality of data that are produced for a given sample (Table 3.1).

No template controls

The most common control used in a PCR is a reaction that is set up with all the components required to produce amplification, except the DNA or RNA template, and is termed the 'no template control' (NTC). These controls should be included in every experimental run and

Table 3.1 Interpreting positive signals in no template controls

Source of positive signal in no template controls		Prevention and troubleshooting
True template	Genomic DNA	One primer designed to span over intron regions of a gene sequence in the absence of known pseudogenes Primers designed in separate exons where the intron is >200 basepairs DNA removal using DNase I or shrimp nuclease Clean laboratory practices
	Amplicon contamination	Addition of dUTP and removal in subsequent reactions using UNG DNA removal using DNase I or shrimp nuclease Clean laboratory practices
Non-specific amplification	Primer dimers	Check for possibility of internal annealing within and between primers Reduce primer concentrations Optimize annealing temperature of primers Adjust concentration of magnesium ions
	Off-targeting	BLAST align primer sequences to determine any potential homology to other transcript and genomic sequences
False positives	Software normalization and corrections	Investigate shape of amplification curve using multicomponent views and comparison to signal in positive controls Run agarose gel electrophoresis/melting analysis to confirm presence of a true product

indeed the recently published MIQE guidelines (minimal information for publication of quantitative real-time PCR experiments) advise that NTCs are an essential part of a PCR to be included each time a reaction is run, and conditions for rejection of data, should they be positive, must be clearly established (Bustin *et al.*, 2009).

A positive signal in a NTC can indicate the presence of specific contamination and it has been suggested that placing multiple NTC's spread across a multiwell plate can help identify the source of this (Bustin and Nolan, 2004). Where possible, for instance when using strip caps to seal a multiwell plate, wells containing the NTC reactions can be sealed before the addition of the test template and others sealed after addition. This will allow the user to distinguish at what point in the protocol the contamination is being introduced. PCR set-up, especially in 96 and 384 multiwell plate formats, involves many different pipetting steps where contaminating template can be introduced from either an environmental source, or from cross-contamination of a test or positive control sample run in the same experiment. In reality, contamination resulting from poor pipetting technique is the simplest problem to resolve and can often be rectified through further practice of setting up the reaction plates and paying careful attention to the individual pipetting steps. Regularly changing the tips and using filtered-tips to prevent aerosols forming and entering pipettes can also help, whilst more complicated set-up of PCR experiments may further benefit from the use of specialized automated instrumentation. However, whatever the solution to the cross-contamination may be, it is still vital to include the NTC reactions in all future experimental runs to satisfy that it has not recurred. A trickier problem to address is when the contaminating template is being introduced at an earlier step of the PCR protocol and in this case could have a number of different sources. For example, any genomic DNA present in the sample could be the source of amplification in a positive NTC reaction and should be considered especially when using RNA as the primary template for amplification (genomic DNA contamination is discussed in the subsequent section on no RT controls). Another common source of contamination can be from the amplicon of a previous PCR that contain sequences to which the primers of the current PCR can anneal (referred to as carry-over contamination). Opening the reaction is not always necessary in PCR (e.g. most applications of real-time PCR), but in some cases it is required for another downstream process to be carried out (e.g. running product on agarose gel or preparation for sequencing reactions). If care is not taken, this could release millions of copies of the PCR product onto the laboratory surfaces and equipment and eventually find its way back into a subsequent experiment. In many of these cases, strict laboratory operating procedures can be used to reduce the incidence of this type of contamination. This can also be overcome by separating rooms for reaction set-up and analysis, using clean hoods, dedicated pipettes and tips can be employed along with clean laboratory practices. However, the mere fact that billons of amplicons are generated in a PCR means that it can still be very difficult to avoid, especially if the same primer set is being used in many reactions, such as a laboratory completing high-throughput testing of samples. One further way to reduce this type of contamination is to include deoxyuridine triphosphate (dUTP) in your reaction set-up to replace (or more often mixed with) deoxythymidine triphosphate (dTTP), thereby replacing thymine bases with uracil. With uracil incorporated into the resulting amplicon, this product can be specifically targeted using the enzyme uracil-DNA glycosylase (UNG), which liberates the uracil residues from the DNA backbone, effectively degrading the amplicon and eliminating it as a potential source of amplification (Longo *et al.*, 1990). By using a thermolabile form of this enzyme, it can be included in an additional step at the beginning of a PCR, to selectively remove only the contaminating products containing uracil, and then the enzyme is deactivated before amplification of the true template commences. Once again most of these sources of contamination are derived from some type of user-error within the PCR work-flow and in many cases can be further avoided by careful attention to pipetting and stricter standard operating procedures.

Another reason for including a NTC reaction is to indicate the presence of any non-specific amplification. Any spurious products that are generated during amplification will consume the components of the PCR, ultimately reducing the efficiency with which the desired product is amplified. Furthermore, certain detection methods of PCR, such as when using a double-stranded DNA binding dye in real-time detection, do not distinguish between the non-specific products and the target amplicon, which could lead to incorrect calculation of the amount of target in a sample. The use of a NTC is important in these situations to indicate the presence of this spurious amplification and then can be used to compare to a true amplification (such as a positive control), using melt curves and gel electrophoresis to distinguish amplicon sizes. Non-specific amplification can be influenced by a number of factors, with the most common culprits being poor primer design leading to primer dimers or priming of other regions of the template (off-targeting), sub-optimal magnesium ion concentration and sub-optimal annealing temperatures (Table 3.1).

One final consideration when using NTC's in PCR is to be aware of false positives especially when using real-time PCR. In some cases automatic calling of a Cq value may not be derived from a true signal, as a result of a mathematical error caused by the instrument normalization and background correction. Examples such as erratic background fluorescence over the first few cycles and anomalies such as FAM creep and ROX drop (caused by alterations in the chemistry of the fluorophore and quenchers over time) can lead to these types of false positive signals. Therefore, it is advisable to use multicomponent views of the amplification plots to determine the validity of these signals. If any doubt remains over its legitimacy, the product can be electrophoresed on an agarose gel and compared to a reaction which is known to produce amplification (see subsequent section on positive controls).

No enzyme controls

When quantifying RNA species it is important to determine that the amplification is specific to the RNA and is not derived from genomic DNA. To control for this, a sample of template RNA is treated identically to the others, except that the reverse transcriptase is omitted from the cDNA synthesis step. This control is normally described as the no-RT control or the no enzyme control (NEC).

The most common method for removal of contaminating genomic DNA from RNA is DNase I treatment (Huang *et al.*, 1996; Klaschik *et al.*, 2002). However, DNase I has some non-specific nuclease activity cleaving double-stranded, single-stranded DNA and hybrid DNA/RNA (Sutton *et al.*, 1997). Therefore, the enzyme cannot be combined with the primers and probes for RT-PCR and all activity of the nuclease must be removed before commencing the reverse transcription reaction. Shrimp nuclease, unlike DNase I, is a nuclease with a high preference for double-stranded DNA only and it is also easily heat inactivated at a relatively low temperature. In some cases it is possible to work around DNA removal by designing the PCR assay to specifically amplify the mRNA, choosing intron-spanning primers. Similarly, in qRT-PCR a detection probe can be designed to hybridize over an exon–exon junction producing a signal for mRNA only. Primers can also be placed in separate exons, but unless the intron is of sufficient size, contaminating genomic DNA in an RNA preparation could still compete for primers and probes and lead to false or imperfect results in qRT-PCR (Bustin, 2002; Freeman *et al.*, 1999). One important consideration when designing primers to get around DNA contamination is that some genes are known to have processed pseudogenes (Mutimer *et al.*, 1998; Raff *et al.*, 1997), which will be amplified identically to the mRNA when genomic DNA is present. In these cases, complete removal of the genomic DNA is the only solution to avoid a positive signal being observed in the negative controls.

Finally, it may be necessary to include a NEC for assessing the absence of DNA when an untested RNA sample and indeed this is one of the essential recommendations of the MIQE guidelines (Bustin *et al.*, 2009), but once a sample has been confirmed to be DNA free, it may be less important to include this control in future PCR, as long as a NTC is always included.

Positive controls

In PCR, a positive control is a reaction which mimics the experimental reactions, but is certain to produce a positive result as long as the reaction conditions are conducive to amplification. The inclusion of positive controls in a PCR experiment enables reliable conclusions to be drawn from negative amplification results, which would by themselves, prove only that amplification has failed (Table 3.2). Considering the numerous causes of PCR failure, positive controls are essential for confirming the nature of negative results and when used in combination with negative controls (see previous section) can aid in the diagnosis of the problem (Table 3.1).

External positive controls

The simplest form of a positive control is a single reaction that is certain to amplify if all the right components have been added to the reaction at the correct concentrations. Such a reaction is an example of an external positive control, as the reaction is run in parallel to those containing the test sample. These basic controls also indicate that the thermal cycling protocol is correct and that the thermal cycler and detection method is functioning properly. It is important to use the same PCR reaction mix in the control reactions as those containing the test template, to ensure that any test reaction failure is not due to an omitted reaction component. Similarly, the test and control reactions must be cycled under the same conditions, using the same consumables and the same thermal cycler, in order to exclude these factors as causes of PCR failure. External positive controls are also useful in the electrophoresis of PCR products, as when run alongside test reactions, they help indicate the correct product size as well as correct functioning of the electrophoresis system (a traditional gel, or automated electrophoresis instrument).

When the starting template for amplification is RNA then two external positive controls can be used; one which includes RNA template and controls for the reverse transcription step and the other which uses DNA in the reaction and controls for the PCR step. The combined results of these two reactions can help to identify at which step the failure has occurred; if the RNA reaction fails, but the DNA reaction produces amplification, then the reverse transcription step is the likely culprit, but if both reactions fail, then the PCR step is the source of failure.

As well as complete failure, it is possible to identify less severe PCR inhibition using external controls. PCR inhibitors can be introduced, either during template isolation and purification, or during the PCR set-up and can include organic compounds such as ethanol, haemoglobin, urea and heparin derived from clinical and forensic samples; proteins, glycogen, fats, and Ca^{2+} (as found in foodstuffs) phenolic compounds (from foodstuffs and environmental samples such as plants), humic acids and heavy metals from plant and soil samples, constituents of bacterial cells, non-target DNA, glove powder, laboratory plastic ware, and cellulose (Wilson, 1997). The yield of a reaction or shift in Cq value in the control reactions of different samples can indicate the presence of PCR inhibitors in a test template. This can be taken a step further with a standard curve prepared from serial dilutions of known concentrations of template and allows more detailed analysis of dynamic range, reproducibility and efficiency to be examined (see subsequent section).

The disadvantages of external positive controls are that they assume that the kinetics of the control reaction is the same as those of the experimental reactions. Therefore, external controls cannot precisely determine the reason for PCR failure due to the very fact that they are not run in the exact same reaction as the test samples! External controls also require additional reagents and consumables to run these separate reactions, consequently increasing experiment costs, and decreasing the sample throughput possible in a single PCR run.

Internal positive controls

An internal positive control (IPC) is used when a target is co amplified within the same reaction as the target of interest. IPCs have the added benefit over external controls, as they can indicate whether a reaction failure is due to a reason intrinsic to the test reaction, without the need for additional reactions to be set-up simultaneously. However, an IPC target must be detectable in such a way that differentiates it from the primary target. This can be achieved in a number of ways including product length discrimination by gel electrophoresis or melting temperature analysis (in qPCR), or by sequence-specific methods such as the use of an oligonucleotide probe labelled with distinct fluorescent markers for each of the control and experimental targets. Generally, amplifying multiple targets in a single reaction poses difficulties, not only because it can be difficult to create and optimize efficient assays which use the same reaction conditions, but also because competition for nucleotides, polymerase and magnesium ions can be detrimental to the amplification of both primary and control targets, becoming more severe with each extra target involved. This competition must be countered by using a reaction mix composition which satisfies the amplification requirements for all targets.

Two mains types of IPC exist which use either a competitive or non-competitive mode of amplification. Competitive IPCs involve amplification of a target from an exogenous template source, using the same set of primers as the target of interest. Examples of this approach include the composite primer method (Siebert and Larrick, 1992) and the substitution of a Herpes simplex virus probe sequence, with a sequence from *Drosophila* (Cone *et al.*, 1992). The most important consideration when using this approach is the concentration of the control template in relation to that of the sample template. The molar ratio between the control and sample template must be optimized so that the control target is not preferentially amplified over the target of interest, as this will be detrimental to the limit of detection and efficiency of the test assay. Generally, the control template should be at the lowest possible concentration that will still yield a reproducible result (Hoorfar *et al.*, 2004). Conversely, using too high a control template concentration will also result in an indiscernible difference in yield if inhibition is low, thus will not indicate any inhibitory effects. Therefore it is prudent to use a control assay that produces a similar Cq value to the test assays, so that any changes observed in the control assay are indicative of what is occurring in the test case. Another important factor to consider for a competitive IPC template is the length of the amplicon compared to the target of interest, as a shorter amplicon will preferentially amplify over a longer target (Sachadyn and Kur, 1998). Therefore, designing the control amplicon length to be longer will ensure that the test sequence out-competes the control target amplification. However, it has been demonstrated that PCR efficiency is not significantly affected below an amplicon length of 500 bp (Abdulmawjood *et al.*, 2002), so in PCR experiments where amplicons are short (such as qPCR), amplicon length is less important when trying to make sure the target of interest is preferentially amplified.

Non-competitive IPC methods use a distinct set of primers for amplification and can be used with either a sequence already contained in the template nucleic acid (endogenous), such as a reference gene, or from the addition of a spiked-in template (exogenous). Sources of exogenous control template can be natural or synthetic nucleic acids and considering that template characteristics (sequence, secondary structure, GC content and length) can affect PCR kinetics, the best option is to choose a natural control template which is similar to the experimental template (e.g. a viral DNA of a different serotype to the test sample). However, because a natural control template is sometimes difficult to find, another convenient option is to use a synthetic nucleic acid, as these can be produced to mimic the experimental amplicon, and is a useful alternative to using a natural template. Synthetic RNA controls can be employed to control for some of the upstream steps in a RT-PCR or qRT-PCR protocol when using RNA as the starting material for the test reactions. For example, this control can be supplemented to the test template either before nucleic acid isolation or before cDNA synthesis in order to provide information on the performance quality of these steps. Due to the propensity of RNA molecules in storage to degrade at a higher rate than DNA molecules, chemically modified synthetic RNA, which is less prone to degradation, can also be employed. However this should be used with careful consideration, as the mere fact that this control is more stable means that the degradation rate will be different to the test sample RNA and thus cannot be used to indicate the RNA quality during storage. Commercially available control assays and kits are widely available, offering the benefit of employing controls without arduous optimization and validation, although the sequences of these assays are often kept proprietary. In many cases it may be advantageous to know the sequences of the control molecules, as this enables the researcher to check for any significant homology between the control and test primers, as this could lead to amplification of non-specific products and/or primer dimers.

Where possible and appropriate, an endogenous IPC is preferable to exogenous IPC, as they mimic the experimental conditions more closely, considering the targets are both present on the same template, the complexity of which can affect the performance of the reaction. The control target must always be present in the test template (e.g. conserved regions of the 16s rRNA for a prokaryotic template). Therefore, endogenous IPCs can only be employed in experiments where the template is always present, thus this approach is only useful for experiments designed to detect particular genes (e.g. virus serotyping), not those experiments designed to detect the presence of template (e.g. virus detection). For the latter type of experiment, or when a suitable endogenous target cannot be found, then an exogenous IPC must be used. The endogenous control target should preferably be present in the same copy number as the target of interest, to ensure that a positive amplification result for the control target implies that the target of interest should also be detectable if it is present. As an endogenous control, reference genes can be useful, as they are known to always be present in the sample at a stable abundance. However, no gene is the perfect 'housekeeper', fulfilling the stability characteristics in all tissues and experimental conditions. Choice of endogenous references is largely dependant on the sample tissue or cell type, its developmental stage, and the experimental conditions employed in the study. In order to make the right choice, it is necessary to test a range of candidate genes which display the desired stability of expression on a study to study basis (Olsvik *et al.*, 2005; Pfaffl *et al.*, 2004; Vandesompele *et al.*, 2002).

In conclusion, a positive control method should be chosen to mimic the experimental amplification as closely as possible, thereby providing the greatest level of confidence in a negative test result. Multiple control reactions used in conjunction can be employed, and if chosen wisely, such a positive control regime can greatly narrow down the possible causes of amplification failure, making troubleshooting simpler. An example would be to perform a basic external control reaction, along with endogenous IPC. This method would indicate whether a test reaction failure was due to PCR inhibition from the template (negative IPC result), incorrect reaction conditions (negative IPC and external control results), or was a genuine negative (both controls produce amplification).

Table 3.2 Interpretation of positive control results

Template	Positive control amplification interpreted as	Without amplification of specific target, this is interpreted as
IPC		
Endogenous	Reagents – OK Consumables – OK Thermal cycling – OK Detection – OK Inhibitors – none/negligible Template – present and sound	Target is absent from sample template
Exogenous	Reagents – OK Consumables – OK Thermal cycling – OK Detection – OK Inhibitors – none/negligible	Test template is absent Target is absent from the sample
EPC		
Endogenous	Reagents – OK Consumables – OK Thermal cycling – OK Detection – OK Template – present and sound Inhibitors in sample– none/negligible	Target is absent from test template Seal failure/reagent evaporation Contamination of reaction/vessel with inhibitor prior to template addition
Exogenous	Reagents – OK Consumables – OK Thermal cycling – OK Detection – OK	Contamination of reaction/vessel with inhibitor prior to template addition; presence of PCR inhibitor in sample template; test template absent; test template is degraded; target is absent from test template; seal failure/reagent evaporation

Standard curves in quantitative PCR

A standard curve, generated by performing qPCR of known template concentrations diluted over several orders of magnitude, provides an excellent tool for determining the initial amount of the target template and for assessing assay efficiency, precision, sensitivity, and working range. The accuracy of gene quantification using qPCR is directly related to the quality of the standard curve and a highly accurate standard curve is required for absolute quantification which is often used for measuring exact copy number (e.g. viral titre). Each concentration of the standard template sample is correlated with a specific Cq value and the exact copy numbers of the unknown target samples can be read directly from the standard curve. Relative quantification, which applies to most gene expression studies focusing on fold change instead of precise copy number, also requires higher level of accuracy when using a gene-specific standard curve, instead of the more commonly used ΔΔCq method for quantification. For accurate gene quantification, it is imperative to have similar RT and PCR efficiencies for the standard and target samples. In addition, a new standard curve should be generated for each new gene being studied, as PCR efficiencies will vary by target.

How to choose and set up standards controls

A standard template can be selected from various sources, for example purified PCR products, plasmid DNA carrying a cloned target sequence or DNA/cDNA from biological or technical controls, as discussed in the previous section. The quality and quantity of starting material is crucial for optimal amplification. In order to minimize variations based on samples, a researcher should always use highly purified template nucleic acid and a standardized extraction procedure throughout the entire study. Furthermore, to avoid any discrepancy for RT-qPCR, ensure that RNA from both standard and unknown samples has been subjected to the same reverse transcription protocol.

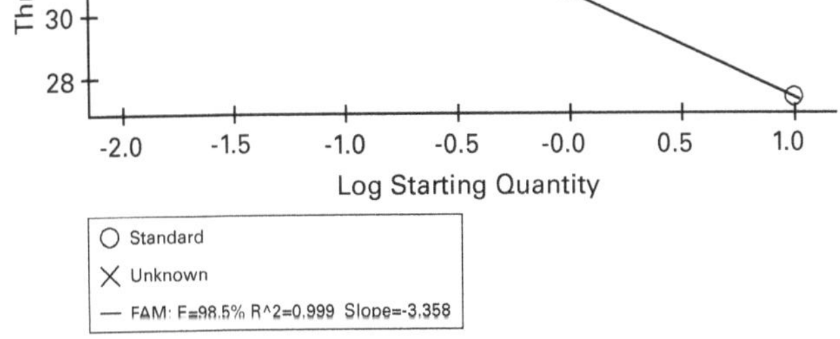

Figure 3.1 A screenshot from BioRad CFX software showing example of a good standard curve. PCR efficiency across four 10-fold template dilution series was measured as 98.5%. The slope of the curve was calculated as –3.358 and the R^2 value as 0.999.

When preparing a standard curve at least four, but preferably six or more, points on the curve should be included using serial dilutions of between 4- and 10-fold of known concentrations of standards. Where possible, each point on the curve should be verified with replicate reactions, ideally using at least three replicates, but no less than two. The intra-assay variation can be calculated from the replicate samples and can indicate the pipetting accuracy during the preparation of the dilution series, as well as the test samples. This simple quality assessment of the data should be performed each time a qPCR is run, and verified before any data analysis can begin. When choosing the dynamic range of a standard curve, plan to extend the range past both the highest and the lowest levels of target expected in unknown test samples. In some situations, the quantity of your unknown samples will not be possible to estimate, or the expectation is that there will be a large fold change difference between the unknowns, and in these cases it is prudent to use a large dynamic range until more information can be gathered. Appropriate negative controls should always be included when running a standard curve, and for SYBR Green dye-based detection a melt curve at the end of protocol can be added to help verify assay specificity. Previously prepared and validated standards stored at 4°C or –20°C can be used for subsequent experiments or in some cases, previously generated data can be imported as external standards on some qPCR instrument software; however, it is strongly recommended to prepare fresh standards each time an experiment is run, to avoid any potential template stability issues and resultant inter-run variation.

How to interpret a standard curve

It is essential to understand and be able to accurately interpret the standard curve in order to get the most use out of this method. Most real-time qPCR instruments already have software that can be set up to automatically prepare a standard curve and to calculate efficiency (Fig. 3.1). However, if that feature is not available, it can easily be done by hand using an

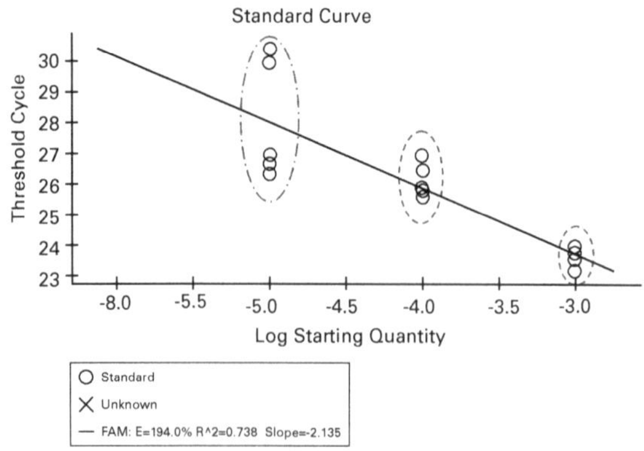

Figure 3.2 A screenshot from BioRad CFX software showing example of a bad standard curve. PCR efficiency was measured as 194.5% across three out of four 10-fold template dilutions. The slope of the curve was calculated as –2.133 and the R^2 value as 0.722. Circles are drawn to highlight replicate samples giving inconsistent Cq values, a potential outcome of inaccurate pipetting.

Excel spreadsheet by plotting a graph of the Cq values versus the log of nucleic acid input level and performing a linear regression. The PCR efficiency can then be calculated by using the gradient (or slope) of the standard curve plot in the following equation (Pfaffl, 2001):

$$\text{Efficiency} = 10^{(-1/\text{slope})-1}$$

A 100% efficient PCR reaction, equivalent to a slope of –3.32, depicts that the amount of PCR product is effectively doubling with each cycle during exponential amplification. A slope calculated to be between –3.58 and –3.10, corresponds to efficiency values of 90% to 110% respectively and is generally considered an acceptable range for a standard curve in qPCR. However, to obtain accurate and reproducible results, reactions should have efficiency as close to 100% as possible. The correlation coefficient of the line or R^2 value is a measure of how well the data fit the standard curve (Fig. 3.1), which should ideally be equal to 1, but in most cases demonstrates sufficient repeatability of a standard curve if it is greater than 0.985.

How to troubleshoot

PCR efficiency is < 90% or > 110%

If the reaction is not efficient (Fig. 3.2), then the quantity of target input, determined from the standard curve, may not be accurate. The use of non-optimal reagent concentrations and poor enzyme quality can result in efficiencies below 90%. In addition, poorly designed primers tend to generate inefficient amplification of the target sequence and so it may be sensible to redesign a new set of primers for this assay. It is good practice to try at least two or three primer pairs when initially optimizing a PCR assay. Another reason for <90% PCR efficiency could be suboptimal primer annealing conditions, for example, the temperature at which primers bind to the template is too high or the annealing time is not long enough. Optimize the thermal protocol by increasing annealing temperature in 2°C increments and annealing time in 3 s increments. Similarly, the polymerase may not be able to make PCR product efficiently because the extension time is too short, which can be solved by increasing the extension time in 5 s increments. Efficiencies greater than 110% may indicate the presence of PCR inhibitors in the reaction mix, in which case an alternative template purification strategy is recommended.

Experimental factors such as length, secondary structure, and GC content of the amplicon can also influence PCR efficiency. To improve efficiency, optimize primer and probe concentrations, use low GC rich amplicon which are unable to form secondary structures, eliminate any potential source of inhibitors and always use good quality and robust reagents.

R^2 value is ≤ 0.985

R^2 value is influenced by the pipetting accuracy and by the range of the assay. A value of ≤ 0.985 raises questions about assay reliability. If one or more points at the lowest levels of input nucleic acid are shifted away from the linear region of the plot, it is likely that this dilution is reaching the sensitivity level for an assay (Fig. 3.2). This may be due to reaching the natural limit of detection of an assay, but could equally be indicative of non-optimal conditions of the qPCR, which is likely to be the case when the calculated efficiency is also out of acceptable range. Similarly, if one or more points at the highest levels of input nucleic acid are shifted away from the linear region of the plot, it is likely that the reaction is saturated and that the level of target exceeds the useful assay range. To address this situation, add less or dilute the sample nucleic acid. Alternatively, if several random points are above or below the line, pipetting accuracy may be a problem. Verify that the pipette tips fit the pipette properly and that the volume dispensed is reproducible. The serial dilutions may not have been calculated properly which means that the line of best fit is drawn using incorrect template concentrations. Repeat the serial dilution using fresh sample material and calculate the concentrations accurately.

References

Abdulmawjood, A., Roth, S., and Bulte, M. (2002). Two methods for construction of internal amplification controls for the detection of Escherichia coli O157 by polymerase chain reaction. Mol. Cell. Probes *16*, 335–339.

Bustin, S.A. (2002). Quantification of mRNA using real-time reverse transcription PCR (RT-PCR): trends and problems. J. Mol. Endocrinol. *29*, 23–39.

Bustin, S.A., Benes, V., Garson, J.A., Hellemans, J., Huggett, J., Kubista, M., Mueller, R., Nolan, T., Pfaffl, M.W., Shipley, G.L., *et al.* (2009). The MIQE Guidelines: minimum information for publication of quantitative real-time PCR experiments. Clin. Chem. *55*, 611–622.

Bustin, S.A., and Nolan, T. (2004). Pitfalls of quantitative real-time reverse-transcription polymerase chain reaction. J. Biomol. Tech. *15*, 155–166.

Cone, R.W., Hobson, A.C., and Huang, M.L. (1992). Coamplified positive control detects inhibition of polymerase chain reactions. J. Clin. Microbiol. *30*, 3185–3189.

Freeman, W.M., Walker, S.J., and Vrana, K.E. (1999). Quantitative RT-PCR: pitfalls and potential. BioTechniques *26*, 112–122, 124–115.

Hoorfar, J., Malorny, B., Abdulmawjood, A., Cook, N., Wagner, M., and Fach, P. (2004). Practical considerations in design of internal amplification controls for diagnostic PCR assays. J. Clin. Microbiol. *42*, 1863–1868.

Huang, Z., Fasco, M.J., and Kaminsky, L.S. (1996). Optimization of Dnase I removal of contaminating DNA from RNA for use in quantitative RNA-PCR. BioTechniques *20*, 1012–1014, 1016, 1018–1020.

Klaschik, S., Lehmann, L.E., Raadts, A., Hoeft, A., and Stuber, F. (2002). Comparison of different decontamination methods for reagents to detect low concentrations of bacterial 16S DNA by real-time-PCR. Mol. Biotechnol. *22*, 231–242.

Longo, M.C., Berninger, M.S., and Hartley, J.L. (1990). Use of uracil DNA glycosylase to control carry-over contamination in polymerase chain reactions. Gene *93*, 125–128.

Mutimer, H., Deacon, N., Crowe, S., and Sonza, S. (1998). Pitfalls of processed pseudogenes in RT-PCR. BioTechniques *24*, 585–588.

Olsvik, P.A., Lie, K.K., Jordal, A.E., Nilsen, T.O., and Hordvik, I. (2005). Evaluation of potential reference genes in real-time RT-PCR studies of Atlantic salmon. BMC Mol. Biol. *6*, 21.

Pfaffl, M.W. (2001). A new mathematical model for relative quantification in real-time RT-PCR. Nucleic Acids Res. *29*, e45.

Pfaffl, M.W., Tichopad, A., Prgomet, C., and Neuvians, T.P. (2004). Determination of stable housekeeping genes, differentially regulated target genes and sample integrity: BestKeeper--EXCEL-based tool using pair-wise correlations. Biotechnol. Lett. *26*, 509–515.

Raff, T., van der Giet, M., Endemann, D., Wiederholt, T., and Paul, M. (1997). Design and testing of beta-actin primers for RT-PCR that do not co-amplify processed pseudogenes. Biotechniques *23*, 456–460.

Sachadyn, P., and Kur, J. (1998). The construction and use of a PCR internal control. Mol. Cell Probes *12*, 259–262.

Siebert, P.D., and Larrick, J.W. (1992). Competitive PCR. Nature *359*, 557–558.

Sutton, D.H., Conn, G.L., Brown, T., and Lane, A.N. (1997). The dependence of DNase I activity on the conformation of oligodeoxynucleotides. Biochem. J. *321*, 481–486.

Vandesompele, J., De Preter, K., Pattyn, F., Poppe, B., Van Roy, N., De Paepe, A., and Speleman, F. (2002). Accurate normalization of real-time quantitative RT-PCR data by geometric averaging of multiple internal control genes. Genome Biol. 3, RESEARCH0034.

Wilson, I.G. (1997). Inhibition and facilitation of nucleic acid amplification. Appl. Environ. Microbiol. *63*, 3741–3751.

Obtaining Maximum PCR Sensitivity and Specificity

4

Cameron N. Gundry and Matthew D. Poulson

Abstract

PCR is a highly sensitive and specific technique used in molecular biology laboratories everywhere. It is able to provide near 100% sensitivity and specificity with appropriately designed assays in controlled situations. However, results do not always match this potential. The most common problems in PCR arise from overlooking basic principles in assay design and optimization. Maximum PCR performance depends on key factors which include (1) choosing an appropriate detection system, (2) using available software for the best primer and probe design, (3) assessing sample quality and controlling inhibitors, (4) avoiding amplicon and environmental contamination, (5) optimizing for reagent quality and concentration, and (6) modifying the thermal cycling protocol for optimal sensitivity and specificity. This chapter will address all of these factors to aid the investigator in designing high quality PCR assays.

Introduction

Although PCR is conceptually straightforward, obtaining the best results requires careful planning and attention to a number of vital factors during experimental design and optimization. One of these factors is the fluorescent detection chemistry, which has a strong effect on the sensitivity and specificity of the assay and should be chosen with care, taking the cost per reaction, level of expertise within the lab, and other factors into consideration. Primer and probe design are equally important and the use of primer and probe design software is strongly recommended. It is also wise to verify primer specificity to the desired organism or gene using BLAST and *in silico* programs when possible. Primer and probe modifications have the potential to enhance sensitivity and increase placement options for primers and probes in difficult targets. These aspects of detection chemistry and primer design will be covered in this chapter with the aim of equipping the investigator to design high quality PCR assays.

Sample and reagent quality are also covered in this chapter. Poor sample quality often disrupts PCR performance and for this reason it is very important to use representative samples bracketing the entire range of variation in quality during assay optimization and validation.

Commercial kits and sample purification systems can improve reproducibility, but never underestimate the damage that can be done from amplicon and other environmental contamination. Buffering reagents can vary significantly in quality. Performing magnesium and other reagent titration experiments are useful for determining the optimal concentrations for these reagents.

We also discuss the importance of ensuring that any designed assay is robust. In less robust assays slight variations in amplification protocols can have a large effect on the results. For instance, lowering the annealing temperature by several degrees can reduce specificity by allowing undesirable amplification products to predominate. Machine-to-machine variability can cause this as well as insidious machine drift.

Finally, the types of controls, and number of replicates are very important in the design of PCR assays. A schematic showing the general process of PCR design and optimization is shown in Fig. 4.1.

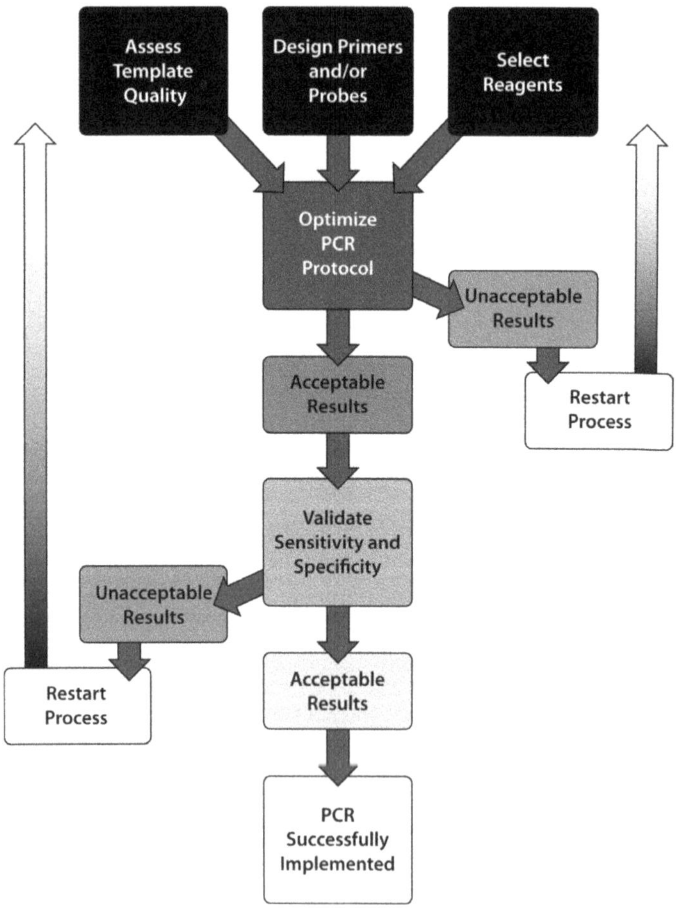

Figure 4.1 Schematic representing key considerations of PCR assay design and optimization. Although the optimization process is illustrated as stepwise, it is also iterative. For example, once an optimal magnesium concentration is identified, it is wise to re-identify an optimal annealing temperature and re-examine the limit of detection.

Definitions of sensitivity and specificity

Before moving on to the design and optimization fundamentals, some brief definitions are in order. Analytical sensitivity for qPCR assays is defined as the minimum number of target copies in a sample that can be detected using a given assay and experimental conditions. This is typically reported as limit of detection (LOD), and is usually expressed as a probability of detection, A probability of 95%, for example, means that there is only a 5% chance of failing to detect the organism or target sequence at the LOD (Bustin *et al.*, 2009). PCR sensitivity can also be reported as the percentage of samples correctly identified. For example, 98% sensitivity means 98% of samples were correctly identified while 2% were incorrectly identified as negative.

Analytical specificity refers to the ability of the assay to identify the desired target, and not undesired targets. Put another way, specificity refers to the proportion of negative samples that are correctly identified as negative, including other organisms or target sequences (or another genotype). For example, in the detection of infectious organisms, specificity would be examined by testing whether the assay detects only the desired species. If the assay provides a positive result with a non-targeted organism, a false positive has occurred. Ninety-six per cent specificity means 4% of the negative samples are false positives.

The situation can become more complicated with qPCR gene expression assays where a false positive may indicate that an empirically obtained difference in expression is not actually present. One way to minimize this risk is to use established data analysis criteria and use multiple normalization genes. Depending on the circumstances, higher sensitivity may be preferred even at the expense of false positives. On the other hand, specificity must be exceptionally high if the PCR assay is diagnostic. Optimization conditions and assay design have a profound effect on PCR sensitivity and specificity.

Choosing an appropriate detection system

Instruments

Detection systems can be defined as the instrument and fluorescent chemistry used to detect the amplified DNA. Instrument choice is a critical factor for good PCR since the quality and type of components determines the quality of data that can be generated and the types of experiment that can be performed. In general, plate-based systems have less temperature precision than air-heated systems but they allow for more replicates and increased throughput. Some instruments have better signal-to-noise ratios through more expensive optics, while factors such as the number of discrete fluorescent channels available and the levels of cross-talk between them vary considerably between different instrument types. Instruments with variable excitation and emission can assist in maximizing the signal for assays in which the choice of fluorescent dyes or probes is inflexible (for example when using commercial assays) and software support for various types of analyses is also important in maximizing assay performance. These issues are treated in more depth in other sections of this book.

Fluorescent detection chemistry

Choosing the best detection chemistry may seem daunting due to the large number of available options. However, there are really only two main classes to consider: probeless and probe-based. Lab expertise, budgetary restrictions and licensing considerations are some of the obvious drivers of the selection, but there are inherent differences between these two main categories that may steer the investigator to probeless or probe-based systems and basic understanding of the two main choices can help to make the best choice. Sensitivity and specificity can be maximized by matching the best detection chemistry with the type of PCR application.

Probeless detection strategies using double-strand DNA binding dyes are often chosen because of the associated low cost and easy assay design. These dyes increase in fluorescence upon binding to dsDNA and as the product accumulates during the PCR reaction, the fluorescence of the sample increases accordingly. The double-strand DNA binding dye ethidium bromide was first used for qPCR (Higuchi *et al.*, 1992) but has been superseded by other dyes, most commonly SYBR® Green I.

A variety of other DNA binding dyes have similar properties, each with its own benefits and drawbacks. Some dyes inherently produce more fluorescence in certain instruments because of a better spectral match between the dye and fluorimeter optics and an example of this is seen in Fig. 4.2. This greater signal improves sensitivity and precision of quantification cycle (Cq) calls in samples with low fluorescence.

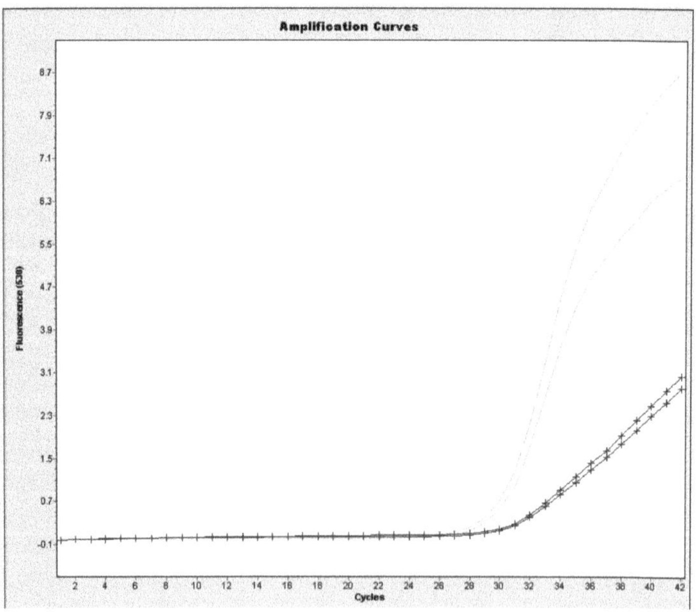

Figure 4.2 Amplification curves on a LightCycler® 1.5, in channel 1. SYBR® Green I provides better signal and earlier Cq values relative to LCGreen® Plus. A 92 basepair product from the human ABL gene was amplified in two pairs of duplicate reactions. SYBR Green I reactions are seen as smooth amplification curves and appear at lower Cq values, whereas LCGreen Plus amplification curves are seen with plus signs and are delayed. The LightCycler optical system is optimized for SYBR Green I and fluorescein emission and is not matched as well to LCGreen Plus.

But DNA binding dyes are not without their disadvantages. Because these dyes bind to any dsDNA, including the non-amplified sample DNA, qPCR signals can be reduced by a large amount of template within the reaction. This effect depends on the type and concentration of dye, as well as the relative amounts of template-to-amplicon DNA, and most often happens if there is a very large amount of sample DNA. Because of the potential to change background levels and alter Cq information due to data normalization and filtering, sensitivity is best when sample concentrations are similar. Gel visualization as the standard method to check for product size is nicely complemented by amplicon melting temperature (T_m) analysis, which is more practical for routine product differentiation (Ririe *et al.*, 1997) because it avoids contamination risks and uses inexpensive dyes. Recently, investigators have demonstrated a high-resolution melting technique as a replacement for gel analysis, even during optimization (Vossen *et al.*, 2009).

In summary, the main advantage of probeless systems is the ability to perform melting curve analysis and to easily obtain a qualitative picture of the PCR products. Positive calls by Cq can be validated with melting curve analysis, dramatically improving specificity.

Fluorescently labelled probe-based detection chemistries offer enhanced specificity due to the presence of sequence-specific oligonucleotide probes. This allows for positive/negative calls to be confidently made without melting curve analysis. Hydrolysis probes work by increasing fluorescence signal of a reporter dye as nucleolytic cleavage of the probe occurs during amplification (Holland *et al.*, 1991). This type of probe is destroyed during amplification and cannot be used for melting curve analysis. Probes are highly specific and provide information derived solely at the location of hybridization, a small section of the amplicon. This is usually seen as an advantage because 'you see only what you want to see' but it can also be a drawback for the same reason. Another disadvantage is that because probes do not provide qualitative information revealing other (undesirable) amplification products, the investigator may be blind to the causes of potentially poor assay performance. This can lead to unexpected sample dropouts if conditions change even slightly. Dual hybridization probes (Wittwer *et al.*, 1997) can be used for both real-time detection and post-PCR melt analysis (Lay *et al.*, 1997, Gundry *et al.*, 1999). With these probes the signal is generated via the fluorescence resonance energy transfer (FRET) mechanism when both probes hybridize within sufficient proximity on the target sequence. This system provides melting curves at the locus of interest but still does not allow for a complete and qualitative picture of all products. Unlabelled probes or LunaProbes™ (Zhou *et al.*, 2004) use a double stranded dye to provide genotyping probe melting profiles. Thus, this system contains added specificity; although it is possible for undesired products such as primer-dimers to masquerade as the probe-to-amplicon hybrid, as it is a one-colour system. High-resolution melting systems and advanced data analysis techniques can mitigate this danger.

Use tools for primer and probe design

Successful PCR begins with good oligonucleotide design. Software should be used for primer selection because it is easy to use and is proven to reduce primer design flaws, saving time and headache in the future. Primer 3 (Rozen and Skaletsky, 2000) is a great web primer selection resource. This software also prevents primers from being selected with significant homologies to known repeat elements within selected genomes. OLIGO 7 (Molecular Biology Insights, Inc., CO, USA), and Primer Premier (Premier Biosoft International, Palo Alto, CA, USA) are good commercial alternatives.

As primer design algorithms continue to improve, PCR will complement other techniques and unlock more regions of the genome such as repeat elements (e.g. Mann *et al.*, 2009). Accurate primer and probe melting temperature (T_m) prediction greatly assists good assay design. A $^{\Delta}T_m$ of 5–10°C is recommended for the most popular probe chemistry options (hydrolysis and dual hybridization). In the case of hydrolysis probes, too low a probe T_m results in failure to generate signal during PCR whereas dual hybridization probes may inhibit PCR if they are too stable relative to primer T_m values. Recently released primer design software programs use updated nearest neighbour thermodynamic parameters for T_m calculations (SantaLucia *et al.*, 1996; Bommarito *et al.*, 2000). Advancements in primer design software are continually being brought to the market. For example, DNA Software (www.dnasoftware.com) has released a suite of primer design programs that uses a secondary stucture folding algorithm, a multistate-coupled equilibrium model, as well as advanced nearest neighbour estimates to provide quantitative estimates of primer binding efficiencies across varying temperatures.

Primer-dimers and primer-probe products result when sufficient sequence homology exists to enable polymerization between two primers and/or probes. These inhibit PCR and can be interpreted as false positives when using double-strand dyes. It is substantially more difficult to eliminate cross-complementarity between oligonucleotides when designing multiplexed reactions so primer design software is even more essential for these types of experiments.

Checking for unwanted binding sites

A well-designed primer set should be checked using available software to see whether it is specific to the intended target. The number of primer homologies (binding sites) within a complex genome is one of the most frequent causes of PCR failure (Andreson *et al.*, 2008). BLAST, and more recently BLASTn, have been widely used to assess the number of homologies by providing primer-sized sequence alignments (in the case of BLASTn) across a huge sequence data repository. If there are many homologies within the organism to be amplified, even if they are not oriented in a fashion to promote exponential amplification, it can create problems. Many primer design programs include quick links to BLAST sites for rapid assessment of the number of bindings within an organism and across organisms.

In silico amplicon prediction to improve primer specificity

Stand-alone *in silico* primer design programs are available on the web and can be used to rapidly double-check primer specificity within a selected genome. These tools check primers in the correct orientation against the user-selected reference genome sequence and display all predicted amplicon sequences. These tools should be used during primer design and before ordering the sequences. University of California Santa Cruz (UCSC) posts a popular *in silico*-PCR tool freely available (with certain restrictions) on the web http://genome.ucsc.edu/cgi-bin/hgPcr?command=start, The National Center for Biotechnology Information (NCBI) posts Primer-Blast http://www.ncbi.nlm.nih.gov/tools/primer-blast/, which combines Primer 3 primer design capabilities with user-selected genome specificity checks and a particularly useful site for prokaryote *in silico* analysis is http://insilico.ehu.es/PCR/ (Bikandi *et al.*, 2004). In Fig 4.3 the results of an acceptable primer set entered into UCSC's *in silico* PCR tool are shown. In Fig 4.4, an unacceptable result is shown where two products

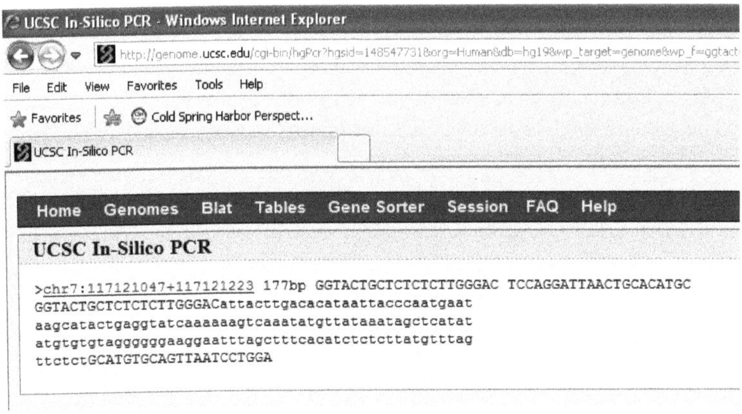

Figure 4.3 UCSC *in silico* amplicon predicts one PCR product. Human chromosome and basepair location of the amplicon is in blue text followed by size (bp) of the amplicon and the primer sequences in capital letters. If only the desired product is predicted PCR is likely to be specific (assuming only one product is desired). If more than one amplicon is predicted, PCR may be compromised. In both cases, other issues such as primer dimers and product secondary structure may affect results, so an appropriate primer design program is also necessary.

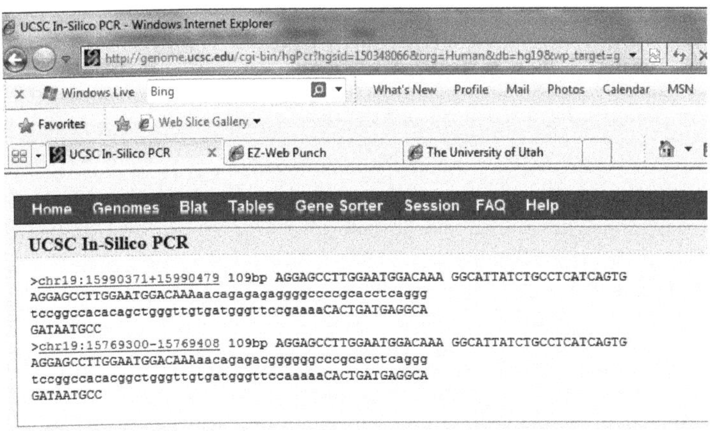

BLAT Search Results

ACTIONS	QUERY	SCORE	START	END	QSIZE	IDENTITY	CHRO	STRAND	START	END	SPAN
browser details	YourSeq	109	1	109	109	100.0%	19	+	15990371	15990479	109
browser details	YourSeq	99	1	109	109	95.5%	19	-	15769300	15769408	109

Figure 4.4 Results from UCSC *in silico* PCR tool showing dual priming to a desired and an undesired site. Although these primers may be good by other measures, they are unacceptable for an assay targeting one product. The predicted amplicons from 4A were inputted into BLAT in 4B. BLAT analysis verified that there are only two products with high homology (95.5%). This would potentially confound the interpretation of sequencing or PCR-based genotyping experiments.

(one undesired) are amplified with the same primers. Ninety-five per cent sequence similarity exists for the two targets specified by the primer set. Recently, it has been pointed out (Hatchwell, 2010) that this type of occurrence has probably caused incorrect data interpretation (Gottlieb *et al.*, 2009). It is best to eliminate this risk by using the available tools to avoid unwanted primer homologies before primer and probe sequences are synthesized.

Primer and probe modifications to increase assay sensitivity and specificity

Various primer modifications have been shown to reduce Cq cycle thresholds and increase PCR yield. For example, 12 to 25 bp of AT rich flaps on the 5′-end of primers were shown to significantly improve SYBR Green qPCR results from bisulfite-treated DNA for methylation detection (Afonina *et al.*, 2007). M13 and other non-homologous sequences added to the 5′-end of primers can also dramatically increase PCR yield when using degenerate primers to amplify across diverse taxa (Regier and Shi, 2005). In a separate type of primer modification, linkers were attached within primers to create two tethered segments that increased specificity even under suboptimal PCR conditions (Chun *et al.*, 2007).

Locked nucleic acids (LNA) modifications (Koshkin *et al.*, 1998; Obika *et al.*, 1998) show greatly enhanced thermal stability (Kaur *et al.*, 2006), compared with natural DNA nucleotides and can be used to increase primer or probe Tm. One application of this technology is to shorten primer or probe sequences, which can increase sensitivity in situations where there is a great deal of sequence variability near the desired location of the probe or primer. Recently, new guidelines have been suggested in order to improve mismatch discrimination in LNA modified oligonucleotides (You *et al.*, 2006). Many studies have shown improved PCR performance by incorporating LNA into primers and probes.

The tripeptide modification 1,2-dihydro-(3H)-pyrrolo[3,2-e]indole-7-carboxylate (CDPI3), commonly referred to as the minor groove binding (MGB) modification, is conceptually similar to LNA but have an even more dramatic effect on raising T_m in oligonucleotides used in PCR. Eight to ten bp MGB-modified primers were used to amplify viral targets in regions containing a high degree of sequence variability between species (Afonini *et al.*, 1997). MGB modifications have also allowed for very short hydrolysis probes to be designed (Kutyavin *et al.*, 2000). This can lower false positives and negatives depending on the situation. It is worth considering either LNA or MGB modifications with PCR primer and probe design when the sequence is not ideal.

Fluorescently labelled primers have been used for gel separation techniques, fragment analysis, FRET probe chemistry genotyping (Lay and Wittwer, 1997) and high-resolution mutation amplicon scanning (Gundry *et al.*, 2003). Various primer and probe modifications will likely continue to emerge as new applications push the limit of current methods.

Undesired template characteristics

Interactions within the template such as, stable hairpins, AT- or GT-rich template, and looped structures can create problems for PCR. Stable hairpins can interfere with primer binding sites decreasing amplification efficiency. Depending on the relative stabilities of the hairpins and primers this can result in a false-negative result. Raising the annealing or denaturation temperatures can destabilize these structures allowing primers and polymerase to function properly. Some primer design programs have integrated tools to predict template structure and there are various hairpin prediction tools available on the web.

Assess sample quality and consistency

Poor sample quality is often overlooked and lowers sensitivity and reproducibility. Good quality samples have adequate and consistent nucleic acid concentration, lack of extensive DNA or RNA shearing or degradation, and an absence or low amounts of PCR inhibitors. It is important to understand the effects of sample variation on PCR results. Because of its importance, sample variation and quality should be rigorously assessed *during*, not after, assay optimization.

Resolving sample condition and sample inhibition

The simplest way to eliminate PCR inhibition is often to dilute the sample, and thus the inhibitors (Wilson, 1997). However, this is not always possible. Alternatively, sample purification can be performed to simultaneously remove inhibitors and concentrate nucleic acids. Inhibition can be detected several ways including an increase in the Cq cycle, flattening of the slope, or reduction in fluorescence. A hybrid method to assess inhibition has been proposed using qPCR amplification slopes and dilutions (King *et al.*, 2009). Correct use of standard curves and controls for assessing PCR efficiency are described elsewhere in this book.

Many types of sample matrices can cause PCR inhibition. Clinical samples such as blood and stool introduce inhibitors. Plant samples contain high amounts of polysaccharides, polyphenolics, and other compounds that can inhibit PCR. Various methods have been devised to overcome this inhibition (e.g. Wilson, 1997; Kranner, 2006). Components of nucleic acid purification kits can introduce PCR inhibitors such as ethanol, salts and sodium dodecyl sulfate (SDS) among others. Thus, changing the extraction buffer (for instance, increasing the etheylenediaminetetraacetic acid (EDTA) concentration) can suddenly cause false-negative results and reduce PCR efficiency. Standardization is important and any changes in sample purification should be revalidated with the PCR assay.

Many samples come from formalin fixed tissues but the fixation process used to prepare them is known to damage to nucleic acid. The fixatives themselves as well as fixation time both have a significant impact on the nucleic acid recovery from these samples (Greer *et al.*, 1991). Ethanol and HistoChoice® are fixatives that yield superior PCR template than samples preserved with formalin (Gianella *et al.*, 1997; Gillespie *et al.*, 2002). Similarly, preserving RNA samples with commercial solutions (e.g. RNAlater, Ambion, USA) can enhance the results of gene expression assays.

When working with fragmented, old, or dilute nucleic acid samples, the most successful strategy is usually to shorten amplicon length (Sefc *et al.*, 2003). Alternatively, greater amounts of template can be used for increasingly fragmented samples. Average size estimates can be performed by performing PCR on smaller and larger fragments and measuring the ratio of PCR success (Swango *et al.*, 2005) or instruments such as the Bioanalyzer 2100 (Agilent Technologies, Santa Clara CA) can provide direct size estimation to assess RNA fragmentation.

Template concentration

Determining the optimal range of template concentrations will improve the reliability and reproducibility of the assay. Both too much and too little template can create problems in the qPCR assay. A reaction containing too much DNA can result in undesirable product formation while evaluating the qPCR assay with too little template will give inconsistent results because it is beyond the LOD for the assay. If large amounts of sample are required for results it has the potential to introduce equally large amounts of inhibitors, potentially compromising the results and lowering sensitivity.

The most common method for measuring nucleic acid concentration is to determine the absorbance at 260 nm (A_{260}). Typically, the 260/280 absorbance ratio is used as a measure of DNA or RNA purity. This is important for absolute or relative PCR quantification of samples. The disadvantage of using absorbance is that a large volume of a high concentration dsDNA solution is needed for reliable results. However, newer low volume spectrophotometers (e.g. Nanodrop, Thermo Scientific, Nanovue, GE Healthcare) only require 0.5–2 µl of sample to obtain accurate results. Using such low volumes requires care. For example, in arid environments evaporation can concentrate nucleic acids in a few seconds leading to erroneous results. In addition, contaminants found in nucleic acid preparations can affect the accuracy of the results.

PicoGreen® is a fluorescent nucleic acid stain used to quantify dsDNA. The PicoGreen method uses dilutions of a standard dsDNA to create a standard curve based on fluorescence data. Unknown sample concentration is then determined by correlating the fluorescence to the standard curve. PicoGreen is able to quantify as little as 1 ng/ml of dsDNA. Discrepancies have been found between absorbance and fluorescence methods. Even small amounts of cetyldimethylethylammonium bromide (CTAB), which is used in many DNA extraction kits, can alter PicoGreen readings in plant DNA extracts (Holden *et al.*, 2009). Therefore, depending on the sample type one method may be preferable to the other.

Additional DNA from other organisms within the sample, especially when present at high concentrations, can interfere with amplification. For example, an assay designed to detect a pathogen in human blood must be able to amplify a small amount of pathogen nucleic acid against a large background of human genomic nucleic acid. To evaluate the possible effects of human nucleic acids on pathogen detection, 1–6 µg of human DNA can be added into the PCR reaction, and any affect on the LOD can be examined. In Fig 4.5 two different assays are compared for their ability to maintain sensitivity at previously determined LOD levels. Different assays behave differently to excess DNA and some types of detection chemistry are more robust. For instance, hydrolysis probes produce good signals even if substantial amounts of undesirable products are produced. However, the large amount of resource consumption often lowers fluorescence or alters the growth slopes during qPCR.

Template type can also result in differences in assay sensitivity. If plasmid templates are used, the linear form of the template is typically a better substrate than the circular form. In our experience, we have found that for pathogen detection assays, the LOD decreased from 3- to 9-fold for assays when linearized plasmid templates were used. However, for other assays, there was no appreciable improvement in the LOD.

Avoid amplicon and environmental contamination

Because of the high sensitivity inherent to PCR even small amounts of amplicon or other environmental contamination can cause false positives. Monitoring for amplicon contamination can be performed simply by swabbing areas, suspending the swabs in buffer, and using this buffer as a template for PCR. Positive qPCR results indicate contamination

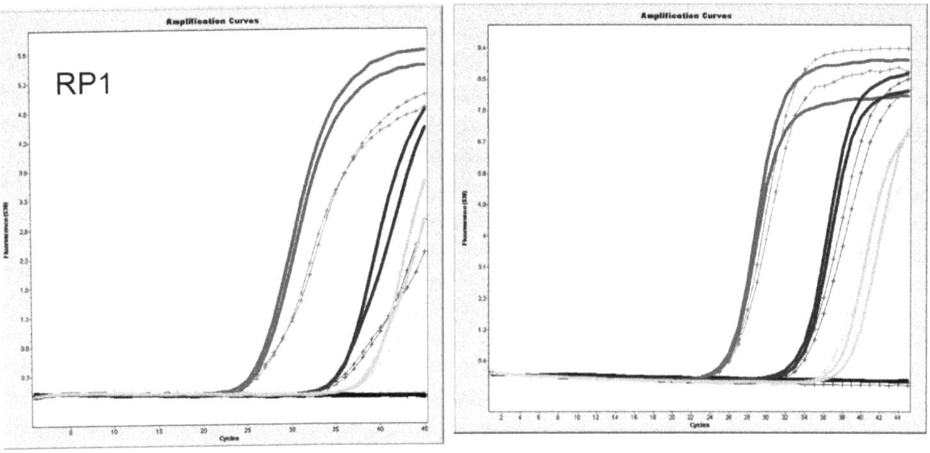

Figure 4.5 Amplification curves of two hydrolysis probe targets. The dark grey amplification curves contained 10,000 genomic equivalents (GE) of microorganism template DNA, black amplification curves contained 50 GE, and light grey curves contained 5 GE. No template controls are also seen as black flat lines. Additional human DNA (1.5 µg) was added in thin-lined curves with plus signs. Each of these two exemplary hydrolysis probe reactions demonstrates varying tolerance to excess non-template DNA. In assay 1 (the upper panel), LOD is compromised with the addition of additional non-template DNA, represented by one negative reaction and one positive. In assay 2 (the lower paned) LOD is not affected by the same additional amount of non-template human DNA.

and the Cq may indicate the level of contamination. The source of contamination must be determined and eliminated and for this purpose, ultraviolet (UV) light is often used to decontaminate PCR work surfaces by dimerization of pyridimine residues (Ou *et al.*, 1991), while commercial acid-based DNA decontamination solutions, as well as bleach are also effective. A 15 min immersion in 3% weight-to-volume (w/v) sodium hypochlorite was found to be required to completely destroy a 500 bp amplicon (Kemp and Smith, 2005). As usual, being proactive is better than reactive. Physical separation between the PCR set-up areas and post-PCR areas is advisable along with appropriate environmental controls (PCR hoods, gloves, gowns, etc.). Amplicon contamination is the primary reason to include negative and positive controls in every run and it is also a good idea to minimize post-PCR manipulations. Any activity involving open manipulation of high copy template materials should be performed in a separate room or hood with minimal foot traffic. Another solution is to use dUTPs instead of dTTPs along with pretreatment of uracil-N-glycosylase (UNG) before PCR begins (Longo *et al.*, 1990).

Select quality reagents and optimize reagent concentration

Using 'hot-start' reagents like anti-*Taq* antibodies or heat activated polymerases, primers, or dNTPs can increase PCR specificity. However, finding optimal concentrations of common reagents such as magnesium, dNTPs, or polymerase can also improve PCR. Magnesium titrations are commonly performed at the beginning of optimization.

Magnesium titration

Cations increase duplex stability by lessening the negatively charged phosphate backbone repulsion between strands of DNA. In addition, magnesium is a required co-factor of *Taq* polymerase and fluctuations in this reagent can greatly affect PCR. Too much magnesium can inhibit PCR and increase the error rate of polymerase (Markoulatos *et al.*, 2002). Optimal magnesium concentrations range from 1 mM to 6 mM, and the final concentrations vary significantly between commercial mastermixes. For qPCR reactions using dual hybridization probes, we have found that optimizing magnesium concentration can alter Cq and maximum fluorescence in interesting ways. For example, magnesium optimization of some assays resulted in an earlier Cq, but the fluorescence remained unchanged. For other assays, magnesium optimization resulted in higher fluorescence, but the same Cq. Still other assays resulted in both earlier Cq and higher maximum fluorescence upon magnesium optimization. Using high magnesium concentrations can lead to increased sensitivity, but typically reduces specificity because the primers have an increased potential to bind to mismatched regions. Therefore, it is important to re-examine the specificity of a reaction when high magnesium concentrations are used.

dNTP concentration

dNTPs are the building blocks of the PCR amplicon. However, they are also a required co-factor of polymerase and an imbalance of dNTPs may reduce the overall fidelity of the enzyme (Kunz *et al.*, 1991). Fig 4.6 shows the results of three different concentrations of dNTPs in a qPCR assay.

Special reagents for difficult templates

Adjuvants such as betaine, dimethyl sulfoxide (DMSO), and polyoxyethylene (20) sorbitan monolaurate (Tween 20) (Henke *et al.*, 1997; Demke *et al.*, 1992) can be used to increase PCR efficiency, especially when amplifying from difficult template such as high GC regions. A special high-pH (9.0) buffer containing DMSO, Tween and betaine, at 1 M, 2%, and 1.5%, respectively, resulted in more reproducible post-PCR high-resolution melting results from bisulfite-treated DNA (Bliek, 2009).

Primer and probe quality and concentration

Primers are essential for PCR. A wide variation in purity can be found among the different post-synthesis purification techniques. Highly purified primers can sometimes have a dramatic effect on assays. In Fig 4.7 two representative melting curves demonstrate the large effect that increased purity can have. Primer concentration also can have a dramatic effect on PCR. While increasing primer concentration may decrease Cq, it also increases primer-dimers and undesirable amplification products.

The sensitivity of PCR reactions using dual hybridization probes can be enhanced by increasing the concentrations of the probes. Although we have not found that increasing probe concentration results in an earlier Cq, it does increase fluorescence, which can increase sensitivity in samples already compromised with inhibition or low template levels.

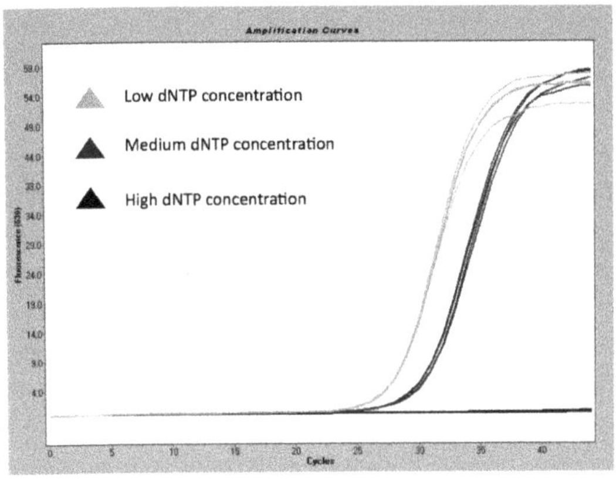

Figure 4.6 Effect of dNTP variation on qPCR. Low (light grey), medium (dark grey), and high dNTP concentrations (black) are shown. In this case, low dNTP concentrations (0.12 mM each dNTP, green) result in earlier Cqs (probably because of more magnesium availability). Medium dNTP concentrations (0.20 mM each dNTP, red) cause slightly delayed Cq values but also slightly higher end fluorescence. High dNTP concentrations (0.28 mM each dNTP, blue) fail to amplify, presumably due to lack of magnesium availability. Total Mg^{2+} concentration is 2mM final. The authors wish to thank Ranae Lems (Idaho Technology) for providing the data for this figure.

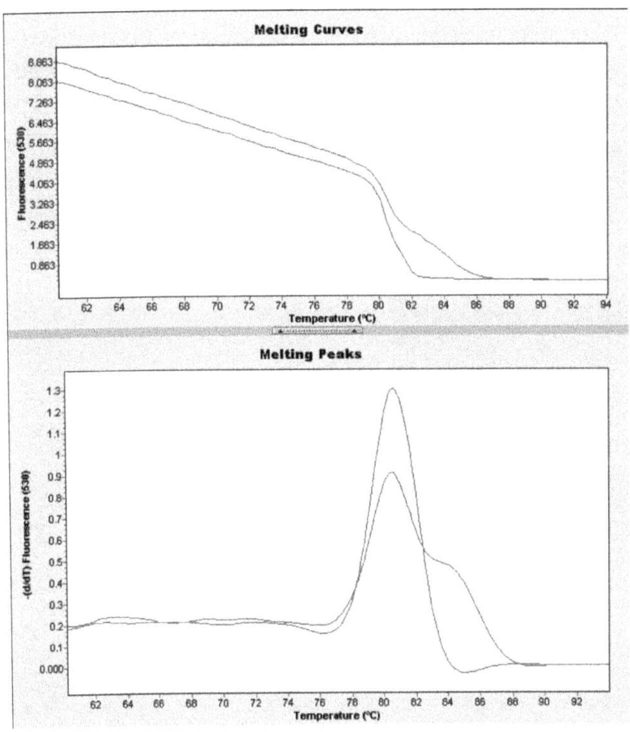

Figure 4.7 Melting curves and normalized derivative melting peaks showing the difference that primer purification can make. HPLC-purified primers were used in the clean peak sample whereas desalted only primers were used in the sample with a high temperature shoulder, which results from additional undesired product generation.

Enzyme selection

Hot-start methods improve PCR performance by preventing dNTP incorporation at the lower less stringent temperatures where primers are capable of hybridization to imperfectly matched template. Although the original hot-start methods were based on physical separation, current commercial hot-start mastermixes usually contain chemically modified (Kebelmann-Betzing *et al.*, 1998) enzymes or anti-*Taq* antibodies (Kellogg *et al.*, 1994). Hot start methods improve low copy reactions, thus increasing sensitivity (Chou, 1992). Most primer dimers are caused at room temperature or low temperatures and can be eliminated by using a hot-start method. However, not all hot-start systems are 100% efficient. Besides this possible limitation, primer dimers may still reduce PCR efficiency even if they are not extendable.

Optimize the PCR protocol

Fast PCR cycling (30 cycles in 10–30 min) with highly accurate thermal control can increase product specificity and yield partly because less time is spent at lower temperatures, where undesirable priming can occur. (Wittwer *et al.*, 1994). Although the annealing temperature is probably the single most important parameter that affects sensitivity (Roux, 1995), denaturation temperatures can also affect the results. For example, too high and too long denaturation damages cytosine residues, leading to deamination and can also damage the polymerase reducing PCR performance. For example, Wittwer and Garling (1991) reported that PCR product yield decreased with denaturation times over 16 seconds .

Even with careful primer design, thermal cycling conditions need to be optimized. An easy first step is to perform an annealing temperature gradient. An example of an annealing gradient PCR for optimization is shown in Fig 4.8. Many of the plate-based thermal cycling instruments are capable of performing temperature gradients. The annealing gradient

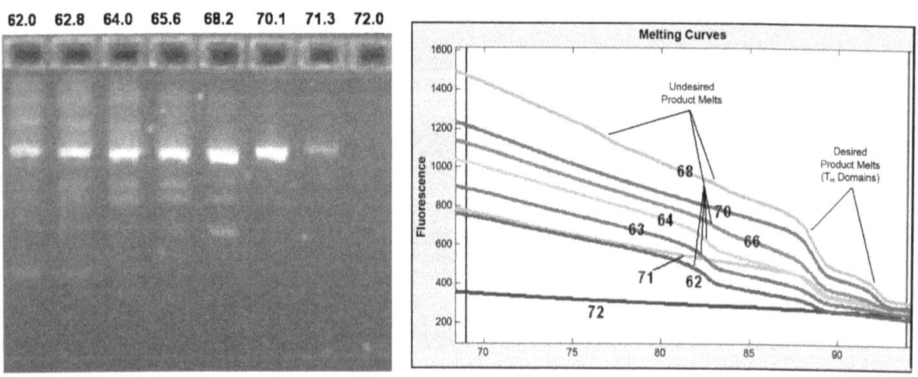

Figure 4.8 Gradient annealing optimization. Optimizing for maximum sensitivity can include gel and melting analysis. The gel image correlates with the melting profiles. Gel analysis enables identification of candidate conditions for PCR. In this example, annealing temperatures of 62–68.2°C show undesired products as visualized by extra bands. 71.3°C and 72°C generate faint single bands, while 70.1°C generates a robust single band, the best annealing temperature for PCR using this primer set. Corresponding melting profiles support the gel results showing undesired (often referred to as non-specific) product at the lower temperatures and the desired product melting profile at 70.1°C.

experiment allows multiple primer pairs to be optimized on a single plate. To set up the annealing gradient, set the gradient to cover ± 10°C of the calculated primer T_m. By varying the annealing temperature in each column or row, you create 8 or 12 discrete annealing temperatures for each primer pair on one plate. SteadySlope® technology (Eppendorf) ensures equivalent heating and cooling temperature ramping rates in order to maximize the chances that results from gradient experiments are reproducible and directly transferable to routine PCR protocols. The best primer sets are those that give clean products over the widest annealing temperature range.

If an assay is being designed for use across multiple instruments, it is important to test the assay's ability to tolerate a variety of annealing temperatures. This is particularly important if the assay will be used in clinical settings. For some assays, small differences in annealing temperature can result in failure to amplify, particularly near the LOD. A simple experiment to examine this possibility is to test annealing temperatures ± 2°C from the ideal temperature. If sensitivity or specificity drops, it would be wise to consider assay redesign. This type of test does not replace annealing gradient optimization but can be used as a valuable check after other parameters have been determined

Special PCR protocols

PCR protocol optimization is crucial and numerous suggestions have been made in the literature. Designing primers with similar stabilities (T_m within one or two degrees) is the most obvious strategy. This allows for a single PCR protocol to be used with multiple primer designs. The step-down or touch-down PCR technique (Don, 1991) is more stringent at the beginning cycles of PCR (Hecker and Roux, 1996), reducing the amplification of undesired products (Hecker *et al.*, 1996). The nested PCR technique has been shown to dramatically increase specificity (Portor-Jordan *et al.*, 1990). Nested multiplex PCR protocols are also attractive for simultaneous detection of multiple organisms. However, transferring the sample between the first and second PCR is not desirable from a contamination standpoint.

Conclusion

Careful attention should be given to each of these key considerations for maximal PCR success. In order to obtain the best PCR sensitivity and specificity the investigator should always strive to make the cleanest PCR products possible, and should guide assay design, optimization, and validation towards that goal.

As more genomes become sequenced, PCR will play an increased role in discovering the complex relationships within living systems. The ability to run more replicates (e.g. using the LightCycler 1536 instrument) will translate into better statistical conclusions from the data, which will allow for a greater number of reference genes to be used in gene expression studies. Better data quality will always drive more accurate conclusions. Better data analysis methods and standardization (Bustin *et al.*, 2009) will accelerate the development of new PCR methods. Finally, complete closed system solutions that integrate sample purification, PCR detection, and data analysis are finding their place in clinical settings. These systems provide high sample throughput (GeneXpert, Cepheid) and massively multiplexed pathogen detection (FilmArray™, Idaho Technology) results to healthcare decision makers. As in the ongoing sequencing revolution, more informative data per run will be expected of PCR.

Techniques and PCR systems that can deliver more meaningful data will continue to lead the way in the discovery and understanding of the role DNA plays in the symphony of life.

Acknowledgements

The authors wish to thank Kent Moyle and Mark Kessler for assistance with Figs 4.1 and 4.8. The authors also thank Haleigh Millward and Irene Ota for assistance with some portions of the writing. The authors also wish to thank Lindsey Cutler, Y. Ranae Lems, and Mikeal Wall for providing data for Figs 4.6 and 4.8. Data for Figs 4.2 and 4.7 were generated by the authors under NIH funding for the grant entitled 'Quantitative RT-PCR: Prediction Lymphoma Transformation' #1R42CA113212-01.

References

Afonina, I., Ankoudinova, I., Mills, A., Lokhov, S., Huynh, P., and Mahoney, W. (2007). Primers with 5′ flaps improve real-time PCR. BioTechniques *43*, 770–774.

Andreson, R., Möls, T., and Remm, M. (2008). Predicting failure rate of PCR in large genomes. Nucleic Acids Res. *36*, e66.

Bikandi, J., San Millán, R., Rementeria, A., and Garaizar, J. (2004). In silico analysis of complete bacterial genomes: PCR, AFLP-PCR, and endonuclease restriction. Bioinformatics *20*, 798–799.

Bliek, Aberrant, J. genomic imprinting in chromosome 11p15-associated congenital growth disorders: consequences for DNA-diagnostics. (2009). PhD dissertation. University of Amsterdam.

Bommarito, S., Peyret, N., and SantaLucia, Jr, J. (2000). Thermodynamic parameters for DNA sequences with dangling ends. Nucleic Acids Res. *28*, 1929–1934.

Bustin, S.A., Benes, V., Garson, J.A., Hellemans, J., Huggett, J., Kubista, M., Mueller, R., Nolan, T., Pfaffl, M.W., Shipley, G.L., Vandesompele, J., and Wittwer, C.T. (2009). The MIQE Guidelines: minimum information for publication of quantitative real-time PCR experiments. Clin. Chem. *55*, 611–622.

Chou, Q., Russell, M., Birch, D.E., Raymond, J., and Bloch, W. (1992). Prevention of pre-PCR mispriming and primer dimerization improves low-copy-number amplifications. Nucleic Acids Res. *20*, 1717–1723.

Chun, J.Y., Kim, K.J., Hwang, I.T., Kim, Y.J., Lee, D.H., Lee, I.K., and Kim, J.K. (2007). Dual priming oligonucleotide system for the multiplex detection of respiratory viruses and SNP genotyping of CYP2C19 gene. Nucleic Acids Res. *35*, e40.

Demeke, T., and Adams, R.P. (1992). The effects of plant polysaccharides and buffer additives on PCR BioTechniques *12*, 332–334.

Don, R., Cox, P., Wainwright, B., Baker, K., and Mattick, J. (1991). Touchdown PCR to circumvent spurious priming during gene amplification. Nucleic Acids Res. *19*, 4008.

Giannella, C., Zito, F.A., Colonna, F., Paradiso, A., Marzullo, F., Alaibac, M., and Schittulli, F. (1997). Comparison of formalin, ethanol, and Histochoice fixation on the PCR amplification from paraffin-embedded breast cancer tissue. Eur. J. Clin. Chem. Clin. Biochem. *35*, 633–635.

Gillespie, J.W., Best, C.J., Bichsel, V.E., Cole, K.A., Greenhut, S.F., Hewitt, S.M.,. Ahram, M., Gathright, Y.B., Merino, M.J., Strausberg, R.L., Epstein, J.I., Hamilton, S.R., Gannot, G., Baibakova, G.V., Calvert, V.S., Flaig, M.J., Chuaqui, R.F., Herring, J.C., Pfeifer, J., Petricoin, E.F., Linehan, W.M., Duray, P.H., Bova, G.S., and Emmert-Buck, M.R. (2002). Evaluation of non-formalin tissue fixation for molecular profiling studies. Am. Pathol. J. *160*, 449–457.

Gottlieb, B., Chalifour, L.E., Mitmaker, B., Sheiner, N., Obrand, D., Abraham, C., Meilleur, M., Sugahara, T., Bkaily, G., and Schweitzer, M. (2009). BAK1 gene variation and abdominal aortic aneurysms. Hum. Mutat. *30*, 1043–1047.

Greer, C.E., Peterson, S.L., Kiviat, N.B., and Manos, M.M. (1991). PCR amplification from paraffin-embedded tissues. Effects of fixative and fixation time. Am. J. Clin. Pathol. *95*, 117–124.

Gundry, C.N., Bernard, P.S., Herrmann, M.G., Reed, G.H., and Wittwer, C.T. (1999). Rapid F508del and F508C assay using fluorescent hybridization probes. Genet. Test. *3*, 365–370.

Gundry, C.N., Vandersteen, J.G., Reed, G.H., Pryor, R.J., Chen, J., and Wittwer, C.T. (2003). Amplicon melting analysis with labeled primers: a closed-tube method for differentiating homozygotes and heterozygotes. Clin. Chem. *49*, 396–406.

Hatchwell, E. (2010). BAK1 gene variation and abdominal aortic aneurysms – Variants are likely due to sequencing of a processed gene on chromosome 20. Hum. Mutat. *31*, 108–109.

Hecker, K., and Roux, K. (1996). High and low annealing temperatures increase both specificity and yield in touchdown and stepdown PCR. BioTechniques *20*, 478–485.

Henke, W., Herdel, K., Jung, K., Schnorr, D., Loening, S.A.(1997). Betaine improves the PCR amplification of GC-rich DNA sequences. Nucleic Acids Res. *25*, 3957–3958.

Higuchi, R., Dollinger, G., Walsh, P.S., and Griffith, R. (1992). Simultaneous amplification and detection of specific DNA sequences. BioTechnology *10*, 413–417.

Holden, M.J., Haynes, R.J., Rabb, S.A., Satija,N., Yang, K., Blasic, Jr, J.R.. (2009). Factors affecting quantification of total DNA by UV spectroscopy and PicoGreen fluorescence. J. Agric. Food Chem. *57*, 7221–7226.

Holland, P.M., Abramson, R.D., Watson, R., Gelfand, D.H. (1991). Detection of specific polymerase chain reaction product by utilizing the 5′→ 3′ exonuclease activity of Thermus aquaticus DNA polymerase. Proc. Natl. Acad. Sci. U.S.A. *88*, 7276–7280.

Kaur, H., Arora, Wengel, A., Maiti, J., S. (2006). Thermodynamic, Counterion, and hydration effects for the incorporation of locked nucleic acid nucleotides into DNA duplexes. Biochemistry *45*, 7347–7355.

Kebelmann-Betzing, C., Seeger, K., Dragon, S., Schmitt, G., Möricke, A., Schild*1*, T.A., Henze, G., and Beyermann, B. (1998). Advantages of a new *Taq* DNA polymerase in multiplex PCR and time-release PCR. BioTechniques *24*, 154–158.

Kellogg, D.E., Rybalkin, I., Chen, S., Mukhamedova, N., Vlasik, T., Siebert, P.D., and Chenchik, A. (1994). *Taq*Start Antibody: hot start PCR facilitated by a neutralizing monoclonal antibody directed against *Taq* DNA polymerase. BioTechniques *16*, 1134–1137.

Kemp, B.M., and Smith, D.G. (2005). Use of bleach to eliminate contaminating DNA from the surface of bones and teeth. Forens. Sci. Int. *154*, 53–61.

Kermekchiev, M.B., Kirilova, L.L., Vail, E.E., and Barnes, W.M. (2009). Mutants of *Taq* DNA polymerase resistant to PCR inhibitors allow DNA amplification from whole blood and crude soil samples. Nucleic Acids Res. *37*, e40.

King, C., Debruyne, R., Kuch, M., Schwarz, C., Poinar, H. (2009). A quantitative approach to detect and overcome PCR inhibition in ancient DNA extracts. BioTechniques *47*, 941–949.

Koshkin, A.A., Singh, S.K., Nielsen, P., Rajwanshi, V.K., Kumar, R., Meldgaard, M., Olsen, C.E., and Wengel, J. (1998). LNA (locked nucleic acids): synthesis of the adenine, cytosine, guanine, 5-methylcytosine, thymine and uracil bicyclonucleoside monomers, oligomerisation, and unprecedented nucleic acid recognition. Tetrahedron 54,3607–30.

Kranner, B.S. (2006). Isolation of high-quality RNA from polyphenol-, polysaccharide- and lipid-rich seeds. Phytochem. Anal. *17*, 144–148.

Kunz, B.A., and Kohalmi, S.E. (1991). Modulation of mutagenesis by deoxyribonucleotide levels. Annu. Rev. Genet. *25*, 339–359.

Kutyavin, I.V., Afonina, I.A., Mills, A., Gorn, V.V., Lukhtanov, E.A., Belousov, E.S., Singer, M.J., Walburger, D.K., Lokhov, S.G., Gall, A.A., Dempcy, R., Reed, M.W., Meyer, R.B., and Hedgpeth, J. (2000). 3′-minor groove binder-DNA probes increase sequence specificity at PCR extension temperatures. Nucleic Acids Res. *28*, 655–661.

Lay, M.J., and Wittwer, C.T. (1997). Real-time fluorescence genotyping of factor V Leiden during rapid-cycle PCR. Clin. Chem. *43*, 2262–2267.

Longo, M.C., Berninger, M.S., and Hartley, J.L. (1990). Use of uracil DNA glycosylase to control carry-over contamination in polymerase chain reactions. Gene *93*, 125–128.

Mann, T., Humbert, T., Dorschner, M., Stamatoyannopoulos, J., and Noble, W.S. (2009). A thermodynamic approach to PCR primer design. Nucleic Acids Res. *37*, e95.

Markoulatos, P., Siafakas, N., and Moncany, M. (2002). Multiplex polymerase chain reaction: a practical approach. J. Clin. Lab. Anal. *16*, 47–51.

Obika, S., Nanbu, D., Hari, Y., Andoh, J.I., Morio, K.I., Doi, T., and Imanishi, T. (1998). Stability and structural features of the duplexes containing nucleoside analogues with a fixed N-type conformation, 2′-O,4′-C-methyleneribonucleosides. Tetrahedron Lett. *39*, 5401–5404.

Ou C.Y., Moore, J.L., and Schochetman, G. (1991). Use of UV irradiation to reduce false positivity in polymerase chain reaction. BioTechniques *10*, 442, 444, 446.

Regier, J.C., and Shi, D. (2005). Increased yield of PCR product from degenerate primers with nondegenerate, nonhomologous 5′ tails. BioTechniques 38, 34–38.

Ririe, K.M., Rasmussen, R.P., and Wittwer, C.T. (1997). Product differentiation by analysis of DNA melting curves during the polymerase chain reaction. Anal. Biochem. *245*, 154–160.

Roux, K.H. (1995). Optimization and troubleshooting in PCR. Genome Res. *4*, S185–S194.

Rozen, S., and Skaletsky, H.J. (2000). Primer3 on the WWW for general users and for biologist programmers. In Bioinformatics Methods and Protocols: Methods in Molecular Biology, Krawetz, S. and Misener, S., eds. (Totowa, NJ, USA: Humana Press), pp. 365–386.

SantaLucia, Jr, J., Allawi, H.T., and Seneviratne, A. (1996). Improved nearest-neighbor parameters for predicting DNA duplex stability. Biochemistry *35*, 3555–3562.

Sefc, K.M., Payne, R.B., and Sorenson, M.D. (2003). Microsatellite amplification from museum feather samples: effects of fragment size and template concentration on genotyping errors. The Auk *120*, 982–989.

Swango, K.L., Timkena, M.D., Chonga, M.D., and Buoncristiania, M.R. (2005). A quantitative PCR assay for the assessment of DNA degradation in forensic samples. Forensic Sci. Int. *158*, 14–26.

Vossen, R.H., Aten, E., Roos, A., den Dunnen, J.T. (2009). High-resolution melting analysis (HRMA): more than just sequence variant screening. Hum. Mutat. *30*, 860–866.

Wilson, I.G. (1997). Inhibition and facilitation of nucleic acid amplification. Appl. Environ. Microbiol. *63*, 3741–3751.

Wittwer, C.T., and Garling, D.J. (1991). Rapid cycle DNA amplification: time and temperature optimization. BioTechniques *10*, 76–83.

Wittwer, C.T., Reed, G.B., and Ririe, K.M. (1994). Rapid cycle DNA amplification. In The Polymerase Chain Reaction, Mullis, K., Ferre, F., and Gibbs, R., eds. (Deerfield Beach, FL: Springer-Verlag), pp. 174–181.

Wittwer, C.T., Ririe, K.M., Andrew, R.V., David, D.A., Gundry, R.A., and Balis, U.J. (1997). The LightCycler: A microvolume multisample fluorimeter with rapid temperature control. BioTechniques *22*, 176–181.

You, Y., Moreira, B.G., Behlke, M.A., and Owczarzy, R. (2006). Design of LNA probes that improve mismatch discrimination. Nucleic Acids Res. *34*, e60.

Zhou, L., Myers, A.N., Vandersteen, J.G., Wang, L., and Wittwer, C.T. (2004). Closed-tube genotyping with unlabeled oligonucleotide probes and a saturating DNA dye. Clin. Chem. *50*, 1328–1335.

RT-PCR Optimization Strategies 5

Martina Reiter and Michael W. Pfaffl

Abstract

PCR technology is based on a simple principle; an enzymatic reaction that increases the amount of nucleic acids initially present in a sample. This powerful method makes it possible to detect specific mRNA transcripts in any biological sample by the application of RT-PCR. The RT-PCR quantitative analysis workflow has several steps, each of which is crucial to the success of the experiment. It starts with a sampling step, followed by nucleic acid extraction and stabilization, cDNA synthesis and finally the qPCR where the mRNA quantification takes place. PCR itself is quite a stable reaction with reproducibility between 2–5% but the number and nature of the pre-PCR steps mean that there are many sources of experimental variance in the workflow. Reliable data can be produced only when the experimental variance is minimized, so the sources of noise must be identified and optimized for each step of each experiment. Typically, the pre-PCR steps are often neglected and optimization is done for PCR reaction only.

In this chapter the optimization of the whole RT-PCR workflow will be discussed and recommendations to reduce experimental variance are shown to produce more reproducible and reliable results are put forward.

Pre-PCR steps and optimization of RNA quality

In RT-qPCR experiments a special focus should be on the pre-PCR steps, because they can be major sources of error, especially during tissue sampling, RNA storage, RNA extraction, and quality assurance procedures. High RNA integrity guarantees valid mRNA and microRNA quantification, but fragmented RNA has a substantial impact on qPCR results and the conclusions drawn from them.

The validity and accuracy of gene expression evaluation by RT-qPCR is known to be profoundly influenced by the quality, the purity and integrity, of the starting RNA (Fleige *et al.*, 2006). Low-quality RNA is highly likely to compromise the results of downstream applications which are often labour-intensive, time-consuming and highly expensive (Raeymarkers, 1993, Imbeaud *et al.*, 2005). Thus, for any taking steps to obtain and preserve high quality, intact and not fragmented RNA, is essential.

This is especially true in clinical applications for which tissues are obtained via surgery or biopsy or after laser micro–dissection and limited sample material is available. It is well known that RNA is sensitive to degradation by inadequate sample handling, stabilization, or storage as well as by postmortem processes (Perez-Novo *et al.*, 2005). RNA extraction from adipose, connecting or collagen-rich tissues often provides samples of a lower yield and quality, which invariably contain partly degraded RNA sub-fractions (our unpublished results). To circumvent this, particular steps during the tissue handling procedure have to be carefully controlled in order to preserve the purity, quality and integrity of the RNA material and reliable RNA quality analysis is necessary prior to quantitative analysis (Bustin and Nolan, 2004). Thus, the ability to quickly assess RNA quality in small samples has become increasingly important (Becker *et al.*, 2010; Fleige *et al.*, 2006).

The quantity and quality of purified RNA varies from sample to sample and, due to the inherent instability of RNA, tends to decline during storage. Especially long mRNAs (up to multiple kilobases) are very sensitive to degradation and will be easily cleaved by nucleases or sheared during the purification process. This can happen through cleavage by endogenous RNAses or exogenous RNAses introduced by handling with tissue samples or RNA extracts (Pfaffl and Fleige, 2006).

All steps of the RNA workflow including tissue sampling and storage, RNA stabilization, storage, extraction, dilution, and quantity and quality checks must be carried out with great care. In order to be considered as 'high quality', RNA samples must fulfil the following criteria (Bustin and Nolan, 2004; Pfaffl, 2005):

- Be free of RNAses.
- Be free of protein (verified at absorbance 260 nm/280 nm).
- Be free of genomic DNA.
- Total RNA should be not fragmented (28S/18S rRNA ratio should be roughly between 1.8 and 2.0) and the RIN/RQI should be at least 5 (Fleige *et al.*, 2006; Becker *et al.*, 2010).
- Be free of enzymatic inhibitors (verified at absorbance 260 nm/230 nm or 260 nm/320 nm) in later RT and PCR reaction. This is strongly dependent on the purification or clean-up methods employed and on the experience of the laboratory personal. Inhibitors can be tested by SPUD assay (Nolan *et al.*, 2006).
- Be free of any substances that complex essential reaction co-factors (e.g. Mg^{2+} or Mn^{2+}, or any nucleases which might degrade RNA).

Further problems can arise during tissue sampling before the tissue is fixed and stored safely in RNase stabilizers [e.g. PAXgene™ (PreAnalytix), LeukoLock (Applied Biosystems), RNALater (Ambion/Qiagen), or PrepProtect (Miltenyi Biotec)]. It is often very challenging to decrease the sampling time to a minimum within the framework of clinical routine sampling to prevent this. RNA quality may also be impaired in samples stored for long periods or under sub-optimal conditions, e.g. in wrong buffer solution or at elevated temperatures (Schoor *et al.*, 2003; Schroeder *et al.*, 2006).

To verify RNA quality, purity and integrity in a standardized manner, we recommend lab-on-chip automated capillary electrophoresis before any RNA based analysis. Various instruments are commercially available to accomplish this, e.g. 2100 Bioanalyzer (Agilent Technologies) and Experion (Bio-Rad). Both of these devices are sensitive, give highly reproducible data and are suitable for reliable quality control of RNAs over a wide concentration and integrity range (Pfaffl *et al.*, 2008).

Overall RNA quality can be determined using an electropherogram and a virtual RNA gel image. From this the ratio of both major ribosomal RNA (rRNA) 28S and 18S subunits can be calculated. Although this serves as a RNA quality marker it is reported, that the 28S/18S rRNA ratio is not a reliable marker for total RNA integrity (Fleige *et al.*, 2006; Pfaffl *et al.*, 2008). Instead the standard RNA integrity and quality markers the RNA integrity number (RIN) and the RNA quality index (RQI) should be used. To simplify the assessment of RNA integrity for the researcher these are typically calculated by the instrument software. A RQI/RIN value of 1 represents completely fragmented and degraded RNA, while a value of 10 represents intact RNA (Schroeder *et al.*, 2006). Previous studies have shown that a RIN/RQI of at least five is sufficient for obtaining reliable PCR results (Fleige and Pfaffl, 2006) and in a recent study, we demonstrated that this threshold value seems also sufficient for quantitative microRNA analysis (Becker *et al.*, 2010).

Significant degradation of target RNA can result in an artificially high quantification cycle (Cq) value, leading to an underestimation of its concentration and the initial copy number. This can be exacerbated by differential sensitivities of mRNAs to degradation and cause an enormous impact on the interpretation of qPCR results since different reference genes appear to be suitable for normalization in degraded versus intact mRNA or microRNA samples. In addition, the degree of dependence can vary between samples, emphasizing the need to understand the relationship between RNA quality and the results in any given study.

MicroRNA is affected to a lesser degree than mRNA, presumably because its small size reduces the ability of RNA nucleases to attack it. Nevertheless, the verification of mRNA and microRNA integrity is recommended prior to any quantitative RNA analysis, although the application of an appropriate normalization method can partly reduce the impact of degradation in qRT-PCR experiments (Fleige *et al.*, 2006; Becker *et al.*, 2010).

PCR detection chemistry

The market for PCR reagents is constantly increasing. Providers now offer a portfolio of chemistries for the whole experimental set-up, from sampling to reverse transcription to final PCR reaction. Chemical composition and concentrations are well established in the offered kits, therefore laborious optimization procedures at the bench can be avoided. But the downside of pre-optimized kits is their limited flexibility in allowing the user to optimize the reaction in order to overcome poor sample quality or eliminate non-specific PCR products. Nevertheless, even with these kits, there are options available for the user to influence the experiment and its performance and these are detailed in this section.

DNA polymerase

The DNA polymerase is the 'workhorse' in the PCR reaction and choice of polymerase type can have a profound effect on the success of a reaction. The first enzyme used for PCR was DNA polymerase I isolated from *E. coli* but today many different DNA polymerases boasting a variety of features, are commercially available.

The basic DNA synthesis process, through the Mg^{2+}-dependent 5′ to 3′ incorporation of free dNTPs to build new complementary second DNA strand, is common to all of the DNA polymerases. But for accurate DNA synthesis, and thus low error-rate PCR amplification, it is essential that the enzyme has a 3′–5′ exonuclease (proofreading) activity, which decreases the rate of incorporation false paired bases.

The first available thermostable DNA polymerase was the *Taq* polymerase isolated from *Thermus aquaticus* which is thermostable but lacks any proofreading activity. For higher conversion rates and long strand synthesis T4 and T7 polymerases from *E. coli* phage T4 and T7 are commonly used, although these are not thermostable. *Pwo* (*Pyrococcus woresei*) and *Pfu* (*Pyrococcus furiosus*), on the other hand, are thermostable polymerases that exhibit good PCR amplification fidelity and proofreading activity, which makes them generally good choices for qRT-PCR. For some special applications like amplifying GC-rich templates, polymerase mixes are often used to combine the positive characteristics of the different enzyme types (Table 5.1) (Al-Soud *et al.*, 1998, Eckert *et al.*, 1991, 1993).

'Hot start' polymerases, which were developed to increase specificity and convenience of the PCR reaction, are another tool worth considering. Hot start polymerases are produced by reversibly blocking the polymerase through chemical modification, a wax-barrier or a *Taq* polymerase-specific antibody. The polymerase is then mostly unblocked at the first high temperature DNA denaturation step at 92–95°C, then further enzyme is activated at the denaturation step of each subsequent cycle. The blocking of the polymerase early on in the reaction prevents the synthesis of non-specific products at lower temperatures during the PCR set-up and early stages. Thus, performing hot-start PCR reduces non-specific amplification and increases specific PCR product target yield (Lebedev *et al.*, 2008).

Polymerase fidelity is described in terms of errors per nucleotide polymerized and varies among the different enzyme types from a rate of 10^{-3} to 10^{-6}. The fidelity of the chosen DNA polymerase under difficult and changing conditions, for example in tissue extracts with varying compositions and inhibitor levels is a very important consideration. In the field of SNP analysis and genotyping the fidelity of amplification is especially important. Thus, in such applications, enzymes with a proofreading activity are highly recommended (Dietrich *et al.*, 2002, Tanguy *et al.*, 2004, Griffith *et al.*, 2007). It has been experimentally shown that low magnesium and dNTP concentrations, a pH 6.0–6.5 and lower cycle

Table 5.1 Overview of different polymerase types

Polymerase type		3′–5′ activity	Temperature optimum	Special features
Taq	*Thermus aquaticus*	no	72–95°C	High fidelity
T4	Coli-Phagen T4	yes		High fidelity
T7	Coli-Phagen T4	yes		High fidelity
Pfu	*Pyroccocus furiosus*	yes	72–74°C	Thermostable; good proofreading
Pwo	*Pyrococcus woesei*	yes	72°C	Thermostable, good proofreading, blunt-End cloning
Tfl	*Thermus flavus*	yes	74°C	thermostable
Phi29	*Bacillus subtilis* phage	yes		Whole genome amplification

numbers improve polymerase fidelity (Eckert *et al.*, 1991, 1998), providing further scope for improving PCR accuracy where needed.

DNA polymerase activity is affected by residual tissue components, such as bile salts, complex polysaccharides (faeces), haemoglobin (blood), proteinases (milk) and urea (urine). Other inhibitors can be retrieved from a contamination out of the extraction procedure by phenols or other protein detergents, which degrade or inhibit any enzyme activity in RT and PCR reactions. Different DNA polymerases show varying sensitivities to these substances so pre-screening various polymerases to identify the optimal choice for a given reaction is recommended (Al-Soud *et al.*, 1998). Special applications need specific enzymes, e.g. GC-rich PCR can be done with an enzyme mix of *Pwo and Taq* polymerase. For long-range PCR special dNTP kits and enzyme mixes exist and for cloning of PCR product a combination of *Taq* and *Pwo* polymerase is helpful.

General recommendations

- Use hot start enzymes for PCR reactions.
- Be aware that RNA integrity and inhibitors can influence polymerase activity.
- Enzymes with proofreading activity are recommended.
- Fidelity is improved with low magnesium and low dNTP concentrations.
- Choose the enzyme depending on your experimental conditions.

Reverse transcription

In the reverse transcription (RT) step the RNA template is reverse transcribed into complementary DNA (cDNA), which will later be amplified in the PCR. The components required in the RT reaction are the reverse transcriptase enzyme, a reverse primer and an optimized buffer containing Mg^{2+} and/or Mn^{2+} ions and free dNTPs. The optimization of the RT step often is disregarded in favour PCR optimization but since the error accumulation in any experimental set-up is higher in the first quantification steps (sampling, extraction and RT) these steps deserve closer attention (Tichopad *et al.*, 2009).

Reverse transcriptase

Reverse transcriptase (RT) is an RNA-dependent DNA polymerase that transcribes single-stranded RNA into cDNA. All known reverse transcriptases are encoded by retroviruses and *in vivo* their role is to copy the viral RNA genome into DNA prior to its integration into host cells. The first isolated RT enzymes were AMV, from avian myeloblastosis virus, and MMLV, from Moloney murine leukaemia virus. A downside to their application in RT-PCR is that, when present at high concentrations, both of these enzymes inhibit PCR due to their RNase H domain which cleaves the DNA–RNA hybrids. However in many commercially available enzymes this domain is often inactivated by a point mutation, called RNAse H^{minus} (Berger *et al.*, 1983; Hawkins *et al.*, 2003). To get an integer and long cDNA strand out of the RT reaction set-up, the use of RNAse H^{minus} is recommended in addition to ensure that the starting RNA is fully intact and exhibits a good RIN or RQI number.

Thanks to the development of thermotolerant RT enzymes that are stable at elevated temperatures (e.g. SuperScript III; Invitrogen), RT is nowadays performed at reaction temperatures of up to 65°C. This melts any secondary stem-and-loop RNA structure, giving the reverse primer better access to template RNA (Stahlberg *et al.*, 2004a).

So which RT enzyme is the best for your RT reaction? Which enzyme will give best RT yield and fully reverse transcribe your starting RNA? Unfortunately, studies show that there is no universal RT enzyme which always shows best results in all mRNA reverse transcribed. All RT enzymes show different RT yields and sensitivities to distinct mRNA secondary structures. Consequently, the results of any quantitative RT-PCR (qRT-PCR) experiment will depend largely on the RT efficiency, and less so on the PCR efficiency (Tichopad *et al.*, 2009). This is especially true for genes that are expected to show a lower expression in the sample need high RT yields (Bustin, 2000; Stahlberg *et al.*, 2004a).

The 'best RT enzyme' gives the highest cDNA yield, fully reverse transcribes all RNAs investigated and introduces no bias or error in the cDNA pool (Stahlberg *et al.*, 2004b). The only way to find the best RT enzyme for a given study is to do extensive testing in the optimization phase.

Nevertheless, some general recommendations for the selection of a good RT enzyme can be made:

- RT enzymes without RNAse H (=RNAse H^-) activity are recommended, especially in samples with low RNA quantity and quality.
- Experimental evaluation of RT reagents in a pre-PCR optimization study is recommended.
- Use high-quality RNA in the RT reaction.
- Use a temperature-stable RT enzyme.
- Heat samples at 60°C before RT to minimize secondary RNA structure.

RT priming strategies

The RT step is one source of variability in a qRT-PCR experiment and for each enzyme the specific reaction conditions has to be optimized. Salt contamination, alcohol, phenol, and other inhibitors carried over from the RNA isolation process can affect the apparent RT efficiency (Wong *et al.*, 1998; Freeman *et al.*, 1996). Another source of variability is the choice of priming method used to initiate cDNA synthesis, which can be either target gene specific or non-specific. In general three different priming strategies exist for the RT step:

- gene-specific priming;
- oligo-dT or poly-dT priming;
- random priming – using hexamer, octamer or decamer primers.

The main differences in the priming strategies are the product family after reverse transcription and the lengths of the resultant cDNAs. Target gene specific priming works well in conjunction with elevated RT-reaction temperatures up to 60°C to eliminate secondary RNA structure and spurious transcripts. Here only one cDNA complementary to the target gene mRNA is synthesized.

Non-specific primers such as random hexamer, octamer or decamer primer mixes can be used to circumvent high inter-assay variations between biological samples during RT. These random primers will anneal at various positions of the single-stranded RNA and give multiple starting points for the reverse transcriptase. Therefore fragments of difference length will be produced from one starting mRNA. The entire RNA sample is effectively reverse transcribed, and thus a total cDNA pool of all present RNA fragments is synthesized, including the highly abundant ribosomal (rRNA) subunits, 5S, 18S and 28S and the tRNA family. Therefore the RT yield using random primers is very high and theoretically 'all genes' can be measured in later PCR.

Similarly, poly-T oligonucleotides (consisting solely of 16–25 deoxythymidine residues) can anneal to the polyadenylated 3′ (poly-A) tail found on most mRNAs (Bustin, 2000). This facilitates the specific reverse transcription of the entire mRNA pool in the sample, although it should be noted that some mRNAs, such as the ubiquitin message, have no poly-A tail so would not be transcribed in such an approach.

cDNA pools synthesized with non-specific or oligo-dT primers can later be split into a number of different target-specific PCR assays. This maximizes the number of genes that can be assayed from a single cDNA pool derived from one total RNA sample. Therefore the gene expression results are comparable between the applied split assays, at least within the same cDNA pool.

It is reported that reverse transcription can take place without any primers, because any RNA fragments present in the extract can be used as primers in the RT reaction (Stahlberg *et al.*, 2004b). However, the yields of such reactions are very low, so the use of RT primers is always recommended.

Gene specific priming is used in one-step RT-PCR, because no separate RT step is required in the procedure. To combine the benefits of the different RT priming strategies, a mix of oligo-dT and random primers is recommended (Table 5.2). Primers can be mixed equally (1:1), so a high specificity of the oligo-dT and the high RT yield of the random primers is guaranteed (Zhang *et al.*, 1999).

Finally, the different priming strategies can be ranked in order of RT efficiency and cDNA yield as follows (unpublished results):

mixture of poly-dT and random primers > random primers > poly-dT primer ≥ gene-specific primer.

Table 5.2 Overview of RT priming strategies

Primers	Processed RNAs	RT yield	Special application
Oligo-dT	mRNA, full-length products	Low	Gene expression analysis
Random primers	mRNA and rRNA and small RNA, fragmented products	High	Multiple applications
Gene specific primers	One RNA = one specific gene sequence	High but only one measurable gene	One-step PCR
Primer mixes	mRNA and rRNA and small RNA, fragmented products	Very high	Gene expression analysis, multiple applications

RT protocol

The variety of RT available protocols, optimized for the sensitivities of the different RT enzymes, provide additional opportunities for the user to optimize their reactions. Generally used parameters are RNA concentrations between 100 and 1000 ng total RNA per reaction, total reaction times of 10–60 min and performance of the RT step at anywhere between 37°C and 60°C, depending on the thermostability of the reverse transcriptase enzyme. A post-reaction denaturation step at 95°C for several minutes is generally included to degrade any remaining template RNA.

For special applications like single-cell expression analysis, where only a few picograms of total RNA are available for each biological sample, a pre-amplification step between the RT and PCR reaction is effective (own unpublished results). Applying 10 cycles of pre-amplification can increase the cDNA by 1000-fold ($2^{10} = 1024$, own results), but is reported to introduce a bias in the generated cDNA pool that skews the quantitative PCR result (Ng *et al.*, 2005; Mestdagh *et al.*, 2008). To avoid this, a very simple method to decrease standard deviation between the samples should be employed. This involves is to setting up the experimental replicates as early as possible; ideally in sampling step but at the latest in the RT step. A very common error is to use one replicate in the RT and take duplicates or triplicates for the PCR reaction, resulting in large deviations in the replicates because the main technical variance belongs to the RT step and not to the qPCR step. Placing replicates already in the RT step can help to increase technical variance of the whole experiment. Choosing the triplicates in the RT step and doing single analysis in the PCR reduces the deviations of the replicates significantly (Stahlberg *et al.*, 2004b; Tichopad *et al.*, 2009; Hammerle-Fickinger *et al.*, 2009).

General recommendations

- RNA should be heated up to 65°C for 5 min before the RT reaction to open secondary RNA structures and allow a better annealing of the primers.
- RT enzyme should be the last substance given to the mastermix (keep the enzyme on ice as long as possible).
- Chose the replicates as early as possible to decrease error accumulation; best at extraction step or in the RT step – not in the PCR step.
- Common RT protocols are based on 100 ng to 1 µg total RNA.

Intercalating fluorophore dyes

Intercalating fluorescent dyes are commonly used for the real-time detection of the newly generated double-stranded PCR product. Sometimes these dyes are called 'non-specific dyes' because they bind to all double stranded DNA (dsDNA) molecules, including unwanted products like primer dimers or genomic DNA. But this name can be misleading because it has nothing to do with the specificity in the PCR reaction. The specificity of the generated PCR product is only determined by the binding abilities of the primer pair. The dye helps to detect the newly synthesized product and to make it visible for the fluorescence acquisition system. During PCR the increase of amplified product is directly proportional to the increase of the fluorescence signal.

The first-generation of dyes used for PCR analysis were fluorophores, such as ethidium bromide (EtBr), YOHO, YOPRO-1, or Hoechst 33258 (Katz *et al.*, 1990), which intercalate with nucleic acids, an effect that is widely used in agarose gel electrophoresis. When exposed to ultraviolet light, EtBr will fluoresce with an orange colour and can be visualized by eye or a CCD camera, allowing the visualization of nucleic acids bands on a gel. EtBr is still used in post-PCR gel analysis because it is one of the cheapest dyes available. EtBr's disadvantages are its high toxicity and low fluorescence enhancement of DNA; high concentrations of nucleic acids (5 ng DNA in single band, 25–50 ng in polydispersed samples) are needed for accurate visualization (Ahn *et al.*, 1996). The second generation of intercalating dyes such as PicoGreen (a DNA binding dye) and RiboGreen (an RNA binding dye) are used in conventional PCR for DNA quantification (Molecular Probes, Invitrogen). These come with the added advantage of having little background noise since the unbound dye has virtually no fluorescence. By the binding to double stranded DNA (dsDNA) the fluorescence is enhanced by 20-fold, therefore low amounts down to 0.25 ng DNA can be measured. This makes these dyes much more sensitive than EtBr (Ahn *et al.*, 1996).

For real-time PCR, the most popular second generation intercalating dye is SYBR Green I (Molecular Probes, Invitrogen). SYBR Green I binds to the minor groove of dsDNA, which results in a 30-fold increase in the fluorescence signal. The free unbound SYBR Green I shows very low background noise. On the downside, high SYBR Green I concentrations inhibit the polymerase reaction and therefore diminish the PCR efficiency (Wittwer *et al.*, 1997).

A third generation of saturated intercalating dyes was introduced to the market in parallel with the development of high-resolution melting (HRM) curve analysis and genotyping. Examples include EvaGreen (Biotium), SYTO 9 and other members of the family of SYTO dyes (Molecular Probes, Invitrogen), LC Green (Idaho Technology), BEBO/BOXTO and Chromofy (TATAA Biocenter, Sweden). These dyes are non-toxic to the polymerase and exhibit greater stability at higher temperatures than earlier dyes, allowing higher concentrations to be used in reaction mixes, resulting in an increase in sensitivity of detection. In addition to their increased sensitivity, the main benefits of these dyes are their low running costs and ease-of-handling in the primer design and assay set-up. Problems can arise with these dyes when primer design and assay quality are poor, in terms of unwanted PCR products or primer dimers. Each double-stranded DNA molecule in the assay will generate a signal, generated by primer dimer formation or genomic DNA. Therefore the PCR product's specificity should be checked by including a melting curve analysis in the PCR protocol, its length verified by agarose gel electrophoresis and its sequence checked by sequencing.

The alternative to dye-based PCR detection chemistries are probe-based systems, which can be separated into single probe and primer-probe based systems.

Primer-probe systems

Primer-probe based systems utilize a specifically designed PCR primer with a covalently attached fluorophore to facilitate product detection (e.g. scorpion primer). The background signal, from the fluorophores of unincorporated primers, is kept to a minimum by a quencher dye. During amplification, the distance between fluorophore and quencher is increased, so that the signal of the reporter fluorophore can be detected. Quencher dyes which themselves are not fluorescent (e.g. black whole quenchers, BHQ) are recommended because the background noise during real-time detection is acceptable low, e.g. as applied in hydrolysis probes. These probes have a very high binding specificity, but detailed attention must be paid to their design. The number of different primer-probe systems supplied by the various vendors are too numerous to list here, but a few prominent examples include LNA probes (Exiqon) and scorpion primer.

Hydrolysis probes

Structurally, probes can be linear or non-linear. Linear probes have no secondary structure so often have better hybridization efficiency and their design is easier because no complementary sequences need to be considered.

The first available, and best known, linear probes are hydrolysis (or nuclease) probes, (e.g. TaqMan probes, Applied Biosystems), which are still the most popular probes used in research and diagnostic applications. TaqMan probes have a fluorophore dye (typically FAM) and a quencher (BHQ) that inhibits the signal of the fluorophore. In the elongation phase the DNA polymerase exonuclease activity separates the fluorophore from the quencher and so the fluorescence signal can be detected.

Hydrolysis probes can be used in both singleplex and in multiplex PCR. For multiplex PCR the probes require different fluorescent labels, to allow the various signals to be distinguished by the instrument (Bustin *et al.*, 2004). Advantages of TaqMan probe systems are their high binding specificity and the multiplexing capabilities. However they are more expensive and primer/probe design is more difficult compared to intercalating dyes. With probes it is also impossible to check for primer dimers or unwanted artefacts in the reaction because the probe only detects the specific product (Shipley, 2006).

Oligo-probes

Second most popular among the linear probes are oligo-probe systems such as HybProbes (Roche Diagnostics), which are often used in multiplex PCR and SNP detection and have become the method of choice for the LightCycler systems (Roche Diagnostics). Such systems consist of two probes that are designed to bind adjacent to one another. One probe, the so called anchor probe, carries the donor dye located at the 3′-end. The second probe, called the reported probe, carries the acceptor dye at the 5′-end. In the annealing phase the energy of the donor dye is transferred to the acceptor dye which is activated by a so called fluorescence energy transfer (FRET) mechanism (Förster, 1948). The signal of the acceptor dye is detected; the donor signal only emits background fluorescence. With HybProbes the generation of melting curves and SNP analysis is possible (Read *et al.*, 2001).

Non-linear probes

Molecular beacons are non-linear probes because of the hairpin structure, with a complementary stem sequence and a loop sequence complementary to the target. A fluorophore and a quencher dye are covalently attached at the 3′- and 5′-ends of the probes (Tyagi *et al.*, 1998). The quencher, 4,4′-dimethylamino-phenylazo-benzene (DABCYL), differs from the description of TaqMan chemistry in that it is non-fluorescent and so does not produce any background signal at all. The tight stem structure of the probe allows a very efficient quenching of the fluorophore dye. In the presence of a complementary sequence, i.e. the primer binding sites, the beacon probe will hybridize to the template and, shift into an open configuration. This increases the distance between the fluorophore and the quencher, allowing the emitted light of the fluorophore to be detected. The occurrence of a mismatch between a beacon probe and its target has a greater destabilizing effect on the duplex than the introduction of an equivalent mismatch between the target and a linear oligo-probe because the hairpin structure provides a highly stable alternate conformation. Therefore, beacons show more specificity than the more common linear oligo-probes making them ideal candidates for detecting SNPs (Tyagi *et al.*, 1998). Molecular beacons can be used in numerous applications like multiplex PCR, SNP analysis, and detection of enzymatic cleavage (Bustin *et al.*, 2004).

Stabilizers or chemical changes in the probe DNA backbone are used to keep melt temperature requirements constant but reduce the required length, in order increase binding affinity and specificity of the probe to the amplicon. Examples for stabilizers are:

- Minor groove-binders (MGB, Applied Biosystems), which are added to one end of the linear probe to increase the binding affinity, for example this is done in TaqMan Probes for SNP detection.
- Locked nucleic acids (LNA, Exiqon; see Fig. 5.2) which are inserted at every 5th–8th base position in the probe sequence and are used for microRNA detection (Exiqon), SNP and allelic discrimination assays. An optimized set of LNA probes is commercially available as Universal Probe Library (Roche Diagnostics). Here primer and probe design is performed automatically according to the selected target and reference gene sequence and pre-optimized kits are used for mRNA quantification.
- Peptide nucleic acids (PNA) (Paulasova and Pellestor, 2004), which have a peptide sequence with an attached fluorescence dye. PNA molecules do not have a negatively charged backbone facilitating stronger annealing to the template strand (Bustin *et al.*, 2000; Shipley *et al.*, 2006).

Comparison of detection chemistries

When comparing signals from the different chemistries, the destruction of nuclease oligo-probes continues despite a plateau in product accumulation whereas intercalating dye fluorescence in the no template control (NTC) generally increases non-specifically during later cycles. This is due to primer dimer which (can) accumulate in late cycles (> cycle 35). Adjacent oligo-probe fluorescence begins to decrease as the rate of collision between the growing numbers of complementary amplicon strands increases favouring the formation of double-stranded DNA over the hybridization of oligo-probe to its target DNA strand. Reported in the literature as 'hook effect' some oligo-probes are consumed by sequence-related nuclease activity of the used polymerases (Mackay *et al.*, 2002).

All three oligo-probe chemistries, SYBR Green I, TaqMan hydrolysis probes and HybProbes are capable of detecting amplified PCR product with approximately the same sensitivity and quantification range. Generally the sensitivity of SYBR Green I assays depends strongly on the limitation by unspecific product formation due to inefficient primer design. The usage of TaqMan hydrolysis probes or HybProbes assays offer higher specificity, at least in the quantification of the generated PCR product (Wittwer *et al.*, 1997).

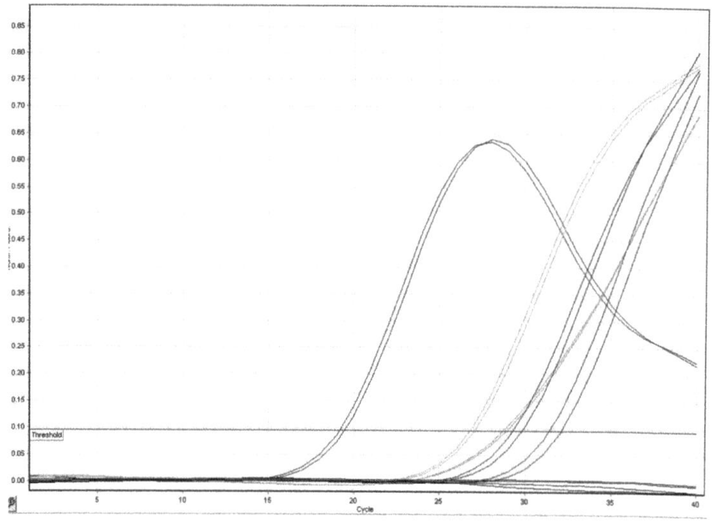

Figure 5.1 Example of hook effect in sample with highest starting amount.

General recommendations

- All detection chemistries are comparable in sensitivity and specificity.
- Use black whole quencher as quencher in probe systems.
- Using intercalating dyes always check specificity of the product by melt curve analysis and gel electrophoresis.
- Saturating dyes give better resolution in melting curve analysis (HRM is possible).

PCR cofactors and additives

PCR buffer and dNTPs

The optimal PCR buffer composition is essential for the fidelity and efficiency of the PCR reaction. Magnesium is the central cofactor for the DNA polymerase so Mg^{2+} ions should be included in the buffer to give a final concentration in the reaction between 1 and 5 mM. Magnesium is a chelating cation which can inhibit the PCR reaction and form complexes that reduce PCR fidelity (Wiedbrauk *et al.*, 1995), therefore excess magnesium (≥ 7.5 mM) should be avoided.

dNTPs are essential for the synthesis of the new dsDNA strain in PCR but since they bind magnesium ions, their total concentration should be lower, ideally 4-fold, than that of the magnesium so dNTP concentrations between 0.2 and 0.8 mM are recommended (Shipley, 2006).

The cations potassium and ammonium are also essential components of the PCR buffer. Pre-optimized kits contain a mix of potassium chloride (KCl) and ammonium sulfate $(NH_4)_2SO_4$ in Tris-HCl. Potassium interacts with the phosphate ion of the single DNA strand and stabilizes the structure for better primer annealing, while ammonium can bind the hydrogen between the bases, decreasing the probability of mispriming and primer dimer formation (Qiagen News, 1996).

PCR enhancers

The amplification capacity of PCR can be significantly decreased by inhibitory compounds found in the sample, such as haem from blood or uric acid from urine or lactoferrin from milk (Al-Soud *et al.*, 2001; Wiedbrauk *et al.*, 2001). Some tissues like fatty tissue or gastro intestinal fluids can themselves cause problems in PCR efficiency (unpublished results).

These inhibitors can often be removed, at least partly, in the sample preparation and extraction steps, but are not always totally eliminated and additional purification steps can be very time-consuming and have potential for loss of target nucleic acid. The influence of PCR inhibitors can be seen in the slope of the real-time PCR fluorescence history plot, in the PCR efficiency or in the variability of the technical replicates in RT and PCR. To check for active inhibitors in the sample extract, we recommend performing inhibition testing assays, e.g. the SPUD assay (Nolan *et al.*, 2006).

The use of PCR enhancers can help to decrease the effect of inhibitors in the reaction by complexing them, annulling their effects and positively influencing the reaction fidelity. Enhancers can have varying effects in different biological samples or tissues and they can be used to increase reaction yield for delicate PCR products or in special PCR applications. Table 5.3 shows some of the most commonly used enhancers and their features.

General recommendations

- Magnesium concentrations have to be higher than dNTP concentrations in the PCR reaction.
- A mix of potassium and ammonium ions in the PCR buffer should be preferred.
- RNA extraction should be performed with care and should contain no inhibitors.
- Working with tissues which have natural inhibitors, PCR enhancers should be used.
- Use a SPUD assay (Nolan *et al.*, 2006) to check for inhibitors.

Optimization of the amplification protocol

In the early days of block PCR amplification the optimization of PCR protocols was a central part of the technology. It was highly correlated with the quality of the result and the length of the annealing and elongation steps which were in the focus of the optimization strategy. Today the PCR protocol is more standardized and reagents for RT and PCR reaction can be bought as ready-to-use kits, with already optimized reagent concentrations and validated temperature profiles. On one hand this saves time for the users because optimization becomes a minor endeavour but on the other, where optimization is required it is more difficult. Nevertheless, increasing PCR specificity and efficiency by changing the protocol is still possible, and potential approaches are described in this chapter.

Original approaches to optimize the PCR protocol centred on obtaining maximal product yield and getting rid of unwanted primer dimer formations. A typical PCR protocol starts with a DNA denaturation step at 95°C for some minutes to fully denature the starting DNA followed by 30–40 iterative PCR cycles, each with a short denaturation step (at least 5 s at 95°C), a primer annealing step at a primer-specific annealing temperature (between 55 and 60°C) and a step at 68–75°C for the elongation of the new strand.

Today, however, two-step protocols (denaturation and annealing step), three-step (denaturation, annealing and elongation step) or four-step protocols (denaturation, annealing, elongation step and elevated temperature acquisition step at 80–85°C) are

Table 5.3 List of enhancers for PCR reaction

Enhancers	Function	Applications	Concentration	References
BSA	Reduces sample binding to reaction vessel Binds inhibitors	Light Cycler systems LC 1.0, LC 1.5 and LC 2.0	1–400 ng/µl	Al-Soud *et al.* (2001), Kreader (1996)
DMSO	Reduces temperature differences between primer DNA	GC-rich and long-range PCR	2–10%	
Betaine	Reduces sequence-dependent temperature differences		0.5–2 M	
Glycerol	Stabilizes DNA polymerase			
TMAC	Eliminates non-specific priming Increases yield and specificity of PCR		15–100 mM	Wood *et al.*, (1985), Chevet *et al.* (1995),
Formamide	Improve the specificity of PCR		1–5%	Gelfand (1989), Sarkar *et al.* (1990)
Triton X-40 Triton X-100 NP-40	Prevents aggregation of polymerase May increase yield but also increase non-specific amplification		0.01–1%	
Tween 20	Reduces non-specific interactions			
T4 Gen 32 Protein	Reduces temperature differences between primer-DNA		100–150 ng/µl	Kreader (1996)

performed. Two- and three-step protocols, either with or without an additional elongation step, are the most popular and commonly used for probe based detection systems. The longer annealing step performed at 60°C should always be as short as possible, to reduce formation of primer dimers, but as long as necessary to make a specific annealing possible. In older publications four-step PCR protocols with fluorescence acquisition at elevated temperatures (80–85°C) were performed to circumvent any detection of generated primer dimers (Roche Diagnostics, 1999). Primer-dimers will still be in the reaction and possibly affect whole assay efficiency. In the early days of qPCR using SYBR Green I detection an additional fourth step was included in the protocol to circumvent the quantification of unwanted primer dimers. Existing primer dimers were melted at elevated temperature acquisition 80–85°C and therefore become invisible in quantification (Pfaffl, 2001).

Today it is very common to perform fast, and yet still quantitative RT-qPCR assays. Fast PCR is the combination between a reliable fast processing polymerase mix and suitable instruments and consumables. In the past, reaction speed was limited by the performance of the Peltier elements used for block heating and cooling. But with the development of new fast PCR block technologies, using either silver block (realplex Eppendorf), honey comb block (Bio-Rad CFX series) or new conductive materials in LightCycler 480 (Roche Diagnostics) or 7500 and 7900 fast cyclers (Applied Biosystems) they are also applicable on block cyclers. The platforms exhibit heating rates up to 6°C/s and cooling rates 4°C/s and show good block homogeneity. But the real fast systems are the good old fashioned rotary systems, e.g. LightCycler family (1.0, 1.5 and 2.0) or the Rotor-Gene family (Corbett Research Rotor-Gene 3000 or 6000; Qiagen Rotor-Gene Q). These show optimal temperature uniformity over all samples, very quick heating and cooling up to 10°C/s and make fast PCR possible since they have been introduced to the market (Swerdlow *et al.*, 1997; own findings). For more information on instruments, see Chapter 6.

Today the focus of the protocol optimization is directed towards the primer annealing step. The annealing temperature depends primarily on the primer sequence and has to be optimized for every analysed primer pair. Generally the predicted melt temperature calculated by the primer design program is not the optimal run temperature. Therefore it is recommended that the annealing temperature should be optimized in a gradient PCR

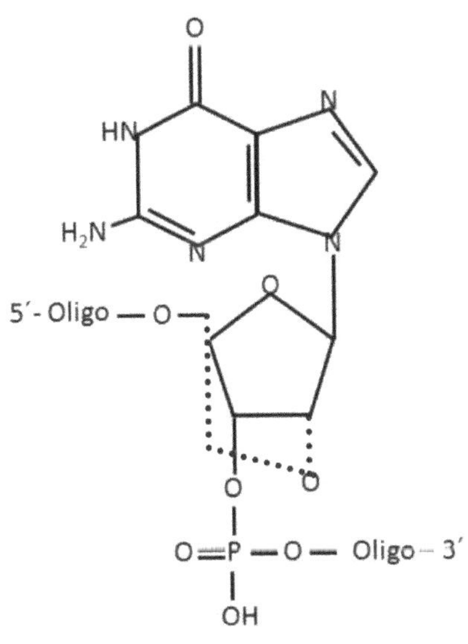

Figure 5.2 LNA probe with 'locked' ring (dotted lines).

experiment (see next section on 'optimization of primers'). The primer annealing also depends on the composition and ionic strength of the buffer system and on the binding affinities of the employed consumables. Whether a three-step protocol is really needed depends on the dye used and on the amplicon length. Second generation dyes were processed with an extra elongation step to allow inclusion of the dye in the new synthesized dsDNA. Nowadays, third generation intercalating dyes can be used with a two-step protocol, including a longer annealing phase, that was already standard for probe assays. Long-range PCR with an amplification of 5–40 kb fragments still needs an additional elongation phase; between 10–30 min so that DNA polymerase has the opportunity to synthesize the new strand (Roche Diagnostics, 2006).

Optimizing the temperature profile can be done by using a RT-PCR, a combined RT and PCR reaction, or by separate RT and PCR reactions. Combined RT-PCR is often used in gene expression studies, to measure various biological samples, and because the RT step is part of the PCR protocol, the possible error accumulation of the RT reaction is definitely reduced (Tichopad *et al.*, 2009). Here PCR primers are also taken in the RT reaction step, so no additional RT primers are necessary. This is cost savings but also means that only one single gene expression can be measured in each PCR run.

General recommendations

- Often two-step protocols are sufficient even with intercalating dyes.
- Perform a stringent protocol or perform fast PCR: annealing temperature as high as possible and the protocol as fast as possible.
- Samples with low starting amounts, e.g. single cells, often need more than 35 cycles for adequate amplification.

Optimization of primers

Primer design requires the sequence of the gene of interest and suitable primer design software that can calculate possible primer pairs for the detection system. In theory this is simple but often the calculated primer pairs do not work in PCR. Therefore optimization of the primer design or the amplification protocol is necessary.

Primer design is the most critical step in PCR analysis, because it influences the specificity, efficiency and accuracy of the assay so it should be done carefully to avoid any primer dimers or unwanted reaction artefacts. Initially this may be laborious and time-consuming, but after failure an additional primer pair redesign and ordering becomes very fretful.

First step is to find the correct sequence for the gene to be quantified. Several databases for gene sequence retrieval are available, the most popular being NCBI (http://www.ncbi.nlm.nih.gov/) and ENSEMBL (http://www.ensembl.org). Specialized databases for bacteria and virus sequences are also available (http://www.ebi.ac.uk/genomes/). As more and more gene sequences are identified and published this can become very complex and confusing because each gene of interest my have several sequence variations posted, for example sequences from DNA or RNA, from different isoforms, splice variants, SNPs, mutations, partial sequences (CDS, introns, exons, 5′- or 3′-sequences) or from sequence variations from different breeds. Therefore it can take some time to find the actual sequence that has to be amplified. It helps to know the abbreviation of the gene name, the accession number, and to look for the gene or transcript information.

After the correct sequence is identified it can then be uploaded into primer design software. Software for both primer and probe design are available, either as freeware, or, more commonly, as a proprietary package bundled with an instrument, e.g. PrimerExpress (Applied Biosystems). The most popular freeware package is Primer 3 (Rozen and Skaletsky, 2000) but many more, too numerous to list in this article are available and a summary can be found here http://primers.gene-quantification.info.

All software can calculate primer pairs from a given gene sequence following predefined parameters. Most important parameters are: primer melting temperature, primer GC content, primer length, product length, GC content of product, self priming potential of primer hetero- and homodimers. The calculated melting temperature (T_m) is especially important for PCR because this should also be the optimal annealing temperature of the primer in the reaction. Development of primer binding algorithms to calculate the primer Tm began in the 1960s and various additional models have been built on these original efforts (Breslauer *et al.*, 1968; Owzarzy *et al.*, 2004; Santa Lucia *et al.*, 1996, 2001; Schildkraut *et al.*, 1965). In Primer 3 the user can choose between these models to calculate optimal T_m. A list of the available models is shown in Table 5.4.

It is very common to design primer pairs at the same T_m for different genes to allow amplification of these different genes using identical amplification protocols. Such designs do not relate to the real T_m and annealing behaviour of the primers, but instead represent a best-fit comprise for the group of targets. By definition, at the optimal T_m half of the primers will be annealed, so in these best-fit primer groups, different amounts of primers will be bound to the each target depending on the actual T_m of the individual target sequence. This becomes very important in multiplex real-time PCR set-up where multiple amplicons are compared within one reaction set-up so similar amplification efficiencies for the measured genes are essential to make them fully comparable. Therefore it is essential that all used primer pairs have an equal concentration binding to the DNA template strands. To calculate similar hybridization behaviours it would be more useful to design primers with similar Gibbs free energy ($\Delta G°$) instead of similar T_m (Zucker *et al.*, 1999). The free energy is based on the enthalpy and the entropy of the reaction that describes the annealing behaviour of the primer to the target strand. Freely available software applets for primer design,

Table 5.4 Summary of research groups developing algorithms for T_m calculation

Model	Calculation of T_m	
2 × 4 rule	4°C × (number G/C bases) + 2°C × (number A/T bases)	
Breslauer (1968)	Nearest-neighbor	For short oligos (8 bp), 1 M Na
Santa Lucia (1996)	Nearest neighbour + end interactions	1 M Na
Santa Lucia (1998)	Nearest neighbour + end interactions + salt correction	Salt concentration ~1.5 M Na
Schildkraut (1965)	Salt correction	
Owczarzy (2004)	Nearest neighbour + end interactions	DNA dumbbells and polymers, short oligos, 25 mM Na

such as Netprimer (Premier Biosoft), can calculate the free energy of the designed primer pairs. We recommend that primers should be optimized for both T_m and $\Delta G°$ between the genes of interest in such experiments.

There is no need for the primer concentrations to be balanced, but they should be high enough to allow sufficient sensitivity, specificity and fidelity within the assay. High primer concentrations increase the risk of primer-dimer formation. A range between 200 and 600 nM is recommended as optimal. In multiplex PCR the adjustment of each primer concentration becomes more difficult and it is recommended that the primer concentration of the genes with lower expression levels is increased so that similar PCR efficiencies are reached and a comparison of the multiplexed products is possible (Persson *et al.*, 2005). In any case, following all recommendations it is possible to optimize primers to meet all of the above-mentioned parameters and experience shows that they perform with reproducible and high PCR efficiency (Allawi *et al.*, 1998; Santa Lucia, 2001).

General recommendations

- Select target sequence with care.
- Use the same model for calculation of T_m in primer design software when several genes are designed.
- Avoid secondary structures within the primer sequence.
- Avoid self-complementarity, especially at the 3′ end of primer pairs.
- Try to design primers with similar T_m and $\Delta G°$.

Consumables

Consumables are not a crucial factor in quantification process but still an important one and the correct choice will depend on sample number and the PCR instrumentation employed. Some instruments require specific plastic consumables, others glassware but for most of the qPCR formats, especially with plate based formats, various types of consumables can be used. The consumables have to be taken in combination with the ability of the hardware, e.g. in performing a fast PCR protocol.

A range of PCR consumables is available for PCR analysis to suit the sample number and reaction volumes required. Choices include:

- single tubes with caps;
- strip tubes with strip caps;
- plates with various sealing films;
- special consumables, e.g. LightCycler capillaries (Roche Diagnostics), Rotor-Gene Q reaction tubes (Qiagen), or Smartcycler reaction tubes (Cepheid).

When performing fully quantitative real-time PCR it is always recommended to calibrate the instrument according to different dye spectra with whatever consumables you intend to use in your analysis. Some real-time instruments can even be calibrated for a range of different consumables and calibration plates and standardized kits are in generally available from the instrument manufacturer (Applied Biosystems; Bio-Rad). The calibration of the instrument can be seen as additional optimization and standardization step, to increase specificity and sensitivity of the fluorescence measurement, for real-time PCR data and should be repeated annually, according to the rules of good laboratory practice (GLP or MIQE guidelines; Bustin *et al.*, 2009). In the LightCycler capillary cyclers, a colour compensation kit can be applied to calibrate platform, glass capillaries, fluorescence acquisition unit and software (Roche Diagnostics).

High-volume 0.2 ml tubes are often taken for block PCR cycling with limited number of samples and 8-strips with caps are used when sample number is low (< 96 samples). Tubes and 8-strips need higher reaction volumes, generally between 15–50 µl to avoid the upper dead volume, although new cap styles exist that minimize the free volume and therefore minimize evaporation rate (Eppendorf). Using single tubes and 8-strips saves on consumables because vial number can be optimized according to the sample number.

As well as the standard 96-well plate, high-throughput 384 well-plates, and even 1536-well plates are now available (Roche Diagnostics). Because the dimension of the plate is standardized, the well volume changes in proportion to number of wells. In 96-well plates 10–25 µl, in 384-well 3–10 µl, and in 1536-well only 0.5–2 µl is used as total volume for a single PCR run. This helps to save chemicals and costs.

To avoid evaporation, plates can be closed with caps or, most commonly, self-adhesive or heat-sealing films or foils. Heat-sealing avoids possible reaction mix evaporation during the PCR run, which can happen using sticky adhesive films if they are old and have lost their stickiness or if the instrument lid pressure is insufficient. Problems with lid pressure can also be avoided if new instruments with high lid pressures are used (Bio-Rad CFX series; Roche LC 480). To access the PCR products in heat sealed vessels the film can either be pierced with the pipette tip or, depending on the sealing system, the films can be removed entirely (Reiter and Pfaffl, 2008).

Users can also choose between transparent and white plates. Transparent plates allow better control of the pipetting process and better fluorescence readout (BioNews, 2009, Eppendorf). This is important e.g. when robot pipetting has to be checked or when the PCR product has to be processed after PCR reaction. Transparent consumables are also indispensable in real-time instruments to allow scanning through the tube (from bottom to top) as it is the case in the Rotor-Gene (Corbett Research or Qiagen) and the Light-Cycler system family (Roche Diagnostics). White plates definitely decrease the detection level of the dye because of the higher fluorescence reflection (BioNews, 2008, Eppendorf) and therefore quantification at lower starting concentrations is possible. Furthermore, white plates could decrease the variance of PCR replicates. Significant differences in PCR efficiency between transparent and white plates have not been detected. In high-throughput analysis or samples with low concentration, a better fluorescence signal can help to increase sensitivity and specificity (Reiter and Pfaffl, 2008).

General recommendations

- White plates show higher fluorescence.
- Heat sealing reduces evaporation and border effects.
- Consumables have influence on data variability.
- High-throughput plates help to save reaction mastermix.

Conclusion

The previous decade has been characterized by important methodological advances that have made quantitative PCR more sensitive, less variable and therefore more valid and reliable. As described in the past chapter, most advances were implemented in the optimization of the PCR method itself, but pre-PCR steps are still highly variable and introduce a noise in the quantification results. Clearly, we are still only half way in terms of having the optimal reaction composition or the optimal quantification work flow (Pfaffl, 2010). The developed MIQE guidelines will help to improve future experiments, but researchers must apply these instructions to get more valid and 'true' quantification results (Bustin *et al.*, 2009).

References

Ahn, S.J., Costa, J., and Emanuel, J.R. (1996). PicoGreen quantitation of DNA: effective evaluation of samples pre- or post-PCR. Nucleic Acid Res. *24*, 2623–2625.

Allawi, H.T., SantaLucia, Jr, J. (1998). Nearest-neighbor thermodynamics of internal AC mismatches in DNA: sequence dependence and pH effects. Biochemistry *30;37*, 9435–9444.

Al-Soud, A., and Radström, P. (1998). Capacity of nine thermostable DNA polymerases to mediate DNA amplification in the presence of PCR-inhibiting samples. Appl. Environ. Microbiol. 3784–3753.

Al-Soud, A., and Radström, P. (2001). Purification and characterization of PCR-Inhibitory components in blood cells. J. Clin. Microbiol. 39, 485–493.

Becker C., Hammerle-Fickinger A., Riedmaier, I., and Pfaffl, M.W. (2010). mRNA and microRNA quality control for qRT-PCR analysis. Methods *50*, 237–243.

Berger, S.L., Wallace, D.M., Puskas, R.S., and Eschenfeldt, W.H. (1983). Reverse transcriptase and its associated ribonuclease H: interplay of two enzyme activities controls the yield of single-stranded complementary deoxyribonucleic acid. Biochemistry *22*, 2365–2372.

BioNews, Eppendorf issue number *28*, 2008.

BioNews, Eppendorf issue number *31*, 2009.

Breslauer, K.J., Frank, R., Blöcker, H., and Mary, L.A. (1986). Predicting DNA duplex stability from the base sequence. Proc. Natl. Acad. Sci. U.S.A. *83*, 3746–3750.

Bustin, S.A. (2000). Absolute quantification of mRNA using real-time reverse transcription polymerase chain reaction assays. J. Mol. Endocrinol. *25*, 169–193.

Bustin, S.A., and Nolan, T. (2004). Chemistries. In: A–Z of Quantitative PCR, Bustin, S.A. ed. (IUL Biotechnology Series), pp. 215–279.

Bustin, S.A., Benes, V., Garson, J.A., Hellemans, J., Hugget, J., Kubista, M., Mueller, R., Nolan, T., Pfaffl, M.W., Shipley, G.L., *et al.* (2009). The MIQE guidelines: minimum information for publication of quantitative real-time PCR experiments. Clin. Chem. *55*, 611–622.

Chevet E., Lemaître, G., and Katinka, M.D. (1995). Low concentrations of tetramethylammonium chloride increase yield and specificity of PCR. Nucleic Acids Res. *23*, 3343–2244.

Dietrich, J., Schmitt, P., Ziegler, M., Preve, B., Rolland, J.L., Chaabihi, H., and Gueguen, Y. (2002). PCR performance of the highly thermostable proofreading B-type DANN polymerase from *Pyrococcus abyssi*. FEMS Microbiol. Lett. 19; *217*, 89–94.

Eckert, K.A., and Kunkel, T.A. (1991). DNA polymerase fidelity and the polymerase chain reaction. Genome Res. *1*, 17–24.

Eckert, K.A., and Kunkel, T.A. (1993). Fidelity of DNA synthesis catalyzed by human DNA polymerase *α* and HIV-1 reverse transcriptase: effect of reaction pH. Nucleic Acids Res. *21*, 22.

Fleige, S., Walf, V., Huch, S., Prgomet, C., Sehm, J., and Pfaffl, M. (2006). Comparison of relative mRNA quantification models and the impact of RNA integrity in real-time qRT-PCR. Biotechnol. Lett. *28*, 1601–1613.

Förster T. (1948). Zwischenmolekulare energiewanderung und fluoreszenz. Ann. Phys. *6*, 55–75.

Frackman, S., Kobs, G., Simpson, D., and Storts, D. (1998). Betaine and DMSO: enhancing agents for PCR. Promega Notes *65*, 27.

Freeman, W.M., Vrana, S.L., and Vrana, K.E. (1996). Use of elevated reverse transcription reaction temperatures in RT-PCR. BioTechniques *20*, 782–783.

Gelfand, D.H. 1988. In PCR Technology, Erlich, H.A., ed. (New York, Stockton Press), p.17.

Goodman, M.F., and Fygenson, D.K. (1998). DNA polymerase fidelity: from genetics towards a biochemical understanding. Genetics *148*, 1475–1482.

Griffith, K., Nayak, S., Park, K., Mandelman, D., Modrell, B., Lee, J., Ng, B., Gibbs, M.D., and Bergquist, P.L. (2007). New high fidelity polymerases from Thermococcus species. Protein Expr. Purif. *52*, 19–30.

Hammerle-Fickinger A., Riedmaier, I., Becker, C., Meyer, H.H.D., Pfaffl, M.W., and Ulbrich, S.E. (2009). Validation of extraction methods for total RNA and miRNA from bovine blood prior to quantitative gene expression analyses. Biotechnol. Lett. *32*, 35–44.

Hawkins, P.R., Jin, P., and Fu, G.U. (2003). Full-length cDNA synthesis for long-distance RT-PCR of large mRNA transcripts. BioTechniques 34(4); 768–773.

Imbeaud, S., Graudens, E., Boulanger, V., Barlet, X., Zaborski, P., Eveno, E., Mueller, O., Schroeder, A., and Auffray, C., (2005). Toward standardization of RNA quality assessment using user-independent classifiers of microcapillary electrophoresis traces. Nucleic Acids Res. *33*, e56.

Katz E.D., Haff, L.A., and Eksteen, R. (1990). Rapid separation, quantitation and purification of products of polymerase chain reaction by liquid chromatography. J. Chromatogr. *20, 512*, 433–444.

Kreader C.A. (1996). Relief of amplification inhibition in PCR with bovine serum albumin or T4 gene 32 protein. Appl. Environ. Microbiol. *62*, 1102–1106.

Lebedev, A.V., Paul, N., Yee, J., Timoshchuk, V.A., Shum, J., Miyagi, K., Kellum, J., Hogrefe, R.I., and Zon, G. (2008). Hot start PCR with heat-activatable primers: a novel approach for improved PCR performance. Nucleic Acids Res. *36*, e131.

Mackay I.M., Arden, K.E., and Nitsche, A. (2002). Real-time PCR in virology. Nucleic Acids Res. *30*, 1292–1305.

Mestdagh P., Feys, T., Bernard, N., Guenther, S., Chen, C., Speleman, F., and Vandesompele, J. (2008). High-throughput stem-loop RT-qPCR miRNA expression profiling using min amounts of input RNA. Nucleic Acids Res. *36*, e143.

Nolan T., Hands, R.E., Ogunkolade, W., and Bustin, S.A. (2006). SPUD: a quantitative PCR assay for the detection of inhibitors in nucleic acid preparations. Anal. Biochem. *351*, 308–310.

Ng G., Roberts, I., and Coleman, N. (2005). Evaluation of 3 methods of whole-genome amplification for subsequent metaphase comparative genomic hybridization. Diagn. Mol. Pathol. *14*, 203–212.

Owczarzy R., You, Y., Moreira, B.G., Manthey, J.A., Huang, L., Behlke, M.A., and Walder, J.A. (2004). Effects of sodium ions on DNA duplex oligomers: improved predictions of melting temperatures. Biochemistry 43, 3537–3554.

Paulasova P., and Pellestor, F. (2004). The peptide nucleic acids (PNAs): a new generation of probes for genetic and cytogenetic analyses. Ann. Genet. *47*, 349–358.

Persson, L., Hamby, K., and Ugozzoli, L.A. (2005). Four-color multiplex reverse transcription polymerase chain reaction-pvercoming its limitations. Anal. Biochem. *344*, 33–42.

Perez-Novo, C.A., Claeys, C., Speleman, F., Cauwenberge, P.V., Bachert, C., Vandesompele, J., (2005). Impact of RNA quality on reference gene expression stability. BioTechniques 39, 52–56.

Pfaffl M.W. (2001). Development and validation of an externally standardised quantitative Insulin like growth factor-1 (IGF-1) RT-PCR using LightCycler SYBR® Green I technology In: Rapid Cycle Real-time PCR, Methods and Applications, Meuer, S., Wittwer, C. and K. Nakagawara, eds. (Springer Press, Heidelberg) pp. 281–291.

Pfaffl, M.W., (2005). Nucleic acids: mRNA identification and quantification. Nucleic Acids, Encyclopedia of Analytical Science, 2nd edn. pp. 417–426.

Pfaffl, M.W., Fleige, S., and Riedmaier, I. (2008). Validation of lab-on-chip capillary electrophoresis systems for total RNA quality and quantity control. Biotechnol. Biotechnol. Eq. *22*, 829–834.

Pfaffl, M.W. (2010). The ongoing evolution of qPCR. Methods *50*, 215–216.

Qiagen News. Effect of P.CR buffer on multiplex In, P.C.R. : Critical factors for successful P.C.R. (2009).

Raeymarkers, L., (1993). Quantitative PCR: theoretical consideration with practical implication. Anal. Biochem. *214*, 582–585.

Read, S.J., Mitchell, J.L., and Fink, C.G. (2001). LightCycler multiplex PCR for the laboratory diagnosis of common viral infections of the central nervous system. Clin. J. Microbiol. 39, 3056–3059.

Reiter M., and Pfaffl, M.W. (2008). Effects of plate position, plate type and sealing systems on real-time PCR results. Biotechnol. Biotechnol. Eq. 22, 3.

Roche (2006). PCR Application Manual, 3rd edn. (Indianapolis, IN: Roche Applied Science).

Roche Diagnostics Technical Note No. LightCycler 1/1999; How to Reduce Primer Dimers in a LightCycler PCR.

Rozen, S., and Skaletsky, H.J. (2000). Primer3 on the WWW for general users and for biologist programmers. In Bioinformatics Methods and Protocols: Methods in Molecular Biology. Krawertz, S., Misener, S., eds. (Totowa, NJ, Humana Press) pp. 365–386.

Santa Lucia, J. (1998). A unified view of polymer, dumbbell, and oligonucleotide DNA nearest-neighbor thermodynamics. Proc. Natl. Acad. Sci. U.S.A. *95*, 1460–1465.

Santa Lucia, J. (2001). Physical principles and visual-OMP software for optimal PCR design. In: PCR primer design, Yuryev, A. ed. (Methods molecular biology), pp. 3–34.

Sarkar G., Kapelner, S., and Sommer, S.S. (1990). Formamide can dramatically improve the specificity of PCR. Nucleic Acids Res. *24*, 7465.

Schildkraut, C. (1965). Dependence of melt temperature on salt concentration. Biopolymers *3*, 195–208.

Schoor, O., Weinschenk, T., Hennenlotter, J., Corvin, S., Stenzl, A., Rammensee, H.-G., and Stevanovic, S., (2003). Moderate degradation does not preclude microarray analysis of small amounts of RNA. BioTechniques *35*, 1192–1201.

Schroeder, A., Mueller, O., Stocker, S., Salwosky, R., Leiber, M., Gassmann, M., Lightfood, W., Menzel, W., Granzow, M., and Ragg, T. (2006). The RIN: an RNA integrity number for assigning integrity values to RNA measurements. BMC Mol. Biol. *7*, 3.

Shipley, G.L. (2006). An introduction to real-time PCR. In Real-time PCR, Tefik, Dorak, M., ed. (Bios advanced methods), pp. 1–37.

Stahlberg A., Kubista, M., and Pfaffl, M.W. (2004a). Comparison of reverse transcriptases in gene expression analysis. Clin. Chem. *9*, 50.

Stahlberg, A., Hakansson, J., Xian, X., Semb, H., and Kubista, M. (2004b). Properties of the reverse transcription reaction in mRNA quantification. Clin. Chem. *50*, 509–515.

Swerdlow, H., Jones, B.J., and Wittwer, C.T. (1997). Fully automated DNA reaction and analysis in a fluidic capillary instrument. Anal. Chem. *69*, 848–855.

Tanguy Le Gac, N., Delagoutte, E., Germain, M., and Villani, G. (2004). Inactivation of the 3′–5′exonuclease of the replicative T4 DNA polymerase allows translesion DNA synthesis at an abasic site. J. Mol. Biol. *336*, 102–134.

Tichopad A., Kitchen, R., Riedmaier, I., Becker, C., Ståhlberg, A., and Kubista, M. (2009). Design and optimization of reverse-transcription quantitative PCR experiments. Clin. Chem. *55*, 1816–1823.

Tyagi, S., Bratu, D.P., and Kramer, F.R. (1998). Multicolor molecular beacons for allele discrimination. Nat. Biotechnol. *16*, 49–53.

Wiedbrauk, D.L., Werner, J.C., and Drevon, A.M. (2001). Inhibition of PCR by aqueous and vitreous fluids. J. Clin. Microbiol. 2643–2646.

Wittwer, C.T., Herrmann, M.G., Moss, A.A., and Rasmussen, R.P. (1997). Continuous fluorescence monitoring of rapid cycle DNA amplification. BioTechniques *1*, 130–138.

Wittwer, C.T., Ririe, K.M., Andrew, R.V., David, D.A., Gundry, R.A., and Balis, U.J. (1997). The LightCycler a microvolume multisample fluorimeter with rapid temperature control. BioTechniques, *22*, 176–181.

Wong, L., Pearson, H., Fletcher, A., Marquis, C.P., and Mahler, S. (1998). Comparison of the efficiency of Moloney murine leukaemia virus (M-MuLV) reverse transcriptase, RNase H-M-MuLV reverse transcriptase and avian myeloblastoma leukaemia virus (AMV) reverse transcriptase for the amplification of human immunoglobulin genes. BioTechniques *12*, 485–489.

Wood, W.I., Gitschier, J., Lasky, L.A., and Lawn, R.M. (1985). Base composition-independent hybridization in tetramethylammonium chloride: a method for oligonucleotide screening of highly complex gene libraries. Proc. Natl. Acad. Sci. U.S.A. *82*, 1585–1588.

Zhang, J., and Byrne, C.D. (1999). Differential priming of RNA templates during cDNA synthesis markedly affects both accuracy and reproducibility of quantitative competitive reverse-transcriptase PCR. Biochem. J. *15*, *337*, 231–241.

Zuker, M., Mathews, D.H., and Turner, D.H. (1999). Algorithms and Thermodynamics for R.N.A Secondary Structure Prediction: A Practical Guide In R.N.A Biochemistry and Biotechnology, Barciszewski, J. and Clark, B.F.C., eds. (N.TO A.S.I Series, Kluwer Academic Publishers) 1999.

Real-Time PCR Instrumentation: an Instrument Selection Guide

6

Sandrine Javorschi-Miller and Ivan Delgado Orlic

Abstract

A paper from 2008 mentions that quantitative PCR is 25 years old (VanGuilder *et al.*, 2008) but routine use of this technology has only taken off during the past 14 years. The first commercial real-time PCR instrument, the ABI Prism 7700, was introduced to researchers in 1996 by Applied Biosystems (Gibson *et al.*, 1996; Heid *et al.*, 1996). Since then, over 40 additional real-time PCR instruments have been developed by more than a dozen vendors. Because there are so many qPCR instruments available utilizing a wide range of technologies, scientists face a daunting selection task. The space includes everything from entry level (single-colour detection, small number of samples, low cost) to more complex (over five channels for multiplexing detection, thousands of samples processed in each run, and expensive system). In this chapter we highlight some key features that differentiate real-time PCR instruments, with the goal of simplifying the criteria needed to select the instrument that best fit a specific scientist's research needs.

Essential components and considerations for a real-time PCR instrument system

A real-time PCR instrument system is the merger of three components that together make it possible to quantify a specific gene of interest:

1. A thermal system that automates the temperature cycling required by the PCR process, from higher temperatures for the melting of double stranded DNA to lower temperatures for primer annealing and extension.
2. An optical system with the dual function of emitting the light energy necessary to activate the fluorophore(s) present in the PCR reaction and capturing the fluorescence generated.
3. A software to control the instrument operation, that provides a user interface to set-up the plate layout and thermal profile, and enables the various functions and algorithms necessary to collect and analyse the data generated.

It is the perfect integration of these disparate components into a single instrument that makes real-time PCR possible.

There are many technical solutions to fulfil the requirements of each component. Most thermal systems are block-based and regulate temperature through the Peltier effect, while others use heated air to achieve thermal cycling. Optical systems vary significantly, with most using lamps or LEDs as the source of excitation energy and PMTs (photomultiplier tubes), CCD (charge-coupled device) cameras, or photodiodes to detect the fluorescence generated.

The most differentiating component, which varies significantly from instrument to instrument, is the software. Nevertheless, there are a number of applications that are shared by most instruments, including gene expression analysis, absolute quantification, relative quantification, allelic discrimination, melt curve analysis, and multiplex analysis.

Understanding these three components will enable you to make an educated choice when considering the purchase of a new qPCR instrument.

Thermal systems

The first PCR instrument (thermocycler, without real-time capabilities) capable of PCR was the Perkin Elmer-Cetus Instruments (PECI) DNA Thermal Cycler 480. It became commercially available in 1987 (Sharrer, 2006). This device established the basic thermal requirement for a device that automates the pre-programmed uniform and rapid thermal cycling required for the PCR process to occur.

Over the years technical variations have been introduced, including the use of heated air (Roche's LightCycler® 2.0), or a resistive heater with air-cooling and centrifugation (Qiagen's Rotor-Gene® Q). However the majority of real-time PCR instruments use blocks that rely on the Peltier effect to actively cycle heat in and out of PCR reactions (Table 6.1a–d). The Peltier effect allows heat to be exchanged by the flow of electrons in a semi-conductor. Fig. 6.1 illustrates how this effect occurs in more detail.

Uniformity

The uniformity of the thermal system is critical to obtaining high-quality and reliable real-time PCR results for the following reasons:

- When running experiments that span multiple plates, a high thermal uniformity will typically equate with higher reproducibility within experiments.
- When running quantification experiments, a high thermal uniformity will positively impact the efficiency of the real-time PCR assay and the precision of the data obtained within a given experiment.
- High thermal uniformity is required for high-resolution melt (HRM) analysis, a novel real-time PCR technique capable of detecting almost any polymorphism, mutation, or epigenetic modification found in DNA samples (Dufresne *et al.*, 2006) based on the precise temperature at which the products melt.

For demanding applications, such as detection of the single basepair polymorphism found in A-to-T Class 4 SNPs, an instrument must be able to distinguish a temperature difference of 0.2°C (Kwok *et al.*, 2007) making uniformity critical. Interestingly, most instruments have thermal uniformities between ± 0.25°C and ± 0.5°C (Table 6.1). Notable exceptions are the Qiagen Rotor-Gene® Q, which achieves a uniformity of ± 0.01°C thanks to its air-cooling and centrifugation format, and the Illumina Eco™ Real-Time PCR System, whose proprietary thermal block achieves a best-in-class uniformity of ± 0.1°C for a block-based instrument.

The variation in thermal uniformity observed in block-based systems is typically due to limitations of the heating element in controlling the thermal edge effect inherent in solid block-based designs. Peltier elements normally only cover a fraction of the bottom surface of the block, which can lead to a temperature gradient from the centre to the edges of the block (Herrmann *et al.*, 2007). Compounding this edge effect is the need to match the multiple Peltier elements needed to cover the surface of a typical block, a task that is far from trivial. Some of these effects can be attenuated by the use of a passive reference, such as ROX, yet ultimately the user must carefully assess the uniformity requirements of his or her application and the limitations inherent in most thermal systems. Alternatively, a unique thermal block design such as the one used by the Illumina Eco system can be used. It is based on a hollow silver block filled with a thermal conductive fluid that is agitated thanks to electromagnetic paddles. This results in perfect thermal uniformity across the block without edge effect.

Speed

Another aspect of the thermal system is speed, which is dependent on the ramping rate of the unit. Conventional Peltier blocks are capable of heating a sample at 1–1.5°C/s, with cooling times generally slightly slower. Improvements like silver and gold plated blocks (Illumina Eco™, using a thermal conductive fluid in its silver hollow block and ABI 7500 FAST, in combination with the use of FAST real-time PCR chemistries) or Therma-Base technology (an efficient heat-equalizing layer between the heat block and the cooling element, found in the Roche LightCycler® 480) have increase sample ramp rates to 2.5–8°C/s. High-thermal conductivity ceramic heating plates (Cepheid SmartCycler®) and resistive heating with air-cooling and centrifugation (Qiagen Rotor-Gene® Q) can achieve rates as high as 10–20°C/s (Table 6.1). This translates to a standard 40-cycle real-time PCR run taking around 90 min, with the fastest instruments performing the same run in under 40 min.

Reaction format

Inherent to the thermal system is the format in which the real-time PCR reactions are enclosed. The most prevalent format is the 96-well plate which is an industry standard 5 inch wide by 3.4 inch deep rectangular plate that holds 8 rows by 12 columns of 0.2 ml wells. Within this same footprint, instrument manufacturers have introduced 384 wells (ABI 7900 FAST HT, Roche LightCycler® 480 and BioRad CFX384™) and even 1536 wells (Roche LightCycler® 1536). For lower throughput, 96-well instruments can hold individual tubes or tube-strips. Recently a number of 48-well instruments (ABI StepOne™, BioRad MiniOpticon™, and Illumina Eco™ Real-Time PCR System) have been introduced that provide a convenient and cost-effective solution between handling individual tubes or tube-strips and using a portion of a 96-well plate per experiment.

Other instruments use glass capillaries (Roche LightCycler® 2.0), SmartTubes (Cepheid SmartCycler®), or Rotor-Discs (Qiagen Rotor-Gene® Q) as alternatives to the standard plate format. These atypical formats offer the advantage of rapid cycling but may require specific additives to work efficiently due to chemistry incompatibilities between vessel and reaction mix.

Table 6.1a Summary of specifications of available real-time PCR systems (Applied Biosystems)

Company	Applied Biosystems				
Model	ABI 7900HT/ABI 7900HT FAST	ABI 7300	ABI 7500/ABI 7500 FAST	StepOnePlus	StepOne
Thermal system	Peltier element block				
Average ramp rate (°C/s)	1.6/3.0	1.6	1.6/3.5	±1.6/±2.2	
Accuracy (°C)	± 0.25 (35–95°C) (3 min after clock start)				
Uniformity (°C)	± 0.50°C (30 s after clock start)				
Time (40 cycles) (min)	<120/33–52	<120	<120/~30	<120/<40	
Temperature range (°C)	4–100				
Optical system	Argon-ion laser	Tungsten halogen lamp		LEDs	
Excitation spectrum	488 nm	1 filter	5 filters	480 nm	480 nm
Detector	CCD + spectrograph	CCD camera		Photodiodes	Photodiodes
Emission filters/channels	4 colours: 500–660 nm	4	5	4	3
Compatible dyes[1]	S, F, V, J, N, R, TR, T	S, F, V, J, N, A, R	S, F, V, J, N, A, R, TR, 3, 5	S, F, V, J, N, A, R	S, F, V, J, R

User dye calibration	Calibrated at install			None	
Passive reference	ROX				
HRM capable	Yes[2]	No	Yes[2]	No	
Format	96-well/384-well plates	96-well plates			48-well plates
Tubes (ml)	0.1/0.05	0.2	0.2/0.1	0.1	
Supported volumes (μl)	10–100/5–20	20–100	20–100/5–30	10–30	
Dimensions (W×D×H) (in)	28 × 33 × 25	13.4 × 17.7 × 19.3	14 × 17.7 × 19.3	9.7 × 20.2 × 20.2	9.7 × 20.2 × 16.8
Footprint (in^2)/volume (ft^3)	924/13.37	237/2.65	247.8/2.77	196/2.29	196 /1.9
Weight (lbs/kg)	180/82	64/29	75/34	53/24	52/23.6

[1]Compatible dyes: S, SYBR Green; F, FAM; V, VIC; J, JOE; N, NED; H, Hex; A, TAMRA; R, ROX; TR, Texas Red; 3, Cy3; 5, Cy5; T, TET; C, Cal Gold 540; CR, Cal Red 610; Q6, Quasar 670; Q7, Quasar 705. AF, Alexa Fluor 350; A5, Alexa 532; A6, Alexa 637.

[2]Only FAST can be upgraded, and it is an option that can be obtained at additional expense to the user.

LED, light-emitting diode; CCD, charge-couple device.

Table 6.1b Summary of specifications of available real-time PCR systems (BioRad)

Company	BioRad				
Model	CFX384	CFX96	iQ5	MyiQ2	MiniOpticon
Thermal system	Peltier element block				
Ramp rate (°C/s)	2–2.5[1]	3.3–5[1]	2–3.3[2]		0.1–2.5
Accuracy (°C)	±0.2 (at 90°C)		±0.3		±0.4
Uniformity (°C)	±0.4 (with 10 s of arrival at 90°C)				
Time (40 cycles) (min)	48/<30 (fast mode)		68/<40 (fast mode)		<40
Temperature range (°C)	0–100		4–100		0–99
Optical system	5 LEDs	6 LEDs	Tungsten halogen lamp		48 LEDs
Excitation spectrum	450–650 nm	450–684 nm	475–645 nm	470–545 nm	470–500 nm
Detector	Photodiodes	Photodiodes	CCD camera		2 photodiodes
Emission filters/channels	5	6	5 channels	2	2
Compatible dyes[3]	S, F, V, H, R, TR, 5, T, C, CR, Q6, Q7		S, F, H, A, TR, 5	S, F/S, F, H, J, T	S, F, V, J

User dye calibration	None		Calibrated at install		None
Passive reference	Not required		Fluorescein		Not required
HRM capable	Yes		No		
Format	384-well plates	96-well plates			48-well plates
Tubes (ml)	0.05	0.2			
Supported volumes (μl)	1–30	1–50	15–100	10–100	10–100
Dimensions (W×D×H) (in)	13 × 18 × 14		11.4 × 22.8 × 15.4	11.5 × 23 × 15.2	7 × 12.5 × 13
Footprint (in²)/Volume (ft³)	235/1.93		261/2.32	264.5/2.32	89.3/0.67
Weight (lbs/kg)	47/21.4		39/17.6/17.9	39/17.6	15/6.8

[1]Average to maximum value.
[2]Average sample rates for cooling and heating for MyiQ2.
[3]Compatible dyes: S, SYBR Green; F, FAM; V, VIC; J, JOE; N, NED; H, Hex; A, TAMRA; R, ROX; TR, Texas Red; 3, Cy3; 5, Cy5; T, TET; C, Cal Gold 540; CR, Cal Red 610; Q6, Quasar 670; Q7, Quasar 705. AF, Alexa Fluor 350; A5, Alexa 532; A6, Alexa 637
Abbreviations: LED, light-emitting diode; CCD, charge-couple device.

Table 6.1c Summary of specifications of available real-time PCR systems (Roche, Agilent)

Company	Roche			Agilent	
Model	LightCycler 1536	LightCycler 480	LightCycler 2.0	Mx4000	Mx3005P
Thermal system	Peltier-based Therma-Base™		Heated air	Hybrid block	Solid-state block
Average ramp rate (°C/s)	2.2–4.8 (cooling and heating rates)		20	±2.2	2.5 (max)
Accuracy (°C)	Not available	±0.2	±0.4	±0.25	
Uniformity (°C)	Not available	Not available	±0.2	±0.25	
Time (40 cycles) (min)	<50	<60/<40 (FAST)	60/30 (FAST)	Not available	90/52 (FAST)
Temperature range (°C)	37–95	Ambient-99			Ambient-95
Optical system	Xenon lamp		LED	Quartz tungsten halogen lamp	
Excitation spectrum	465 nm, 533 nm	5 filters	470 ± 10 nm	350–750 nm	
Detector	CCD camera		Photodiodes	Photomultiplier tubes	Photomultiplier tubes
Emission filters/channels	2	6	6 channels	4 FW: 350–830 nm	5 FW: 350–700 nm
Compatible dyes[1]	Not available	S, F, V, H, 5, LC dyes, Cy 5.5		S, F, V, J, H, A, R, TR, 3, 5, T, AF	

User dye calibration	Not available	Any dye compatible with filters		User defined	
Passive reference	Not available	Not required		ROX (optional)	
HRM capable	Not available	Yes		No	
Format	1536-well plates	96-well/384-well plates	32 capillaries	96-well plates	
Tubes (ml)	0.002	0.1/0.05	0.02–0.1	0.2	
Supported volumes (μl)	0.5–2	10–100/5–20	20–100	10–50	
Dimensions (W × D × H) (in)	22.6 × 23.1 × 19.6	24 × 24 × 21.5	11 × 11.2 × 19.9	30 × 18.1 × 20	13 × 18.1 × 16.9
Footprint (in^2)/volume (ft^3)	522/5.92	576/7.17	123.6/1.42	543/6.28	235.3/2.3
Weight (lbs/kg)	123/55.6	123/55.6	48.5/22	110/50	45/20.4

[1]Compatible dyes: S, SYBR Green; F, FAM; V, VIC; J, JOE; N, NED; H, Hex; A, TAMRA; R, ROX; TR, Texas Red; 3, Cy3; 5, Cy5; T, TET; C, Cal Gold 540; CR, Cal Red 610; Q6, Quasar 670; Q7, Quasar 705. AF, Alexa Fluor 350; A5, Alexa 532; A6, Alexa 637
Abbreviations: LED, light-emitting diode; CCD, charge-couple device.

Table 6.1d Summary of specifications of available real-time PCR systems (Cepheid, Qiagen, Eppendorf, Illumina)

Company	Cepheid	Qiagen	Eppendorf		Illumina
Model	SmartCycler	Rotor-Gene Q	Realplex[2]	Realplex[4]	Eco™
Thermal system	**Ceramic plates**	**Air-cooled**	**Gradient peltier**		**Peltier block**
Average ramp rate (°C/s)	2.5–10	15–20	3–4		5
Accuracy (°C)	±0.5	±0.25	±0.2		±0.1
Uniformity (°C)	±0.5	±0.01	±0.3–0.4		±0.1
Time (40 cycles) (min)	20–40	40–60	<30		<40
Temperature range (°C)	Ambient-95	Ambient-99	4–99		Ambient-100
Optical system	**4 LEDs**	**Up to 6 LEDs**	**96 LEDs**		**2 panels of 48 LEDs**
Excitation spectrum	510–750 nm	365–680 nm	520–550 nm	520–605 nm	350–700 nm
Detector	Photodetectors	PMT	Photomultiplier tubes	Photomultiplier tubes	CCD camera
Emission filters/channels	4	Up to 6	2	4	4

Compatible dyes[1]	F, T, TR, 3, 5, A5, A6	F, V, J, H, R, TR, 5, T, C, CR, Q7, AF, and others	All formats except hybridization probes	S, F, H, R, 5
User dye calibration	None	None	Calibrated at install	None
Passive reference	Not required	None	Not required	ROX (optional)
HRM capable	No	Yes	No	Yes
Format	16 tubes	Rotor disk 100/72/36	96-well plates	48-well plates
Tubes (ml)	Proprietary tubes	0.1/0.2	0.2	0.2
Supported volumes (μl)	25–100	10–50	10–50	5–30
Dimensions (W × D × H) (in)	12 × 12 × 10	14.6 × 16.5 × 11	10.2 × 16.1 × 15.6	13.6 × 12.2 × 12.6
Footprint (in^2)/volume (ft^3)	144/0.83	241/1.51	165.3/1.5	165.9/1.19
Weight (lbs/kg)	22/10	31/14	52.9/24	30/13.6

[1]Compatible dyes: S, SYBR Green; F, FAM; V, VIC; J, JOE; N, NED; H, Hex; A, TAMRA; R, ROX; TR, Texas Red; 3, Cy3; 5, Cy5; T, TET; C, Cal Gold 540; CR, Cal Red 610; Q6, Quasar 670; Q7, Quasar 705. AF, Alexa Fluor 350; A5, Alexa 532; A6, Alexa 637

Abbreviations: LED, light-emitting diode; CCD, charge-couple device.

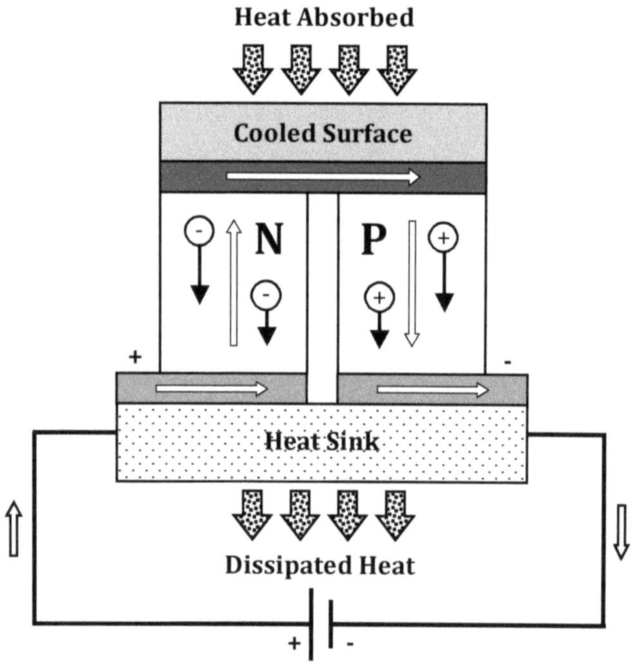

Figure 6.1 Peltier effect schematic. The Peltier effect corresponds to the heat transfer from one end of a conductor to the other (thick arrows) caused by the passage of an electrical current. When a voltage is applied across a semiconductor thermocouple, electrons in the N type material migrate towards the positive terminal. Similarly holes in the P type material migrate towards the negative terminal. The charge movement is accelerated thanks to an electric field (thin arrows) created by applying a voltage between the terminals of the device and this result into heat being dissipated through a heat sink.

Gradient

Some real-time PCR instruments (Bio-Rad instruments and the Eppendorf Mastercycler® ep Realplex) have gradient capability, which means that the user can specify a range of temperatures across the heat block. This capability can be useful and time saving because one can determine the optimal annealing temperature for a given qPCR assay in a single run. However, improved chemistries have all but eliminated the need for extensive optimization of annealing temperatures, limiting the need for gradients.

Optical systems

The addition of an optical system capable of detecting the amplification of DNA as it takes place at every cycle of PCR, to a thermal system provides the basis of a real-time PCR instrument. Many current real-time PCR instruments were not built from the ground up but instead were built on top of pre-existing thermal platforms. For example the ABI 7300 was built over a GeneAmp 9600 thermocycler, the ABI 7500 over a GeneAmp 9700, the BioRad CFX96™ and CFX384™ over a C1000, the BioRad iQ™5 and MyiQ™ over an iCycler, the BioRad MiniOpticon™ over an MJ Mini Cycler, and the Eppendorf Mastercycler® ep Realplex over a Mastercycler®. While these instruments are capable of running real-time PCR experiments, their thermal units were not designed with the extreme thermal uniformity required for optimal real-time PCR performance.

Optical systems are made up of two main components: an excitation light source and a fluorescence emission detector.

Light source

Light sources can be subgrouped into two types: narrow or broad spectrum.

Almost all broad-spectrum instruments utilize incandescent tungsten–halogen or xenon bulbs as their light source (Table 6.1), which generates high intensity light energy in the lower part of the visible range (475–645 nm). One of the main advantages of a broad-spectrum light source, when paired with a set of filters, is that it provides the widest choice of detectable fluorophores. For example, the light emitted by the tungsten–halogen lamp of the ABI 7500 goes through five excitation filters, enabling this instrument to detect up to five different fluorophores in the same reaction (known as multiplexed RT-PCR). The main disadvantage of incandescent bulbs is that they need to be replaced (as often as every 6 months) and decay over time.

Most narrow-spectrum light sources found in real-time PCR instruments are light emitting diodes (LEDs) with the exception of the ABI 7900HT, which uses an argon-ion laser. Unlike halogen lamps, which emit light spanning up to 200 nm, LEDs emit light of a very narrow wavelength range. For example a standard blue-green LED has a peak emission of 470 nm, which means that it can only excite a limited number of fluorophores (e.g. FAM, SYBR® Green). To get around this limitation some instruments use multiple LEDs (BioRad CFX96™ and CFX384™, Qiagen Rotor-Gene® Q, Cepheid SmartCycler®, and Helixis PIXO™), each emitting light of different wavelengths, essentially recreating the multiplexing capabilities of a broad-spectrum light source. The main advantage of LEDs is their lifespan; they last up to 40,000 hours so they never need to be changed. LEDs are expected to outlive the real-time PCR instrument that they are installed in.

Emission detector

Once the narrow- or broad-spectrum light source stimulates the emission of fluorescence by the fluorophore(s), this energy needs to be collected by a fluorescence emission detector. Detectors come in many styles, including CCD cameras, photodiodes, and photomultiplier tubes (PMTs). Table 6.1 summarizes the type of detectors found on the different commercially available instruments. Each one of these detectors has advantages and disadvantages:

- Photodiodes have a great linearity of output current as a function of light, generate little noise, are cheap, and have a long lifetime.
- PMTs are more sensitive and have a quicker response time than photodiodes, but they are more expensive.
- CCD cameras have high quantum efficiency and very linear outputs. They can image all wells simultaneously. They do not need to be moved from sample to sample (scanning), and have the advantage of needing minimum moving parts leading to a more robust and cost-efficient mechanical design.

- Ultimately the benefit of using these different detectors has to be considered in the context of the optical system as a whole.

Dye calibration

Finally, another consideration is the requirement, or not, for dye calibration of the instrument. While some instruments need to be calibrated before they can be used or as part of a regular maintenance schedule (ABI 7300 and 7500, BioRad MyiQ™ and iQ™5, and Eppendorf Mastercycler® ep Realplex), to ensure best optical accuracy, most instruments come factory calibrated and ready to be used (Table 6.1). Dye calibration specifies which pure dyes were used to teach the instrument how to read the emission and excitation spectra of those particular dyes. While experiments can be run using other dyes with very similar spectra to that of the calibrating dyes, issues such as lower signal detection and bleed-through to other channels need to be considered before choosing non-standard dyes (Fig. 6.2).

Software

An often overlooked, but crucial component of a real-time PCR instrument is the software. Software can be broken down into two components: the back-end and the front-end.

The back-end

The back-end consists of all the processes that occur beyond the user's control, like the communication between a circuit board and the thermal block, or the algorithms used to translate the fluorescent signal generated into output data. The minimum requirements

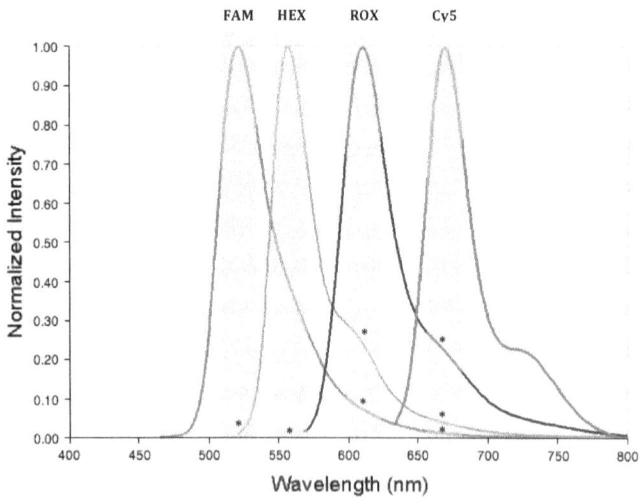

Figure 6.2 Dye calibration and deconvolution principles. Emission spectra, and spectral overlap, of four commonly used fluorescent dyes (FAM, HEX, ROX, and Cy5). To deconvolve the signal of multiple dyes within a single channel, dye calibration information is used to remove all fluorescent-signal present in a given channel that does not belong to the dye of interest. In short, dye calibration information is used to subtract, using simultaneuous equations, any dye bleed-through emissions (denoted by * above) from the emission signal generated by the dye of interest.

of all real-time instruments are: signal detection, normalization, and Cq calculation. For instruments with multiplexing capabilities, dye de-convolution is also necessary.

Dye de-convolution is achieved through algorithms capable of separating signals from different channels and minimizing crosstalk between them. This is required because the emission spectra of different fluorescent dyes often overlap (Fig. 6.2). To analyse only the desired fluorescence in a given channel in the presence of multiple fluorescent sources, the instrument needs to know which dyes are present in the reaction. By providing this information, during assay set-up, the software will be able to separate (deconvolve) the various fluorescent signals present in different channels and only analyse the signal from the desired dye. This is achieved mathematically through a linear equation that uses dye calibration matrices stored in the instrument to remove any fluorescence present in a given channel that does not belong to the dye of interest.

The front end (user interface)

The front end is the user interface (UI). The software must provide the controls necessary for the user to specify their plate set-up and the type of experiment to be run. Typically, software is capable of the following experiments:

- gene expression analysis: absolute or relative quantification;
- allelic discrimination;
- melting curve analysis.

The software can automatically perform these analyses or, alternatively, once the instrument captures the fluorescent signal of an experiment, many of these calculations can be performed manually using standard programs like Microsoft Excel.

High-resolution melting (HRM) analysis, a novel cost-efficient method of doing gene scanning or genotyping is enabled by the latest generation UIs coupled with the high thermal uniformity of the newest instruments. HRM can distinguish almost any genetic variation (SNPs, methylation and mutations) in PCR amplicons based on melting temperature.

To date, two real-time PCR instruments (LightCycler® 480, and Illumina Eco™) have native HRM capabilities. A software/hardware upgrade is available for purchase to make the ABI 7500 and 7900 HT FAST, Qiagen Rotor-Gene® Q models, or the BioRad CFX96™ and CFX384™ instruments, amenable to HRM. It is important to understand that even if the software is capable of HRM, thermal and optical systems capable of very high uniformity and accuracy are still a must.

The UI also enables plate set-up. Setting up an experiment can be a very time-consuming step in the UI when preparing a real-time PCR experiment. During this step the user needs to specify:

- the identity of each sample in each well;
- the assay to be run on each well;
- the optical channel each assay will use;
- the inclusion of all necessary controls;
- thermal profile.

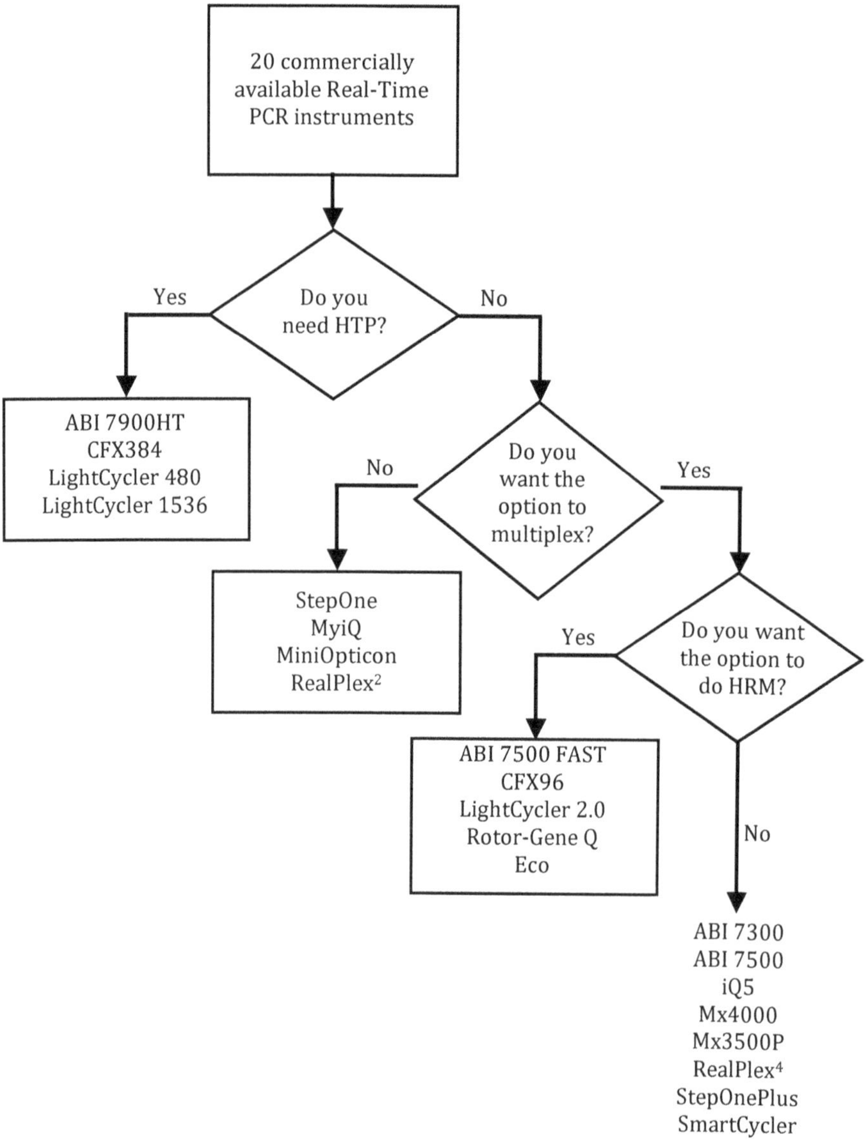

Figure 6.3 Instrument selection. Selection of a qPCR instrument based on throughput, multiplexing technology, and HRM capabilities.

A critical part of the process of setting up a real-time PCR experiment is specifying which optical channels to collect data from. If the input of these settings is done incorrectly, the whole experiment may be lost. Some instruments get around this possibility by collecting data from all channels in every experiment (ABI 7300, StepOne™ and StepOnePlus™, Illumina Eco™). This makes it impossible to lose data by choosing the wrong channels for data collection.

Commercially available real-time PCR instruments

We will now highlight and categorize the main real-time PCR instruments available on the market today. Instruments can be broadly categorized according to the following:

- throughput (small lab use vs. large core facility/shared instrument/large projects);
- multiplexing features (capable of detecting up to four targets in a single reaction);
- ability to run HRM experiments;
- sample format (plate versus tubes/capillaries);
- instrument size.

With many instruments tailored to fit almost every research need, the challenge is in identifying which platform fits your current and future research needs. The decision tree in Fig. 6.3 should help you find the perfect instrument for your needs based on throughput, multiplexing and HRM requirements.

Throughput

A first step towards narrowing down the selection of a real-time PCR instrument is to separate them based on throughput. High-throughput instruments include the ABI 7900HT, the CFX384™, and the LightCycler® 480 and 1536. If your research needs require the running of hundreds or thousands of qPCR reactions per day, these are the instruments of choice.

Multiplexing features

The next step is to determine whether you need an entry-level system or one with full multiplexing capabilities, capable of detecting four or more colours in a single reaction. Under this criterion, the StepOne™, MyiQ™, MiniOpticon™, and RealPlex (Gibson *et al.*, 1996) are all entry-level instruments (Fig 6.4). When multiplexing is not a major requirement these instruments can perform real-time PCR applications satisfactorily.

Among entry-level systems, another consideration is whether they are open or closed systems. The MyiQ™ and the MiniOpticon™ require the use of BioRad's mastermixes because of their fluorescein normalization requirement, reducing the choice of available reagents for some applications.

High-resolution melt capabilities

The ability to run HRM experiments is becoming an important selection criterion. The ABI 7500 FAST, CFX384™, CFX96™, LightCycler® 2.0, LightCycler® 480, Rotor-Gene® Q, and Eco™ are the only HRM-capable instruments (Fig. 6.3). When considering HRM-capable instruments, it is important to note that many systems require a costly hardware and software upgrade to enable this feature. Instruments coming with HRM capabilities by default will be a more cost-efficient and have a higher performance situation.

Sample format

Another selection criteria are the well formats each instrument supports. There are advantages of plate-based reaction vessels as they are readily amenable to the use of standard multi-channel pipetting equipment or robotics for high-throughput applications.

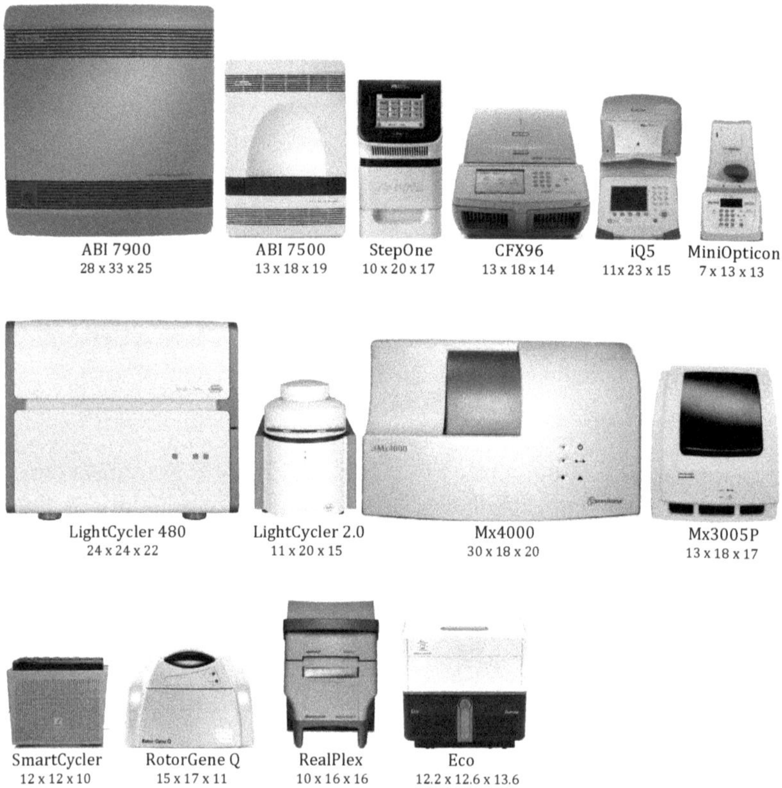

Figure 6.4 Real-time PCR Instruments. Images of real-time PCR instruments. Front views of the various real-time PCR instruments are shown, to scale relative to each other.

The instruments that use the industry-standard 0.2-ml, 96-well plate format are: the ABI 7300, ABI 7500, CFX96, ™ iQ™5, MyiQ™, Mx4000®, Mx3005P®, Mastercycler® ep RealPlex^2, and Mastercycler® ep Realplex (Sharrer, 2006). Other plate-based systems that utilize smaller reaction volumes (ABI 7900HT 384-well block, CFX384, LC 1536, LC 480 384-well block, and PIXO) are compatible with multi-channel pipets or robotics designed around this smaller format. Alternative formats, such as the proprietary tubes of the SmartCycler and the capillary tubes of the LightCycler 2.0 provide unique advantages, but lead to increased consumable expenses and require increased processing time per sample.

Instrument size

Size may be an important part of your decision. Some instruments take over a large chunk of bench real estate and can require dedicated resources. These include the ABI 7900HT, the LightCycler® 1536 and 480, and the Mx4000®. Other instruments are small enough to fit almost anywhere, including the MiniOpticon™, SmartCycler®, and Eco™. The rest vary considerably in size, but overall are large enough to require special space considerations (Fig. 6.4).

Price

A final consideration is price. Real-time PCR instruments range in price from $13,900 to more than $85,000. Typically the price reflects the capabilities of the instrument, with more affordable instrument having fewer capabilities than the more expensive ones, yet this is not always the case. With economic pressure, new platforms are breaking this established rule. For example, Illumina Eco™ real-time PCR instrument, which can perform multiplexing and HRM even while being the least expensive instrument available.

Once you have narrowed down the list of possible real-time PCR instruments, it is strongly recommended to demo instruments before making a decision. No list of specifications or published comparisons can substitute for hands-on, real-life testing of an instrument. One of the tested instruments will typically better satisfy your specific experimental needs than the others and become the instrument of choice for your purchase.

Conclusions/general considerations

In this chapter we tried to provide a comprehensive overview of the different real-time PCR instruments available today. In addition we presented some guidelines that can be used during the purchase decision process. It must be noted that our focus was on the instruments, and to a great extent we left out other considerations that need to be taken into account, including:

- available chemistries (instrument-specific versus universal mastermixes);
- consumables (instrument specific versus industry standard);
- technical and customer support.

The adoption of real-time PCR continues to grow, not just in traditional academic settings but also in more specialized and regulated environments. As technology advances and competition intensifies, real-time PCR is becoming more powerful and affordable. With 15 years of real-time PCR utilization behind us, we can look forward to a future full of innovation and increased accessibility.

References

Dufresne S.D., Belloni, D.R., Wells, W.A., and Tsongalis, G.J. (2006). BRCA1 and BRCA2 mutation screening using SmartCycler II high-resolution melt curve analysis. Arch. Pathol. Lab. Med. *130*, 185–187.

Gibson U.E., Heid, C.A., and Williams, P.M. (1996). A novel method for real time quantitative RT-PCR. Genome Res. *6*, 995–1001.

Heid C.A., Stevens, J., Livak, K.J., and Williams, P.M. (1996). Real time quantitative PCR. Genome Res. *6*, 986–994.

Herrmann M.G., Durtschi, J.D., Wittwer, C.T., and Voelkerding, K.V. (2007). Expanded Instrument Comparison of Amplicon DNA Melting Analysis for Mutation Scanning and Genotyping. Clin. Chem. *53*, 1544–1548).

Kwok A., Bassam, B., and Reja, V. (2007). Genotyping a Class 4 SNP by high-resolution melt (HRM) using SYBR Green I. Corbett Research. Research Report 080701.

Sharrer T. (2006). Mr. Cycle: An 'Automated' PCR Prototype. The Scientist *20*, 92.

VanGuilder H.D., Vrana, K.E., and Freeman, W.M. (2008). Twenty-five years of quantitative PCR for gene expression analysis. BioTechniques *44*, 619–626.

7

Quantitative PCR Data Analysis – Unlocking the Secret to Successful Results

Jan Hellemans and Jo Vandesompele

Abstract

Real-time quantitative PCR (qPCR) is the gold standard for fast, accurate, sensitive and cost-efficient gene expression analysis. Despite its conceptual simplicity and ease of use, the multistep qPCR workflow contains many potential pitfalls. An intelligent experiment design and set-up, high-quality reagents and assays, quality controls in each step of the workflow, proper quantification models and appropriate biostatistical analyses pave the way to successful gene expression results. This chapter will cover all data analysis aspects from the evaluation of pilot studies and quality controls, through universally applicable quantification models and biostatistics, to the reporting of experiment results.

Analysing pilot studies

Prior to starting a qPCR based gene expression profiling, three types of pilot experiments should be performed. In the first type, newly designed assays are validated, while a second type of pilot study has to be performed if you start working on a new sample type or treatment condition to determine the best set of reference genes for normalization. The third pilot experiment is needed to assess the variability and magnitude of the observed differential gene expression, enabling a power study to determine how many samples are required to reach statistical significance. The search for stably expressed reference genes and the power analysis may be combined in one experiment.

Assay quality controls

High quality assays are one of the key elements required in order to obtain precise and accurate results. The design and *in silico* evaluation of new assays is not the subject of this chapter. More information on this topic can be found in Chapters 4 and 5. Once new assays have been ordered they need to be empirically tested in a pilot experiment, which requires three kinds of quality control: specificity determined by melt curve analysis (when using a double-stranded DNA specific binding dye), specificity determined by product size analysis and amplification efficiency calculated from a dilution series (also called standard curve). Importantly, commercially purchased predesigned assays should also undergo this empirical quality control to demonstrate performance in your laboratory setting.

Melt curve analysis (Fig. 7.1) is possible if an intercalating dye such as SYBR Green I is used to monitor the qPCR reaction and a melt step is included after the amplification

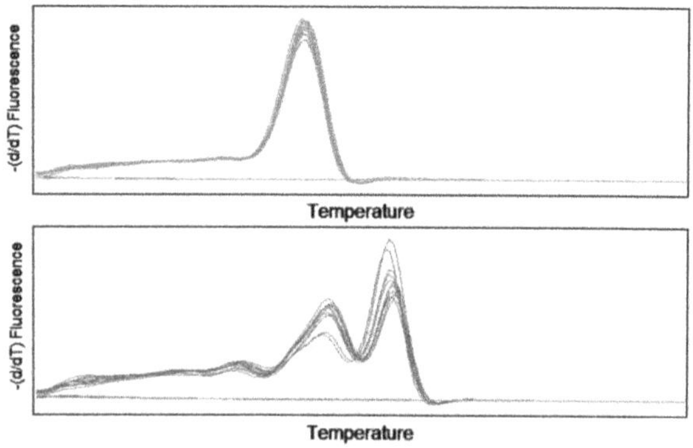

Figure 7.1 Melt curve analysis. The upper plot reveals a unique melt peak, suggestive for specific amplification. The plot at the bottom contains two clearly distinct melt peaks, indicative for non-specific amplification. A colour version of this figure is located in the plate section at the back of the book.

process in the qPCR program. A good assay should reveal a single peak in the plot depicting the negative first derivative of fluorescence over temperature ($-\mathrm{d}F/\mathrm{d}T$) while additional peaks usually indicate the presence of non-specific amplification products. A lower more flattened peak typically points to primer dimer formation and in some rare events a single specific PCR product may show two peaks as a result of a non-random distribution of GC content across the amplicon, especially if a long amplicon is melted.

Confidence in assay specificity can be increased by performing a one time only product size analysis. This is typically done using agarose gel electrophoresis but recently more convenient, microfluidic based alternatives (such as Caliper's LabChip GX and Shimadzu's MultiNA) have become available. Specific assays should show a single band (peak) in such analyses.

Although alternatives are being investigated (Ruijter *et al.*, 2009; Rutledge and Stewart, 2008; Zhao and Fernald, 2005), dilution series are still the gold standard for assessment of amplification efficiencies. A detailed protocol for the creation of proper dilution series can be found at http://medgen.ugent.be/CMGG/onderzoek.php?topic_id=11. A typical dilution series contains six 4-fold dilution points for a representative reference sample (e.g. a mixture of all samples to be analysed), and a no-template control. The (relative) quantities and corresponding Cq values are used to create a standard curve in which Cq values are plotted against log transformed quantities and a linear trend line is fit to the data. Consequently, the slope of the trendline is used to estimate the PCR efficiency. An optimal efficiency (100%), reflecting a doubling of the PCR product each cycle, is characterized by a slope of –3.32. In general, assays with amplification efficiencies between 90% and 110% are considered as acceptable. The R^2 and the error on the estimated amplification efficiency can be used to evaluate the quality of the efficiency determination (Fig 7.2). The mathematical formulas to determine amplification efficiencies and their corresponding errors have been described by Hellemans *et al.* (2007). Most qPCR instrument software packages include basic tools to calculate efficiency values from standard curves and a comprehensive analysis of the PCR efficiency including the error on the estimated efficiency is available in Biogazelle's qbase^{PLUS} software (http://www.qbaseplus.com).

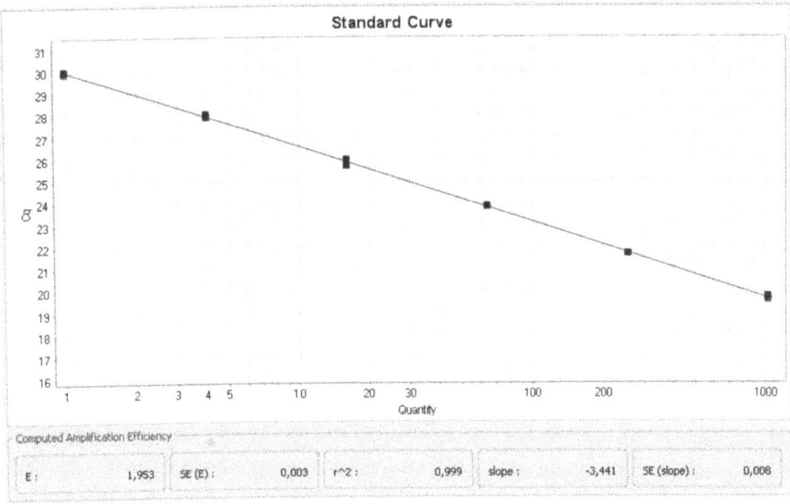

Figure 7.2 Typical standard curve obtained from a 6-point 4-fold dilution series for an assay with a good amplification efficiency ($E = 1.953$, efficiency = 95.3%) as calculated by qbase^{PLUS}. A high correlation coefficient ($r^2 = 0.999$) and low error on the estimated efficiency (SE(E) = 0.3%) indicate a high-quality standard curve from which reliable efficiency estimations can be drawn. A colour version of this figure is located in the plate section at the back of the book.

Good assays should pass all three types of QC, i.e. have a single melt peak, a unique band of the expected size and an amplification efficiency in the 90–110% range. Assay QC is best to be repeated whenever experiment conditions are modified or a fresh batch of primers is ordered. Experimental conditions are known to have an effect on specificity, whereas the oligonucleotide provider appears to affect amplification efficiency (S. Derveaux, unpublished).

Reference gene selection

It is generally recognized that reliable expression results can only be obtained if multiple stably expressed reference genes are used to normalize the data. Identification of the best set of reference genes is typically done in a pilot geNorm experiment (Vandesompele *et al.*, 2002). In such an experiment a number of candidate reference genes (typically 10) are measured in a representative set of samples. The candidate reference genes should not contain genes that are co-regulated (e.g. genes from the same pathway or members of the same gene family) because such genes may falsely appear to be stably expressed. If you plan to analyse different groups of samples (e.g. patient vs control, or treated vs untreated), you should have an equal representation of samples from all groups to avoid a bias towards reference genes that are stably expressed in the largest group, but not really stable in samples of the other groups. A reliable selection of reference genes can only be made on a sufficiently large set of representative test samples. As a guide, 10 samples are sufficient if no clearly distinct groups are present whereas 12 samples are required if 3 sample groups can be recognized (3×4 samples).

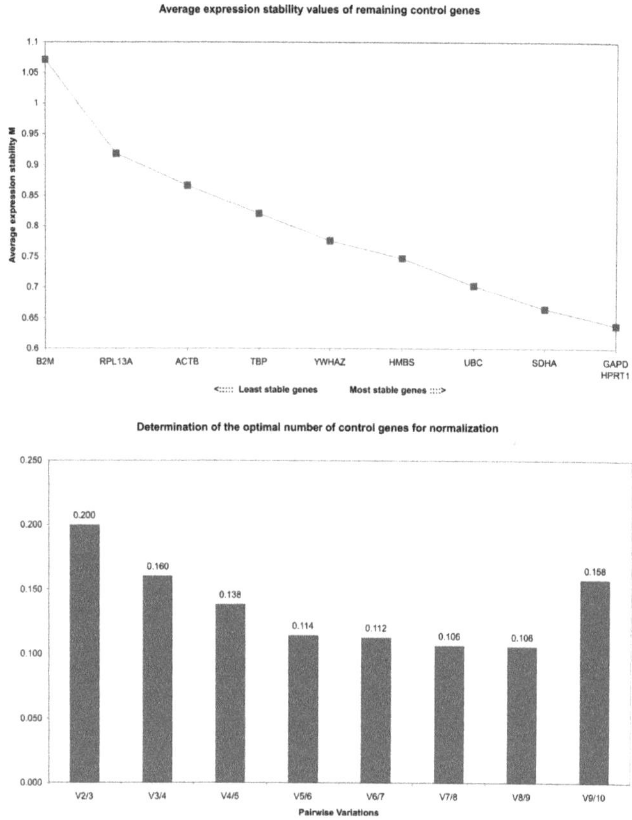

Figure 7.3 Outcome of a geNorm analysis on a heterogeneous cancer biopsy sample set. M values (line graph on top) for candidate reference genes start out rather high (1.35 for B2M at the left), then gradually drop towards the more stably expressed genes at the right (TBP and YWHAZ with an M value of approximately 0.5). The analysis of *V* values (bar chart at the bottom) indicates that 4 reference genes are minimally required for optimal normalization in this heterogeneous sample set (V4/5 is below 0.15, indicating that there is not need to include a 5th gene in the normalization factor). A colour version of this figure is located in the plate section at the back of the book.

A geNorm analysis requires a table with relative quantities (RQ, see below for calculation details) for all sample – target combinations as input. This table should have sample names as row headers, target names as column headers, an empty 'A1' cell and no missing RQ values and the file should be imported into the geNorm package, which can be freely downloaded from (http://medgen.ugent.be/~jvdesomp/genorm/), before start the analysis. Alternatively, the raw Cq values can be input from your qPCR experiment into qbase^{PLUS} for fully automated analysis and interpretation based on an enhanced geNorm algorithm. In a geNorm analysis, two types of graphs are generated. The first graph (Fig 7.3a) ranks the candidate reference genes according to their expression stability in the tested samples. Genes with variable expression (high geNorm M stability measure) are found on the left, the most stably expressed genes (low M value) are located on the right side. The best two reference genes cannot be separated because a geNorm stability analysis requires at least two genes. The second graph (Fig 7.3b) can be used to determine the minimum number of reference genes that are required for reliable normalization. Starting with the two best reference genes, additional genes should be added to the selection until the V-value drops below 0.15 (Vandesompele *et al.*, 2002).

Determination of the statistical power

The objective of a gene expression experiment is to draw a clear-cut conclusion based on the results. This is only possible if the statistical test yielded a *P*-value < 0.05 (significant difference) or a *P*-value > 0.05 with a sufficient 'power' to find a significant difference if there was one. Typically, at least 80% power is required (the higher the power, the higher the chance of finding a significant difference if there is one). Determination of the power of an experiment is done in a so-called power analysis experiment; based on the observed variance and magnitude of differential expression, the minimal sample size is calculated to ensure that statistical significance is reached if there is indeed a true difference with a minimal expected magnitude. There are many free software tools online (see URLs below) to perform a power analysis. The required input info is the accepted false positive rate (alpha value, typically 0.05), the observed variance (standard deviation), the minimal fold change you want to see (typically 2-fold) and the power level (typically between 80% and 99%). Importantly, a power analysis should be performed before the final experiment; it is completely useless to do power analysis on the final experiment. Also, beware of an 'overpowered' study; when the sample size is large enough you can always find a significant (small) difference, but this is often biologically or clinically irrelevant. Finally, for reasons explained later, all input parameters (expected minimal difference and standard deviation) should be on the log scale. Online Software Tool URLs:http://biostatistics.mdanderson.org/SoftwareDownload/SingleSoftware.aspx?Software_Id=41http://www.cs.uiowa.edu/~rlenth/Power/

The quantification model

Any quantification model is based on raw quantification cycle (Cq) values, which can be determined by two different algorithms. The baseline/threshold method first defines the baseline cycle range below which amplification does not occur and then uses a multiple of the standard deviation of the noise levels in the baseline to establish a threshold line which, upon crossing the amplification curves, generates Cq values. Other methods are based on identification of a unique point in the sigmoidal amplification curve (e.g. the first maximum of the second derivative of the fluorescence over time. These curve shape based methods are generally more objective and less sensitive to user based interpretation of the data. In our hands, they also result in smaller standard deviation of the PCR replicates).

Different quantification models for relative quantification have been proposed since the invention of the qPCR technology (Higuchi *et al.*, 1993). The first model, known as the delta-delta-Cq method (equation 1), is still very popular despite its known limitations: normalization with a single reference gene only and assuming 100% amplification efficiency (doubling in PCR product amount in every cycle, base of exponential amplification = 2) for both the gene of interest and the reference gene (Livak and Schmittgen, 2001). An improved quantification model (equation 2) taking gene specific amplification efficiencies into account was developed soon afterwards by Pfaffl (2001). A universally applicable quantification model (equation 3) with multigene normalization possibilities was finally described in detail in 2007 (Hellemans *et al.*, 2007). This publication includes full mathematical details and a corresponding error propagation approach. The qbase quantification model is not fundamentally different from its predecessors as it can be simplified to the Pfaffl model if a single reference gene is used, and to the Δ-Δ-Cq model if also 100% amplification is assumed.

The polymerase chain reaction is an exponential process whereby the specifically amplified product ideally doubles each cycle. The Cq values derived from the amplification plot are logarithmic values that need to be converted into linear relative quantities (RQ), the exponential function that transforms ΔCq values into RQ values is given in equation 4. In this formula, ΔCq is defined as the difference between a reference Cq value (Cq value of a calibrator sample (e.g. untreated control), or preferably the average Cq across all samples to minimize overall errors if error propagation is done (Hellemans *et al.*, 2007)) and the Cq value for the sample of interest. If repeated qPCR measurements are included in the experiment, Cq values may be replaced by average Cq values.

It is important to realize that the calculated relative quantities represent the measured difference between samples and not necessarily the differences in gene expression level between samples because experimentally induced variations also contribute to the observed difference (Fig 7.4). Such variations include both differences in cDNA reaction input amounts as well as the often-underestimated run-to-run variation that occurs if measurements for a given assay are spread across multiple runs. Both types of variation can be corrected for, provided that appropriate controls are included in the experiment.

The variation in cDNA input amount is corrected for by means of a normalization procedure (equation 5, step 1 in Fig 7.4). Normalized relative quantities (NRQ), reflecting gene expression differences (and possibly technical run-to-run variation) are obtained by dividing RQ values by a sample specific normalization factor (NF), determination of which is dependent upon the experiment set-up. Traditionally, this normalization factor equals the relative quantity of a single reference gene in that sample. If multiple reference genes are included in the experiment, the NF is defined as the geometric mean of the RQ values for all reference genes measured in that sample (Vandesompele *et al.*, 2002). Therefore, both approaches require the inclusion of at least one, but preferably multiple validated reference genes. As recently proposed, normalization factors may also be based on the RQ values for all genes, provided that the screened set of genes is sufficiently large and not biased towards a specific pathway or a set of genes that have been implicated in a related condition (Mestdagh *et al.*, 2009).

Although technical run-to-run variation can often be avoided by using the preferred sample maximization approach (Hellemans *et al.*, 2007) or by using a sufficiently large measurement platforms (e.g. a 384-well instead of a 96-well instrument), it still may be part of the observed variation in some experiments. This type of variation can be minimized by standardizing the operating procedure as much as possible (e.g. using the same reagents and Cq determination method, to name a few). However without the inclusion of one (or preferably more) inter-run calibrator(s) (IRC, i.e. a given sample measured in all runs for all genes) one cannot determine whether the variation is acceptably low. The inclusion of IRC samples in all runs (and for each gene) allows a calibration factor (CF) to be determined for every gene–run combination. This CF is defined as the geometric mean of the NRQ values of all the IRCs in a given gene–run combination, and allows NRQ values to be converted into calibrated normalized relative quantities (CNRQ) (equation 6, step 2, in Fig 7.4). By

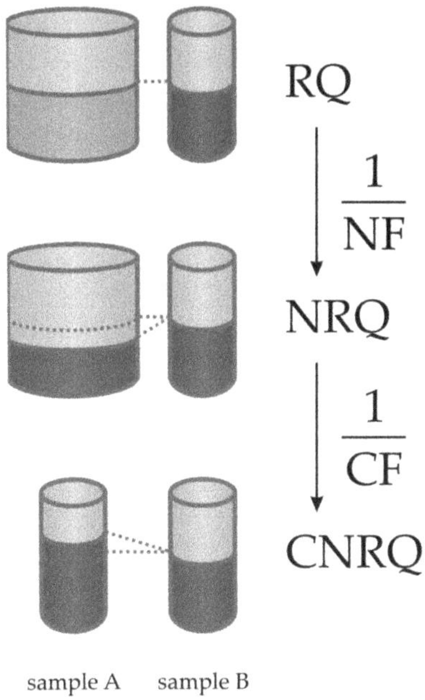

Figure 7.4 Calculation workflow. Identical relative quantities (RQ) of a given target between two samples (A and B) do not necessarily imply equal expression levels of the measured target by the applied assay. Differences in final cDNA concentration (represented by a different shade) or technical run-to-run variation (represented by a variable width) may obscure the picture of true differential expression. The normalization procedure removes the variation resulting from the use of different cDNA concentrations, thus yielding normalized relative quantities (NRQ) in which the expression level is only affected by possible run-to-run variation. This remaining variation can be removed by calibrating NRQs into calibrated normalized relative quantities (CNRQ) that truly reflect the differences in gene expression between the two samples. From a mathematical perspective normalization and inter-run calibration are equivalent procedures and are effected by division by a normalization factor (NF, geometric mean of reference gene NRQ values) or a calibration factor (CF, geometric mean of IRC sample NRQ values), respectively. A colour version of this figure is located in the plate section at the back of the book.

then, all the experimentally induced variation has been corrected for and the obtained CNRQ values truly represent the differences in gene expression.

The equations in this chapter are represented in simplified form, full mathematical details have been described elsewhere (Hellemans *et al.*, 2007). The correct calculation of CNRQ values, along with proper error propagation of all the contributing errors is non trivial and time-consuming when performed manually in a spreadsheet. Most qPCR instrument software packages include basic analysis tools but usually fail in some of the aspects described above (e.g. multigene normalization, correction for run-to-run variation or proper error propagation). Professional qPCR data analysis software packages such as qbase^{PLUS} (http://www.qbaseplus.com) offer the entire range of quantification tools, irrespective of the instrument that has been used to generate the Cq values. In addition to this, a set of important quality controls (see below) are included to ensure the highest quality of results by excluding bad repeated measures or non-stably expressed reference genes.

Data quality control

Data quality control (QC) is an essential aspect of qPCR experiments, not only in the pre-analytical steps (sample and assay QC) but also during the data analysis phase. A first type of quality control is the inspection of melt curves for all measurements to verify that assays have the same specificity as determined initially in the pilot experiment. Additional types of QC inspect Cq values (controls and replicates) or intermediate results (reference gene stability or consistency of inter-run calibrators).

Several types of controls can be included in an experiment. The most widely used control is the so-called no template control (NTC), a reaction to which no (c)DNA has been added, serving as a control for contamination. An NTC should be included for every assay and every time a new run is set-up. Ideally, NTCs should not result in any amplification and, as a result, have no Cq value. However, observing a Cq value for a no template control does not necessarily mean that the results for that assay should be discarded. If the difference in Cq value between the NTC and the sample with the lowest expression (highest Cq value) is sufficiently large (e.g. > 5 cycles), the level of potential contamination can be neglected (only 2^{-5} or about 3% for a 5 cycles difference). The same reasoning can be applied to all other types of negative controls such as the no amplification and the so-called minus-RT control (used to assess genomic co-amplification). Positive controls on the other hand should have a clear Cq value, preferably within a predetermined Cq interval.

PCR replicates may improve accuracy, but are mainly included to detect aberrant reactions that result in outlying Cq values and incorrect estimation of relative expression levels. With three or more PCR replicates, outliers can be detected and excluded from further calculations but very often only two repeated measurements are included in a study. This is sufficient to evaluate the overall quality of the measurements and allows the identification of bad replicates. Replicate QC is then based on the evaluation of the difference in Cq value between the replicates with the highest and lowest Cq values. A threshold for the maximum allowed difference in Cq value between the PCR replicates [typically 0.5 cycles (corresponds to a standard deviation of 0.35), preferably 0.3 cycles for experiments in which small differences in expression are expected] can be used to evaluate which and how many repeated measurements pass this QC. Software solutions such as qbase^{PLUS} are convenient in that they provide users with a summary of the overall quality of PCR replicates, and flag those that fail this quality control. There are different ways to deal with bad duplicates: (1) exclude both measurements because they cannot be trusted, (2) keep both measurements, acknowledging that there will be a higher uncertainty on the results (reflected by larger error bars) or (3) remove one of the replicates. The last approach can only be used if there is additional information pinpointing the bad replicate, e.g. a known pipetting error, abnormal melting curve or an abnormally high Cq value for that assay.

In addition to performing a pilot experiment in which the best set of reference genes is determined by means of a geNorm analysis, it is good practice to re-evaluate the stability of the selected reference genes in the actual experiment. One or more of the included reference genes may appear to be no longer stably expressed due to (for example) modifications in experimental procedures or the inclusion of samples with a deviating gene expression profile. Two types of stability parameters have been proposed (Hellemans *et al.*, 2007): the geNorm M value on the relative quantities and the coefficient of variation (CV) on the normalized relative quantities of a reference gene. Lower values for these parameters indicate more stably expressed reference genes. Reference values for acceptable reference gene stability measures are given in Table 7.1. If all reference genes are of poor quality (high M and CV values), one should repeat the geNorm pilot experiment on a representative set of samples to find reference genes that are sufficiently stably expressed. In situations where only one out of three or more reference genes has substantial higher M and CV values than the proposed reference values, the inappropriate reference gene should be excluded from the normalization procedures. As explained in the original geNorm paper, there is a trade-off between cost-effectiveness (fewer reference genes) and accuracy (more reference genes). The program calculates the minimal number of genes required for reliable normalization.

Statistics

It is beyond the scope of this chapter to review all statistical tests to determine significance of a difference in gene expression between 2 or more groups, to identify a diagnostic or prognostic RNA marker with high confidence, to find correlations between gene expression patterns or samples, or to identify relevant pathways or new sample subgroups. However, a few important points can be considered.

First, it is good practice to log transform the final gene expression results (i.e. the normalized relative quantities), in order to make the data distribution symmetrical (as gene expression data is often log normally distributed) (Fig 7.5). Together with the central limit theorem, this allows the use of parametric statistical tests and calculations that rely on a distribution that resembles a normal distribution (e.g. classic *t*-test, confidence intervals, analysis of variance) (Motulski, H. 1995. Intuitive biostatistics. Oxford: Oxford University Press).

Secondly, independent biological replicates are required to draw meaningful and reliable conclusions. The minimum number of such biological replicates depends on the statistical test and on the power one wants to achieve (e.g. for confidence interval analysis, at least three replicates are needed, for a non-parametric paired test (Wilcoxon signed-rank test), at least six pairs are needed). It must be clear that statistics on repeated measurements (e.g. PCR replicates) are absolutely nonsense, as only technical variation is measured.

Third, the statistical test should be selected prior to doing the actual experiment, whereby the choice is based on the question that needs to be addressed, the number of data points, and the distribution of the data. If in doubt, a (bio)statistician should be consulted. Table 7.2 provides an overview of basic statistical tests to assess the significance of differential gene expression. More than 10 years of experience shows us that the majority of gene expression questions can be addressed by the tests in Table 7.2. Three steps need to be taken to select the proper test are (1) log transform the data (see above), (2) evaluate whether a paired test is needed and (3) assess the distribution of the population and select the corresponding test. Pairing is needed when the value of one subject in the first group is

Table 7.1 Reference values for reference gene QC

	M	CV
Homogeneous sample set, e.g. cultures of the same cell type, blood from healthy individuals	0.5	25%
Heterogeneous sample set, e.g. combination of different cell types, cancer biopsies	1	50%

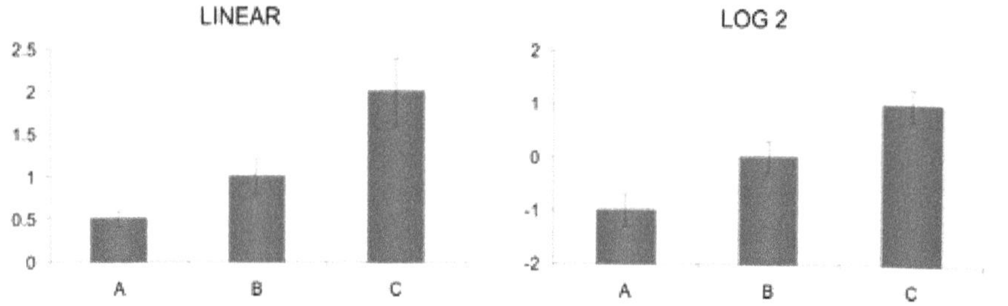

Figure 7.5 The left panel shows the NRQ value of a target gene for three samples (A, B and C) in linear scale. While sample A has a two times lower expression than sample B, sample C has a 2-fold higher expression than sample B. Doing relative quantification, we understand that sample A is as different from sample B as is sample C (i.e. a 2-fold difference), However, in linear scale, sample C appears more distant (1 unit) from sample B as does sample A (0.5 units). This creates a problem when using a statistical test (e.g. *t*-test) that relies on the absolute difference. A simple workaround is to log transform the data (here base-2 logarithm, but any base is fine). In the right panel, sample C and A are equally distant from sample B (i.e. 1 base-2 log unit). Please note that specific error propagation rules are required to go from error bar values in linear scale to error bars on the log transformed bars (formulas in Hellemans *et al.*, 2007).

Table 7.2 Lookup table for statistical tests

	Type of data	
Goal	Measurement from Gaussian population	Rank, score or measurement from non-Gaussian population
Compare two paired groups	Unpaired *t*-test	Mann–Whitney test
Compare two unpaired groups	Paired *t*-test	Wilcoxon test
Compare three or more unmatched groups	One-way ANOVA	Kruskal–Wallis test
Quantify association between variables	Pearson correlation	Spearman correlation

expected to be more similar to a particular subject in the second group (e.g. measurements before and after an intervention). The choice between a parametric or non-parametric test is the most difficult one, but less important when the sample size is large (e.g. > 20 data points per group): the central limit theorem allows use of parametric tests, even if the population is non-Gaussian; non-parametric tests work well with large samples from Gaussian populations. For small groups, one cannot rely on the central limit theorem: the p-value might be inaccurate for data from non-Gaussian populations and non-parametric tests on data from Gaussian populations tend to result in too high p-values (these tests lack statistical power). When dealing with small groups, try to increase the sample size. If this does not work, then it's better to be safe than sorry. Many biostatisticians prefer to be conservative: statistical significance obtained by a non-parametric method is likely to represent a true difference [type I error (false-positive rate) is lower, but type II error (false-negative rate) is higher]. In conclusion, selecting a non-parametric test is always a good and safe choice.

Fourth, important to realize is that the generally accepted threshold (alpha value or type I error) of 0.05 to designate statistical significance is only relevant when applying only one statistical test, to verify one hypothesis only. As soon as you need to test multiple hypothesis (e.g. assessing the differential expression of more than 1 target), the accepted threshold of 0.05 should be modified. In simple words, the more tests you run, the lower the threshold should be to keep the false positive rate under control. When 10 independent hypotheses are tested, there is a 40% probability of one or more p-values being less than 0.05 by chance (false positives). According to the conservative Šidák method for multiple testing correction, a threshold of 0.0051 ($1 - (1–0.05)^{(1/10)}$) should be used to identify significant changes with a false-positive rate of 5%. Less conservative methods (i.e. resulting in fewer false negatives) have been developed and are successfully being used (e.g. Benjamini-Hoghberg, 1995).

Fifth, if the data speak for itself, then let it do so. Sometimes, the difference is so striking or obvious that there is no need to perform a statistical test.

Reporting

A valuable tool to evaluate the procedures applied to a qPCR experiment is the MIQE checklist (http://www.rdml.org/miqe.php). This checklist is part of the Minimum Information for Publication of Quantitative Real-Time PCR Experiments (MIQE) guidelines that have been developed to provide authors, reviewers and editors specifications for the minimum information that must be reported for a qPCR experiment in order to ensure its relevance, accuracy, correct interpretation and repeatability (Bustin *et al.*, 2009). The benefit for the researcher is that adherence to these guidelines, including important sections on validation and data analysis, may help avoid stepping in to any of the pitfalls of qPCR experiments and guarantee the highest quality of data possible.

The last item on the MIQE checklist is the submission of the unprocessed data (Cq values with or without fluorescence based amplification curves) using the RDML format. The Real-time PCR Data Markup Language (RDML) is a recently developed universal standard for the straightforward exchange of qPCR data between different users and different software packages, and for the submission of data along with a publication (e.g. as supplemental file) or into public data repositories (Lefever *et al.*, 2009). Compatibility with RDML is built in qbase^{PLUS} as of version 1.2 and is expected to be implemented in most qPCR instrument software in the near future. This evolution will greatly facilitate data processing and collaboration between researchers operating different qPCR machines.

References

Benjamini, Y., and Hochberg, Y.(1995). Controlling the false discovery rate: a practical and powerful approach to multiple testing. Journal of the Royal Statistical Society, Series B (Methodological) 57, 289–300. MR1325392. http://www.math.tau.ac.il/~ybenja/MyPapers/benjamini_hochberg1995.pdf.

Bustin, S.A., Benes, V., Garson, J.A., Hellemans, J., Huggett, J., Kubista, M., Mueller, R., Nolan, T., Pfaffl, M.W., Shipley, G.L., *et al.* (2009). The MIQE guidelines: minimum information for publication of quantitative real-time PCR experiments. Clin. Chem. *55*, 611–622.

Hellemans, J., Mortier, G., De Paepe, A., Speleman, F., and Vandesompele, J. (2007). qBase relative quantification framework and software for management and automated analysis of real-time quantitative PCR data. Genome Biol. *8*, R19.

Higuchi, R., Fockler, C., Dollinger, G., and Watson, R. (1993). Kinetic PCR analysis: real-time monitoring of DNA amplification reactions. Biotechnology *11*, 1026–1030.

Lefever, S., Hellemans, J., Pattyn, F., Przybylski, D.R., Taylor, C., Geurts, R., Untergasser, A., and Vandesompele, J. (2009). RDML: structured language and reporting guidelines for real-time quantitative PCR data. Nucleic Acids Res. 37, 2065–2069.
Livak, K.J., and Schmittgen, T.D. (2001). Analysis of relative gene expression data using real-time quantitative PCR and the $2(-^{\Delta\Delta}C(T))$ method. Methods 25, 402–408.
Mestdagh, P., Van Vlierberghe, P., De Weer, A., Muth, D., Westermann, F., Speleman, F., and Vandesompele, J. (2009). A novel and universal method for microRNA RT-qPCR data normalization. Genome Biol. *10*, R64.
Pfaffl, M.W. (2001). A new mathematical model for relative quantification in real-time RT-PCR. Nucleic Acids Res. 29, e45.
Ruijter, J.M., Ramakers, C., Hoogaars, W.M., Karlen, Y., Bakker, O., van den Hoff, M.J., and Moorman, A.F. (2009). Amplification efficiency: linking baseline and bias in the analysis of quantitative PCR data. Nucleic Acids Res. 37, e45.
Rutledge, R.G., and Stewart, D. (2008). A kinetic-based sigmoidal model for the polymerase chain reaction and its application to high-capacity absolute quantitative real-time PCR. BMC Biotechnol. 8, 47.
Vandesompele, J., De Preter, K., Pattyn, F., Poppe, B., Van Roy, N., De Paepe, A., and Speleman, F. (2002). Accurate normalization of real-time quantitative RT-PCR data by geometric averaging of multiple internal control genes. Genome Biol. 3, RESEARCH0034.
Zhao, S., and Fernald, R.D. (2005). Comprehensive algorithm for quantitative real-time polymerase chain reaction. J. Comput. Biol. *12*, 1047–1064.

Equations

(1) Relative expression $= 2^{-\Delta\Delta Cq}$

(2) $$\text{Relative expression} = \frac{RQ_{goi}}{RQ_{ref}} = \frac{E_{goi}^{\Delta Cq}}{E_{ref}^{\Delta Cq}}$$

where *goi* = gene of interest and *ref* = reference gene.

(3) $$\text{Relative expression} = = \frac{RQ_{goi}}{geomean\left(RQ_{ref}\right)} = \frac{E_{goi}^{\Delta Cq}}{\sqrt[n]{\Pi E_{ref,i}^{\Delta Cq}}}$$

where *goi* = gene of interest and *ref* = reference gene.

(4) $$RQ = E^{\Delta Cq}$$

(5) $$NRQ = \frac{RQ}{NF} = \frac{RQ}{geomean\left(RQ_{ref}\right)}$$

where *ref* = reference gene.

(6) $$CNRQ = \frac{NRQ}{CF} = \frac{NRQ}{geomean\left(NRQ_{irc}\right)}$$

The MIQE Guidelines Uncloaked 8

Greg Shipley

Abstract

The MIQE (Minimum Information for Publication of Quantitative Real-Time PCR Experiments) guidelines have been presented to serve as a practical guide for authors when publishing experimental data based on real-time qPCR. Each item is presented in tabular form as a checklist within the MIQE manuscript. However, this format has left little room for explanation of precisely what is expected from the items listed and no information on how one might go about assimilating the information requested. This chapter presents an expanded explanation of the guideline items with commentary on how those requirements might be met prior to publication.

Introduction

Led by Dr Stephen Bustin, a number of us with many years of experience in performing qPCR experiments contributed to the content within the MIQE manuscript (Minimum Information for Publication of Quantitative Real-Time PCR Experiment) that outlines the minimal requirements for the publication of real-time qPCR data and provides a checklist for reviewers and authors of new manuscripts containing real-time qPCR data (Bustin *et al.*, 2009). The purpose of this chapter is to help flesh out these requirements and hopefully better explain the thought processes behind the items listed. A copy of the checklist can be seen in Table 8.1. When possible, as much information as possible should be provided either within the manuscript or as supplementary data available on-line for those readers interested in more in-depth information about how the technique was performed.

The performance of a real-time quantitative PCR (qPCR) experiment is not inherently difficult. Having said that, qPCR is a quantitative technique and therefore requires a degree of rigor in how the experiment is performed. Up until this year, qPCR data has been accepted for publication based on the strengths of the technique alone without questioning how carefully the various steps required for the experiment were carried out. The aim of the MIQE paper and guidelines is to provide the reader with all the information required to either repeat the experiment or be able to judge whether the data are sound. I'm sure everyone has experienced the frustration of looking at the Methods section of a paper to find some detail of how a qPCR experiment was performed only to find that there was scant information present. There are examples of research papers being retracted following publication that should never have seen the light of day, as pointed out in the MIQE manuscript (Huang *et al.*, 2005; Bohlenius *et al.*, 2007).

Table 8.1 MIQE checklist for authors, reviewers, and editors

Item to check	Importance	Item to check	Importance
Experimental design		qPCR oligonucleotides	
Definition of experimental and control groups	E	Primer sequences	E
Number within each group	E	RTPrimerDB identification number	D
Assay carried out by the core or investigator's laboratory?	D	Probe sequences	D[d]
Acknowledgment of authors' contributions	D	Location and identity of any modifications	E
Sample		Manufacturer of oligonucleotides	D
Description	E	Purification method	D
Volume/mass of sample processed	D	qPCR protocol	
Microdissection or macrodissection	E	Complete reaction conditions	E
Processing procedure	E	Reaction volume and amount of cDNA/DNA	E
If frozen, how and how quickly?	E	Primer, (probe), Mg^{2+}, and dNTP concentrations	E
If fixed, with what and how quickly?	E	Polymerase identity and concentration	E
Sample storage conditions and duration (especially for FFPE[b] samples)	E	Buffer/kit identity and manufacturer	E
Nucleic acid extraction		Exact chemical composition of the buffer	D
Procedure and/or instrumentation	E	Additives (SYBR Green I, DMSO, and so forth)	E
Name of kit and details of any modifications	E	Manufacturer of plates/tubes and catalog number	D
Source of additional reagents used	D	Complete thermocycling parameters	E
Details of DNase or RNase treatment	E	Reaction setup (manual/robotic)	D
Contamination assessment (DNA or RNA)	E	Manufacturer of qPCR instrument	E
Nucleic acid quantification	E	qPCR validation	
Instrument and method	E	Evidence of optimization (from gradients)	D
Purity (A_{260}/A_{280})	D	Specificity (gel, sequence, melt, or digest)	E
Yield	D	For SYBR Green I, C_q of the NTC	E
RNA integrity: method/instrument	E	Calibration curves with slope and y intercept	E
RIN/RQI or C_q of 3′ and 5′ transcripts	E	PCR efficiency calculated from slope	E
Electrophoresis traces	D	CIs for PCR efficiency or SE	D
Inhibition testing (C_q dilutions, spike, or other)	E	r^2 of calibration curve	E
Reverse transcription		Linear dynamic range	E
Complete reaction conditions	E	C_q variation at LOD	E
Amount of RNA and reaction volume	E	CIs throughout range	D
Priming oligonucleotide (if using GSP) and concentration	E	Evidence for LOD	E
Reverse transcriptase and concentration	E	If multiplex, efficiency and LOD of each assay	E
Temperature and time	E	Data analysis	
Manufacturer of reagents and catalogue numbers	D	qPCR analysis program (source, version)	E
C_qs with and without reverse transcription	D[c]	Method of C_q determination	E
Storage conditions of cDNA	D	Outlier identification and disposition	E
qPCR target information		Results for NTCs	E
Gene symbol	E	Justification of number and choice of reference genes	E
Sequence accession number	E	Description of normalization method	E
Location of amplicon	D	Number and concordance of biological replicates	D
Amplicon length	E	Number and stage (reverse transcription or qPCR) of technical replicates	E
In silico specificity screen (BLAST, and so on)	E	Repeatability (intraassay variation)	E
Pseudogenes, retropseudogenes, or other homologs?	D	Reproducibility (interassay variation, CV)	D
Sequence alignment	D	Power analysis	D
Secondary structure analysis of amplicon	D	Statistical methods for results significance	E
Location of each primer by exon or intron (if applicable)	E	Software (source, version)	E
What splice variants are targeted?	E	C_q or raw data submission with RDML	D

[a] All essential information (E) must be submitted with the manuscript. Desirable information (D) should be submitted if available. If primers are from RTPrimerDB, information on qPCR target, oligonucleotides, protocols, and validation is available from that source.
[b] FFPE, formalin-fixed, paraffin-embedded; RIN, RNA integrity number; RQI, RNA quality indicator; GSP, gene-specific priming; dNTP, deoxynucleoside triphosphate.
[c] Assessing the absence of DNA with a no–reverse transcription assay is essential when first extracting RNA. Once the sample has been validated as DNA free, inclusion of a no–reverse transcription control is desirable but no longer essential.
[d] Disclosure of the probe sequence is highly desirable and strongly encouraged; however, because not all vendors of commercial predesigned assays provide this information, it cannot be an essential requirement. Use of such assays is discouraged.

It is beyond the purview of this chapter to cover all possible contingencies that might arise during a real-time qPCR experiment. However, an attempt will be made to offer some specifics or suggestions when possible for the various requirements as set out in the MIQE manuscript. I have avoided mentioning specific products, for the most part, as most of the reagents on the market today work quite well. However, there are a few cases where there is only one option and those products have been referenced. Another consideration in this decision is the observation that while the folks in laboratory A will have great success with product X over Y, those in laboratory B down the hall will swear that product Y is much better than product X. This illustrates what is most likely the most undocumented part of any research project, the effect of experience and the quality of the personnel involved on the outcome of the experiment. Since this factor can not be quantified in any reliable way, we have to depend upon the parameters outlined in the MIQE checklist as a measure of their success.. Unfortunately, there are many other examples of qPCR data in the literature that have been improperly or inadequately reported.

The following discussion will cover the main topics of the MIQE checklist in order (Table 8.1). Hopefully, my commentary will help flesh out what we (the MIQE authors) had in mind when these items were put on the list.

Experimental design

The most important component of any experiment is in the planning. Like a chess match, looking several moves ahead prior to beginning can save a lot of time, energy and potential heartache down the road. One of the most important considerations is in determining how many members will be required in each experimental group to achieve statistical significance compared to one or more control groups. This is determined by performing a power analysis based on preliminary experimental data from a smaller subset of samples. It is important to define what constitutes an experimental and control group, the number (N) in each group and a clear representation of the data from each group.

As a Core Laboratory director, I insist that publications in which the data presented are generated within our facility mention our involvement in the acknowledgements or methods section of the paper. The 'coin of the realm' for Core Labs is providing data for investigators that lead to publications and grants.

Sample preparation

The term 'sample' in any experiment can have a large range of possible definitions. Therefore, it is critical that the author define what a sample is for each of their experiments. This will include a brief discussion of the origin of the sample (e.g. tissue culture, animal tissue, FFPE material, laser capture, blood, faeces, plant, microbe); how it was obtained (e.g. cell lysis, biopsy); how it was handled (e.g. flash frozen, RNAlater, homogenized immediately, fixation); and how it was stored prior to analysis (e.g. temperature and time prior to nucleic acid isolation). The amount of sample collected can also be critical when evaluating the RNA or DNA isolation procedure and should be reported. The collection of samples is arguably the most important phase of the experiment. If the nucleic acids are not stabilized sufficiently at this step, there is nothing you can do later on to rectify the situation. It has been my observation over the years that it is important for investigators who have little experience with the collection, isolation and purification of nucleic acids to practice all phases of this procedure prior to initiating what may be a very costly set of experiments and ensure that they are technically up to the challenge. This is also true for personnel who have a lot of experience but are obtaining their material from a new source. Never assume that past methods will work with the new material.

Extraction of nucleic acids

Once samples are collected, the nucleic acids therein stabilized and stored appropriately, the next important step is the isolation of RNA, DNA or both. The method used is important and will vary according to sample type. There is no 'one size fits all' for DNA or RNA isolation. Although the isolation of RNA or DNA is done primarily by hand, if one of the instruments available for automated sample preparation was utilized, this should be documented in the Methods. Usually a kit is used for nucleic acid isolation and if so the name, catalogue number and manufacturer of the kit should be mentioned. However, it is possible that homemade or additional reagents were employed during the purification process and these need to be completely described. A normal part of the isolation procedure will be the treatment of RNA with DNase I to remove a potential contamination source in the real-time RT-qPCR. Unlike RNase, *E. coli* DNase I requires Ca^{2+} for enzymatic activity and is stimulated by Mg^{2+}. Further, divalent cations will facilitate the cleavage of RNA with prolonged treatment at high temperature, even with no enzyme present. There are a number of methods and protocols for this step and the one used should be given. Treatment of purified DNA with RNase is common and should also be documented.

Once nucleic acid has been purified, its concentration, purity and quality are three critical parameters that must be measured and reported. The easiest way to measure the concentration is spectrophotometrically with an A_{260} measurement. An A_{260} value of 1 at a 1 cm path length equivalent will be 50 μg/ml for DNA and 40 μg/ml for RNA. However, this assumes that the nucleic acid is 100% pure, which is hard to achieve in practice. For that reason, an A_{260} measurement cannot be used to normalize real-time qPCR data for loading variations from sample to sample. For that purpose, a more specific fluorescence dye-binding assay, such as picogreen for DNA and ribogreen for RNA, is required to determine nucleic acid concentration. Since foreign proteins and other contaminants can cause PCR inhibition, it is critical that you report the purity of your nucleic acid preparations. The easiest way to accomplish this is by measuring ratios of light absorbance at different wavelengths. A ratio of A_{260}/A_{280} provides a measure of how much protein is in your RNA or DNA preparation. An often-cited minimum acceptable ratio is 1.8 up to the maximum of 2.2 for RNA or 1.8 to 2.0 for DNA but depending on the sample source, you may find that the minimum acceptable ratio is lower, although you must be able to show that these samples are working as well as cleaner samples. Another measure of cleanliness is the A_{260}/A_{230} ratio which will pick up salts and other contaminants. A ratio of 1.7 or higher is a good rule of thumb. For publication, a statement on the general range of these ratios for all samples used in the publication would suffice.

Of course, none of the above will tell you anything about potential RT or PCR inhibition but there are a number of ways to determine whether inhibition is a problem for your samples. One is to make 10-fold dilutions of the RNA or DNA and see if the samples dilute properly (roughly a 3 PCR cycle decrement for each 10-fold dilution). You can spike in an *in vitro* transcribed RNA or a DNA oligo spanning the PCR amplicon for a gene or transcript not found in the sample and compare the Cq values when run alone or within a number of samples. For a more thorough analysis on PCR inhibitors see Chapter 2.

The last criterion is the most important, sample integrity. I have seen many samples that were abundant and clean of contamination but were also totally degraded and thus useless. The common method of determining RNA quality is by running the samples in a chip-based micro-electrophoresis chamber such as the Agilent 2100 Bioanalyzer or the Bio-Rad Experion instruments. Sample quality is based on an algorithm that takes the amount of the sample present in the 18S and 28S rRNA peaks as well as other factors to obtain a number that expresses RNA quality for that sample. It is also possible to use real-time qPCR assays for a number of transcripts made to both the 5′- and 3′-ends of each transcript. An ideal result would be if the amount found for the quantity of transcript at the 5′-end were identical to the amount recorded for the 3′-end of the same transcript. Thus, a ratio of 1 for the two values would be ideal but is not always found. However, the samples should have similar ratios within an experiment for each transcript (Nolan *et al.*, 2006a). DNA is much more stable than RNA during isolation so cannot be run on a microchip assay for analysis as total genomic DNA. But looking at DNA integrity on a low percentage agarose gel would be sufficient. No matter how you determine the integrity of your sample, the method used, along with the size range measured must be stated for all samples in the study (Nolan *et al.*, 2006b). A qualitative method that can give some indication of sample integrity is to observe the shape of the amplification curves following real-time qPCR. The amplification curves should all be parallel even when shifted to higher cycle numbers. If some samples have amplification curves that are less acute (upright) than the others or compared to a control template, PCR inhibition should be investigated using one of the methods described.

The reverse transcription step

The next most important step in an RT-qPCR experiment, after sample quality, is the reverse transcription of RNA into cDNA as most of the variability in Cq values will come from this step. Once the cDNA has been made, making replicates for the PCR becomes primarily a pipetting exercise, which is a common source of variability, but replicating the PCR alone does not constitute a true replicate of the RT-PCR. Thus, it is better to have multiple RT reactions, a triplicate RT per sample for example, with a single PCR from each RT reaction than to make a single cDNA reaction with a triplicate PCR.

There are many vendors that sell reverse transcriptase mastermix with two major kinds of reverse transcriptase, AMV and MMLV, and multiple ways to prime a cDNA reaction (assay-specific, random primers, oligo-dT and a mix of both). Thus, the combinatorial possibilities are rather large making it critical that the investigator describe in detail how the reverse transcriptase reaction was performed. The requirements are easily described but I will elaborate for each one.

1 The amount of RNA in micro- or nanograms and the volume used for the RT reaction should be given.
2 The gene-specific primer (GSP) is usually the reverse primer of the PCR assay but can be another primer more 3′ to the PCR amplicon. If the latter, the Tm is usually low enough so it can not participate in the PCR but can anneal at the temperature used for cDNA synthesis. The annealing position within the sequence (5′-base position of the primer) along with the concentration of the primer must be stated.
3 The type of reverse transcriptase used, along with its concentration in units (or volume if units are not known) must be stated. I personally think that every investigator should

know what kind of reverse transcriptase (AMV or MMLV) they are using as each has its own advantages.

4 Temperature and time used for the cDNA reaction simple but essential pieces of information that must be provided. The instrumentation used, whether a thermocycler, heater block and/or water bath, should also be stated..

5 When using a commercial mastermix the name of the mastermix, catalogue number and manufacturer should be stated.

6 The most important control for a successful RT reaction is the -RTase control. It is often questioned whether the Cq value of the –RT control has to be 40 cycles or no signal. The simple answer is the –RT control does not need to be negative. However, the Cq value for this control should be well shifted from the +RT reaction values. This is one of many examples of a situation where each investigator must decide on his or her own personal, hard and fast rule for data analysis and the rule should be reported. My personal rule here is a 6 Cq shift must be present between the mean Cq values of the +RT reactions and the -RT Cq value. Since a 3 Cq shift is roughly 10-fold, a 6 Cq shift equals a 100-fold difference between the RNA and any DNA contaminating signal. Thus, the RNA signal would contain at most 1% DNA contamination in the +RT Cq value. For me, that is acceptable. For you, it may not and you can make another rule. The key is to follow the rule no matter what and to state it in every publication. If data have to be thrown out, so be it.

7 Finally, you must state how the cDNA is stored prior to performing the qPCR. cDNA is fairly stable but it is important you treat it with care nonetheless and communicate how you have done so. At a minimum the storage temperature needs to be recorded and it is also useful to mention whether the sample was aliquoted prior to storage.

Assay template information

Whether you are working with transcripts or genes as primary targets, it is important to document each sample properly so the reader will know exactly to what you are referring. Common names for genes are not reliable and have evolved as more rigorous naming conventions have been employed to encompass all the many different genes, transcripts and variants found for many organisms. The following expands on the checklist requirements.

1 The gene symbol can be found at the NCBI, Ensembl, UCSC and many other databases available over the web. For example, if the Oestrogen receptor alpha was one of your gene targets, the gene symbol is ESR1.

2 Similarly, these sites will also provide an accession number which is a unique identifier for each gene (NC_xxxxxx or NR_xxxxxx), transcript (NM xxxxxxx) or protein (NP_xxxxxx) sequence. The appropriate accession number should be provided in a table with other important components about the target (see Table 8.2).

3 The chromosomal location of the gene may or may not be important for the publication, but if it is, it should be provided, perhaps along with information on neighbouring genes or regulatory sites.

4 The PCR amplicon length (total span of the assay in bases) is useful information to provide. For example, if you are using FFPE samples and your SYBR Green I assays are

Table 8.2 Sample table for submission of assay information for publication

Accession number	NR_003286	NM_001101
Assay name	18S Ribosomal RNA	human beta-Actin
Gene symbol	RN18S1	ACTB
Species	Mammalian	Homo sapiens
Forward primer 5′ base	1234	1040
Forward PRimer sequence	CGGCTTAATTTGACTCAACAC	CCCTGGCACCCAGCAC
Reverse primer 5′ base	1302	1110
Reverse primer sequence	ATCAATCTGTCAATCCTGTCC	GCCGATCCACACGGAGTAC
Probe 5′ base	1258 (+)	1052+
Probe sequence	AAACCTCACCCGGCCCG	ATCAAGATCATTGCTCCTCCTGAGCGC
Reporter/quencher	FAM/BHQ1	FAM/BHQ1
LOD copies	2100	200
PCR amplicon (bases)	68	71
PCR efficiency (%)	95	95
BLAST	1 unique transcript	1 correct transcript, 1 called pseudogene and 3 predicted pseudogenes
Exon location	1	5/6

BLAST: basic local alignment search tool.

There is no splicing in the 18SrRNA transcript. The limit of detection (LOD) for 18SrRNA assay is high due to human DNA contamination in primers/probes from all oligo houses checked to date. Human b-actin: crosses the exon 5/6 junction.

all over 300 bases in length, it would not be a surprise that your ability to detect many transcripts was not very good.

5 If you run your PCR amplicon sequence on a BLAST search against the entire genome or transcriptome of the species in question, how many non-specific hits come up? Is there a reasonable expectation that both primers (and probe, if present) will allow you to detect the target sequence in a total RNA or gDNA population with high specificity?

6 Just because your assay crosses an exon/exon junction does not mean that you can not obtain amplification from contaminating DNA *a priori*. Pseudogenes are spliced copies of transcripts found in alternative places within the genome. Since they are occasionally transcribed they often have mutational differences compared to the true gene. Known pseudogenes will show up on a BLAST search but not all of them are known. For this reason, it is best to keep DNA contamination to a minimum in RNA samples.

7 Prior to assay development, I always run an m-fold analysis to look for long stem structures than can prevent efficient binding of primers (or probes) to the target sequence. I find it best to avoid regions of high secondary structure during real-time qPCR assay development.

8 The location of exon boundaries within a transcript sequence is easily found these days for human, mouse and will be available for other species as progress is made on the analysis of each genome. Not all transcripts are spliced, of course. For easy identification, I find it is best to number each primer with the 5′-nucleotide base position within the refseq sequence and for probes we do the same. I also indicate which strand the probe binds to (+ or −) so the reader will know if the sequence is the same as the reverse complement of the reference sequence.

9 If the assay has been designed to detect one or a subset of splice variants known for a transcript, that information should be stated. If the assay will detect all know splice variants, we call that a generic assay but again this should be stated. Even if this information does not seem important at the time the paper is submitted it may have great relevance at a later date.

Assay components

In manuscripts where real-time qPCR is one of the techniques used, the sequences of the primers (and probe) used are invariably given. However, more detailed information concerning the assay past SYBR Green or probe-based is often missing. For proper review, it is critical that there be complete disclosure of the assay components and the location of the assay within the transcript or gene sequence. Further, comments on whether splice variants, allelic isoforms, SNPs, or other variations within the sequence are or are not detected with the assay should be disclosed if these issues are salient to the data presented within the manuscript. Assay component information is best presented in tabular form as part of the Methods section. Table 8.2 shows the salient column headings and two example assays have been entered for reference.

The manufacturer of the primers/probe and oligo DNA standards, if appropriate, should always be given within the Methods section. Less important is the purification method used for the primers as the standard and free desalting for SYBR and probe-based assay primers is usually sufficient. There are instances, however, where HPLC purified primers are desirable. One example would be for primers used for high-resolution melt (HRM) analysis and this should be mentioned within the Methods section, although the jury is still out as to whether HPLC purified primers are in fact necessary for HRM. There are multiple synthesis and purification options for dual-labelled fluorescent hybridization probes so the method of purification as well as the reporter and quencher moieties should be given along with the manufacturer information. Lastly, there are base modifications that can be added to probes and primers and these need to be detailed (base position(s) and modification, e.g. LNA, inverted bases, etc.) along with the manufacturer information.

qPCR protocol

Describing the reaction conditions for a real-time qPCR experiment would seem to be an obvious addition to any Methods section when qPCR was used in data collection but there are publications where scant information on this topic is provided. Without this critical information it is nearly impossible for a reviewer to determine whether the data presented are credible. This part of the checklist outlines the basic information that must be stated in describing your protocol.

- If a commercial PCR mastermix was used: the name, catalogue number and manufacturer, along with a description of any modifications made by the investigator.
- If a home-made mastermix was used: a complete description of every component, manufacturer and final concentration,.
- The volume, amount of substrate added per reaction, plate type and geometry (48-, 96-, 384-well plates, and tubes for block instruments, capillaries or tubes for rotary instruments), model and manufacturer of the instrument.
- The thermocycling conditions.
- Although most experiments are set up by hand pipetting; if liquid handling robots are used that should be stated along with manufacturer and model.

qPCR validation

This section may be one of the most important in the checklist and yet is most often ignored in any Methods section and is generally given short shrift by investigators anxious to 'get the data' that they forget the data is only as good as the assay used to collect it. As stated earlier, after the quality of nucleic acid preparation, there is nothing more important than the qPCR assay and its' correct implementation in collecting data. All the fancy data analysis in the world can not make up for a poor assay. A good assay is joy to use forever and a poor one is a recurring nightmare.

The first step in new assay development is to acquire the information outlined in section 6 above. Next, determine where within the sequence the assay can and cannot be positioned to satisfy the research question at hand. All of these steps are accomplished *in silico*. This may necessitate aligning related sequences, splice variants, related species, etc. to make this determination. Once the target area(s) have been identified, the actual assay design can commence. Whether the assay will use a probe (hydrolysis, beacon, scorpion, etc.) or use SYBR Green I or one of the many other possible dyes, a software program should be used. A list of assay design software is presented at the end of this chapter.

Two critical factors to keep in mind are that the primers are the most critical part of the assay and potential template secondary structure that can interfere with primer/probe binding. If your software of choice does not include secondary structure analysis as part of the design process, the sequence around the final assay PCR amplicon should be checked by m-fold analysis. Primers or probes that will anneal to strong stem regions within the target sequence will not work well and should be avoided.

Once the assay components have arrived in the lab, a preliminary PCR or RT-PCR should be run on a real-time instrument to see how well the assay is working. More importantly, the PCR product can be used to construct a standard curve. Even if you do not plan to run standard curves for data analysis, it is essential that one be run for assay quality control. The data from this experiment is part of the checklist information. Depending on the Cq value recorded for this first run, dilute the PCR product 100- to 1000-fold in 100 ng/µl *E. coli* or yeast tRNA (nuclease-free, molecular biology grade) in nuclease-free water. A 7-log range in 10-fold decrements from the initial dilution is usually sufficient to cover a high range down to 1–10 copies of template. From this experiment, you will find the dilution at which the assay no longer functions in a linear fashion with the higher template concentrations. The last dilution where the value was still linear (falls on the standard curve) is the LOD or lowest limit of detection. You can make 2-fold dilutions around this value to get a more accurate value if required. The critical consideration is that no Cq value can be reported if it falls below the LOD of the assay. If your assay is not good enough, one solution is to go back to the assay design software and find more primers that can work with the existing ones and try them in all possible combinations. Moving the primer over by 1 base can make a world of difference in how it works with a certain primer partner. I order 4 primers by default and do this for every new assay design. Another solution is to try the primers you have at different concentrations, both symmetric and asymmetric, and with different $MgCl_2$ concentrations and annealing temperatures. Once the results from these experiments gives satisfactory results, you can calculate the PCR amplification and PCR efficiency from the slope of the standard curve (see formulas at the end of this chapter). My QC limits are that the assay must be linear down to 10 template copies and have a PCR efficiency of 93% or better. If your assay does not meet these requirements but works for your experimental requirements, it can still be used. But, you will have to keep the limitations in mind for future work.

Some folks put a lot of stock in the r^2 measurement from the standard curve. I agree it should be near 1 but it is not the end all that some seem to think it is. It merely tells you that the person running the experiment can pipette accurately, which is important, but it is not the assay diagnostic some seem to think it is. A more telling value is the y-intercept but to get an accurate value for this term you have to accurately calculate the number of copies in a given mass of standard. When using PCR products, this can be hard to determine using an A_{260} measurement which is not all that accurate and is a sum of all components that absorb light at 260 nm. Using a value of 2×10^7 copies or molecules for a Cq value of 13 will get you close to the desired y-intercept value. The y-intercept is a value that sums up all the important aspects of the assay (standard curve) in one value. A y-intercept value of 37–38 is within the desired range. A statement should be made in the Methods or Results section that all Cq values of unknowns fall within the linear quantifiable range of all the assays used.

If you have designed a multiplex assay (two or more assays within the same reaction), you must present data for each component assay in the multiplex as described above. Instruments have different ways of limiting spill over from the emission of one reporter dye into a second reporter dye that emits in near proximity (e.g. FAM and VIC or HEX) and this method should be mentioned in the Methods section.

Data analysis

The Data Analysis section primarily deals with how the final post-run data set was analysed. However, data analysis can be divided into two parts, immediate post-run analysis and post-run project data analysis. Immediate post-run analysis is queried under 'Cq Method'. Most instrument software depends on the user performing an initial analysis based on a thresholding process. That is a determination of the proper baseline and threshold to use to get the final Cq values for the run.

All software programs will do this automatically but it has been my experience, and one found in an unpublished ABRF study by the nucleic acids research group, that the best data analysis for all of the instruments in use at that time (2005) was when the end user set these two variables manually verses using any combination of manual and automatic or automatic analysis mode. I would encourage you to look at the manual settings for these two variables and see if you can make your data just a little better by setting these two variables yourself. It's easiest to determine this when running a standard curve over 6 or 7 logs but the proper settings will have an effect on how the Cq values of similar replicates appear as well. Having control of the threshold can be important when you want to compare samples across plates, for example. You should set both of these variables to the same value for all plates being compared, but you will find that the instrument software will not set them exactly the same and instead try to optimize them for each plate run. Not all instrument software works the same so it is impossible to give a generic recommendation here. Having said that, it is important that the baseline setting not include cycles with positive fluorescent signal from any of the samples and that the threshold is set high enough to avoid spurious low level PCR signals but not set so high as to being in the linear, verses the geometric amplification region of the PCR curve. Recently, Roche has introduced the 2nd derivative method of determining Cq values for their LC480 real-time instrument. This method eliminates the need for the investigator to intervene in determining Cq values. The new Bio-Rad CFX and Qiagen Rotor-GeneQ instruments use a similar algorithm in their software. I mention this issue here because I fear most investigators do not know enough about how their real-time qPCR software works to have investigated these settings for themselves. However, if the initial Cq values are not the best they can be, succeeding data analysis can not rescue the data set.

The remaining parts of this section deal with the aspects that first come to mind when you think of data analysis. Although most of us have used Excel and some free-ware for the analysis of real-time qPCR data, there are some excellent programs on the market now that can provide a number of different analyses and save time to boot. The best of these are now commercial products but there are some free ones available as well. A website that lists all the available choices is one managed by Michael Pfaffl's group in Freising, Germany (http://www.gene-quantification.de/download.html), and an in-depth discussion of data analysis is found in Chapter 7. Regardless of which one you use, it should be referenced in the manuscript. The commercial software is also useful in determining when data points are statistically legitimate outliers. If you are performing an Excel -based manual analysis, the method for determining outliers must be given.

I think it goes without saying that data normalization is one of the most difficult parts of any real-time qPCR experiment. When a new experimental protocol is used without prior information on possible transcripts that could be used for data normalization, an empirical approach should be taken to determine possible candidates. One of the goals of the MIQE paper was to eliminate the use of the term 'housekeeping gene' whose very definition was 'a gene that was constitutively transcribed and invariant in concentration regardless of circumstances'. We now know that there are no genes that fall into that category and that any transcript, including the main ribosomal genes 18S and 28S rRNA, can vary depending on the experimental conditions. Why do the data need to be normalized, you might ask? The simple answer is that absorbance estimations of nucleic acid concentration are not sufficiently accurate to be used for data normalization. Thus, when you think you have added the same amount of RNA by absorbance assay for 20 samples into an RT-qPCR experiment, in reality you will find that there will be about a 10-fold spread in the actual measured amounts for any 'control or normalizer' transcript, even if all the samples are from control biological replicates. This just emphasizes the variability in determining macromolecule concentrations using an absorbance assay. Thus, it is critical that something besides absorbance be used to normalize for the amount of sample going into each RT-qPCR or qPCR. Following the seminal publication by Vandesompele *et al.* (2002), it was clearly shown that normalization of RT-qPCR data was best done using multiple relatively invariant transcripts rather than a single one. In practice, any transcript can be used for data normalization if it is relatively invariant in all samples within that experiment. The converse of that has been seen as well where transcripts that one thought would be used for data normalization turned out to be used as experimental ones due to the changes in that transcript within the experiment. It is truly the case that you do not know until you know what will work for data normalization and it is critical that this determination be made early during the experimental development phase. There are programs that can help identify the best transcripts or genes to use for data normalization from empirical data. Software that is dedicated to data normalization is geNorm from the Vandesompele lab (http://medgen.ugent.be/~jvdesomp/genorm/) or NormFinder at the Nordic Centre of Excellence in Molecular Medicine (http://www.mdl.dk/publicationsnormfinder.htm). A number of vendors sell premade plates with multiple potential normalizers on them for rapidly screening and determining which ones to use for a new experimental condition.

It is also true that it is not always possible to rely on any transcript or gene for data normalization. Some examples I have encountered are actinomycin D-treated cells for transcript half-life experiments, cells undergoing apoptosis, and genes or transcripts that originate outside the context of a cell such as serum, plasma, saliva, urine, pus, faeces or other biological fluids. I am sure there are other good examples from microbiological or ecological studies as well. In these cases the best option is to measure the total nucleic acid in the sample using a fluorescence assay such as picogreen for dsDNA, oligreen for ssDNA and ribogreen for RNA (Molecular Probes/Life Technologies, Carlsbad, CA, USA). The final data would be expressed as molecules or copies/ng total RNA or DNA. Copy number can be interpolated from a standard curve or relative quantities can be calculated from Cq values using the PCR amplification value for the assay in the formula $(\text{amplification})^{-\text{dCt}}$, where an ideal real-time qPCR assay would have the formula, $2^{-\text{dCt}}$.

Another item that is important is the difference between a technical replicate and a biological replicate. Technical replicates are repeated analyses of the same sample. That is multiple PCRs from the same cDNA pool or performing replicate RTs from a single sample and a single PCR from each RT reaction (preferred). In both cases the final values will be averaged and a standard deviation calculated (if triplicates or above are determined) but the value will still represent a statistical $N = 1$ for that data point. Technical replicates give you more accuracy in measuring the Cq value for a single transcript from a single sample. A biological replicate is a collection of different samples, all treated the same. So, if you have five control samples, you might run a triplicate technical replicate for each one but at the end of the day, you would still have an $N = 5$, one mean value for each sample which can then be averaged to give the final mean for all the samples in that experimental group. Biological replicates give you more confidence that the data point you end up with is as close as possible to the 'real' value for that experimental treatment within the total cellular or organismal population. Bottom line, it is important to report the number of technical and biological replicates used for data analysis. From this data set, it is important to report the intra-assay variability for each transcript or gene determination. This is best done by showing mean, standard deviation and %CV for each assay from a single plate or run determination. If multiple plates or runs are used to obtain the complete data set for each transcript or gene over the entire set of samples, inter-assay variability should be presented as well.

Finally, the method used to determine statistical significance for your data set needs to be reported. If you performed a power analysis to determine how many members would be required to obtain significant differences in each experimental group, that information should be provided as well along with the software or formulas used for these calculations.

Final comments

The overarching goal of the MIQE requirements is the full disclosure of how a real-time qPCR experiment was performed so that the reader can either repeat the experiment(s), if desired, or at least know exactly how they were performed. Once assays are completely disclosed in the literature and the procedures outlined for an investigator, it should be possible to refer back to the seminal paper in future manuscripts if, and only if, no changes in the assays and procedures have been made in succeeding experiments. If some modification has been made, a description of the modified portion of the experimental protocol should be reported.

It has become common practice to pare the Methods section down to collections of partial and in some cases bare bones information that is insufficient if one wishes to repeat the experiment or just be able to determine if it was done properly. This was brought on by pressures to keep the number of pages to a minimum due to the main factor that rules most things, cost. With the advent of the web, we can now put the missing detailed information that has heretofore been lacking in supplemental information sections that are readily accessible by the reader and yet still keeping the economic realities of paper publication in check.

Satisfying these requirements may seem to be rather daunting but in fact they are quite manageable and more importantly just plain good science. Hopefully the information in this chapter has answered questions you may have had about the publication of real-time qPCR data. As with any experimental protocol, there are many wrong but few right ways to perform the technique. With the goal of having solid real-time qPCR data you want to share with the rest of the research community, if you follow the MIQE guidelines you will succeed.

Primer design programs

Product	Company/source	Website
Commercial		
Beacon Designer	Premier Biosoft	www.premierbiosoft.com/
AlleleID	Premier Biosoft	www.premierbiosoft.com/
3Visual OMP™	DNAsoftware	http://dnasoftware.com/
Primer Express	Applied Biosystems	www.appliedbiosystems.com
Free		
RealTimeDesign	Biosearch Technology	http://www.biosearchtech.com/realtimedesign.aspx
Primer3	MIT	http://frodo.wi.mit.edu/primer3/
PCR Design Tool	IDT	www.idtdna.com/scitools/applications/realtimePCR

Real-time qPCR analysis programs

Product	Source	Website
GenEx	MultiD	http://www.multid.se/genex.html
qBasePlus	Biogazelle	http://www.biogazelle.com/products/qbaseplus
RealTime Statminer	Applied Biosystems	http://www.appliedbiosystems.com/absite/us/en/home.html
RealTime	Applied Biosystems	http://www.appliedbiosystems.com/absite/us/en/home.html
REST-2008	Gene-Quantification	http://www.gene-quantification.de/download.html

Useful websites

NCBI (http://www.ncbi.nlm.nih.gov/)
m-fold (http://mfold.bioinfo.rpi.edu/)
ensembl (http://www.ensembl.org/index.html)
UCSC genome (http://genome.ucsc.edu/)

Useful real-time qPCR formulas

$$\text{Assay amplification} = 10^{-(1/\text{slope})}$$

The ideal value here is 2, for one template converted to two within one cycle.

$$\text{Assay PCR efficiency} = (\text{assay amplification} - 1) \times 100$$

The ideal here is 100%: $(2 - 1) \times 100 = 100\%$.

$$\text{Relative quantity for a sample} = (\text{amplification})^{-\text{dCt}}$$

$$\text{Relative quantity for a sample in relation to another sample} = (\text{amplification})^{-\text{ddCt}}$$

Molecules in 2 pg of ssDNA template = $\{2 \times 10^{-12}/[\text{DNA length in bases} \times 330\ [\text{average MW of dNTP})]\} \times 6.023\text{E}23$

In 2 pg of a 78-base ssDNA template there would be 4.68×10^7 copies

Molecules in 2 pg of dsDNA template = $[2 \times 10^{-12}/(\text{DNA length in bases} \times 660)] \times 6.023\text{E}23$

Molecules in 2 pg of an RNA template = $[2 \times 10^{-12}/(\text{RNA length in bases} \times 340)] \times 6.023\text{E}23$

References

Bohlenius, H., Eriksson, S., Parcy, F., and Nilsson, O. (2007). Retraction. Science *316*, 367.

Bustin, S.A., Benes, V., Garson, J.A., Hellemans, J., Huggett, J., Kubista, M., Mueller, R., Nolan, T., Pfaffl, M.W., Shipley, G.L., Vandesompele, J., and Wittwer, C.T. (2009). The MIQE guidelines: minimum information for publication of quantitative real-time PCR experiments. Clin. Chem 55, 611–622.

Huang, T., Bohlenius, H., Eriksson, S., Parcy, F., and Nilsson, O. (2005). The mRNA of the Arabidopsis gene FT moves from leaf to shoot apex and induces flowering. Science *309*, 1694–1696.

Nolan, T., Hands, R.E., and Bustin, S. (2006a). Quantifiation of mRNA using Real Time RT-PCR. Nat. Protoc. *1*, 1559–1582.

Nolan, T., Hands, R.E., Ogunkolade, W., and Bustin, S.A. (2006b). SPUD: a quantitative PCR assay for the detection of inhibitors in nucleic acid preparations. Anal. Biochem. *351*, 308–310.

Vandesompele, J., De Preter, K., Pattyn, F., Poppe, B., Van Roy, N., De Paepe, A., and Speleman, F. (2002). Accurate normalization of real-time quantitative RT-PCR data by geometric averaging of multiple internal control genes. Genome Biol. *3*, RESEARCH0034.

PCR Applications for Epigenetics Research

9

Gavin Meredith, Miroslav Dudas, Mark Landers, Vasiliki Anest, Jonathan Wang, Caifu Chen, Peter Jozsi and Christopher Adams

Abstract

The field of epigenetics transcends traditional genetics, genomics, and molecular biology, and is poised to revolutionize the field of medical research and healthcare. It is a diverse field that encompasses the study of nuclear components such as chromatin structure, including histone modifications, protein/DNA interactions, protein/RNA interactions, and how these factors influence gene function. It also includes the study of DNA methylation and the role that non-coding RNAs play in influencing DNA methylation patterns, chromatin structure and ultimately regulating gene expression. Just as the field of epigenetics is broad and complex, so is the molecular technology of the polymerase chain reaction (PCR). For every question one would like to address in any of these areas of epigenetics, there is a PCR application and instrumentation suitable to address it. For example there are numerous PCR-based approaches to look at DNA methylation patterns, densities, and even the methylation status of individual cytosine residues by PCR. Additionally, there are PCR methods to survey ncRNA expression and identify regions of the genome where proteins and RNA interact or where certain functional histone marks are located. This chapter provides an overview of these methodologies with a focus on the advantages and disadvantages of each approach.

Introduction

It is now accepted that genetic information is not only encoded by the linear sequence of DNA, but also by various chemical modifications to DNA and chromatin associated proteins collectively known as epigenetic marks. Much of today's epigenetics research is focused on the study of covalent and non-covalent modifications to DNA and histone proteins, the mechanisms that establish them and the mechanisms by which such modifications influence chromatin structure and transcription. Since many classes of non-coding RNAs (ncRNAs), like miRNAs, siRNAs, PASRs and TASRs (Taft *et al.*, 2009; Hawkins *et al.*, 2009; Affymetrix ENCODE Transcriptome Project, 2009) are known to play a role in establishing these epigenetic marks, they are also by definition considered epigenetic regulators and are part of the field of epigenetic research. When it comes to studying epigenetic marks, ncRNAs and their involvement in gene regulation, no existing methodology plays a more important role than the molecular workhorse known as the PCR.

PCR is probably the most versatile and widely used technique in molecular biology-based research. Although many different experimental approaches have been developed for interrogating either genome-wide or specific epigenetic changes in a genome, PCR is often a critical component of these experimental workflows or the actual method used. No other methodology in molecular biology has been modified, customized or tweaked more than PCR. This chapter focuses on historical and emerging PCR methods developed for characterizing and quantifying epigenetic changes throughout a genome. The basic principles, benefits and limitations of each PCR application are presented.

PCR methods for the analysis of DNA methylation

In all mammals, including humans, DNA methylation occurs almost exclusively at the 5 position of deoxycytidine bases (creating 5-methyldeoxycytidine or '5-me C') in the context of CG (also known as CpG) dinucleotides. The patterns of DNA methylation are complex because CpG dinucleotides are non-randomly distributed throughout the genome and the frequency of methylation of any given CpG varies with tissue type, differentiation status, cell state and dynamically in response to environmental changes (Metivier *et al.*, 2008, Kangaspeska *et al.*, 2008). Genome-wide, while CpG dinucleotides occur less frequently than expected based on total C/G content, they tend to occur in clustered regions denoted 'CpG islands'. While DNA methylation in promoter regions and particularly in promoter CpG islands correlates with transcriptional silencing, it is becoming clear that CpG methylation in regions of lower CpG density and distal to promoters can also correlate with gene expression (Irizarry *et al.*, 2009, Schmidl *et al.*, 2009).

Since the initial observations that aberrant methylation is a key alteration in cancer cells, researchers have been developing modified versions of end-point and quantitative PCR methods to investigate methylation patterns. DNA methylation can be analysed quickly and easily by qualitative or quantitative PCR-based methods that include methylation specific PCR (MSP, Herman *et al.*, 1996), combined bisulfite restriction enzyme analysis (COBRA, Xiong *et al.*, 1996), methylation sensitive single nucleotide primer extension (Ms-SNuPE, Gonzalgo *et al.*, 2002), quantitative real-time Multiplex MSP (Fackler *et al.*, 2004) and a combination of methylated-DNA precipitation and methylation-sensitive restriction enzymes (COMPARE-MS, Yegnasubramanian *et al.*, 2006). Moreover, a variety of quantitative MSP methods have been developed and patented over the last decade that permit the analysis of individual CpGs and the relative levels of methylation at specific loci (reviewed in DNA Methylation: Methods and Protocols; 2nd edn, vol. 507; 2009; Humana Press). Although a few of these approaches rely on differential DNA digestion with methylation-sensitive or methylation-dependent restriction enzymes, e.g. methylation-dependent fragment separation (Boyd *et al.*, 2006) and methylation-sensitive restriction fingerprinting (Huang *et al.*, 1997), the majority of approaches involve chemical conversion of the DNA by bisulfite treatment (Frommer *et al.*, 1992). Bisulfite treatment of DNA causes conversion of non-methylated cytosine residues to uracil through a stepwise deamination/desulfonation reaction while leaving methylated cytosines unchanged. This sequence alteration has been the basis for the development of numerous PCR and qPCR approaches to analyse methylation patterns and densities. Examples of these approaches include Methylight (Eads *et al.*, 2000), MethylHeavy (Cottrell *et al.*, 2004) and MethylQuant (Thomassin *et al.*, 2004).

While each of these methods has its own advantages and disadvantages, some generalities exist. Generally, the advantages are ease of use, low cost per assay, a high degree of specificity and unparalleled sensitivity. Often, nanograms and even picograms of converted DNA input are required to analyse a specific genetic locus like a gene promoter. The common disadvantage of these methods is that they only permit the analysis of a handful of CpG sites within a particular target locus. These methods have been around for years and have been thoroughly described by others (Fraga and Esteller, 2002), so will not be discussed in detail here.

DNA methylation: patterns yet to emerge

Affinity-based methods for methylated DNA enrichment

In order to obtain purified DNA sequences for methylation analysis, affinity methods have been developed to enrich methylated sequences from fragmented whole genomic DNA. Affinity purification of methylated sequences avoids the sequence biases and incomplete genome sampling that result from alternative methods for enriching methylated sequences that utilize methylation-sensitive and methylation-specific endonucleases. The affinity methods utilize either methyl-CpG binding domains (MBDs) from methyl-CpG binding proteins, pioneered with the protein MeCP2 in the laboratory of Adrian Bird, or monoclonal antibodies raised against 5-methylcytosine and invariably involve capture on a solid phase support such as agarose, Sepharose™, or magnetizable beads (Reynaud *et al.*, 1992; Cross *et al.*, 1994). The use of anti-5-methylcytosine to purify methylated sequences has been termed 'MeDIP' for 'methylated DNA immunoprecipitation' and has become widely used due in part to the commercial availability of the antibody and protein A or G beads (Weber *et al.*, 2005). MBD-based strategies have emerged that utilize the mammalian MBD2 protein since among methyl-CpG binding proteins tested it has the highest affinity and specificity for symmetrically methylated CpGs in double-stranded DNA (Hendrich *et al.*, 1998; Fraga *et al.*, 2003). Two such systems are MIRA (methylated-CpG island recovery assay) which utilizes the MBD2 and MBD3L proteins bound to a Sepharose™ matrix and MCIp (methyl-CpG immunoprecipitation) which uses the MBD from MBD2 fused to a human IgG Fc domain that can be precipitated with protein A or G beads (Rauch *et al.*, 2005; Gebhard *et al.*, 2006). More recently commercial kits have become available that use the Methyl-binding domain (MBD) of the human MBD2 protein tightly coupled to paramagnetic beads as their capture medium. One such kit is called the MethylMiner™ Methylated DNA Enrichment kit (Invitrogen Corp.), which fuses the MBD2 DNA binding domain to Dynabeads® M-280 Streptavidin via a biotin linker.

The binding interactions in MeDIP and MBD-based methods between the methylated DNA and the capture molecules differ fundamentally and these differences necessitate differing protocols as illustrated in Fig. 9.1. Since the monoclonal antibody used for MeDIP was selected to bind a fully accessible methylated cytosine base, it can only bind ssDNA that contains 5-methyldeoxycytidine (Reynaud *et al.*, 1992). Furthermore, this constraint may reduce capture efficiency since fragments may reanneal during the binding reaction. To compensate for this, antibody–DNA binding is often performed for several hours to overnight at 4°C. After washing the immunoprecipitation beads, the DNA can only be recovered by digestion with proteinase K, usually with heat. This necessitates further

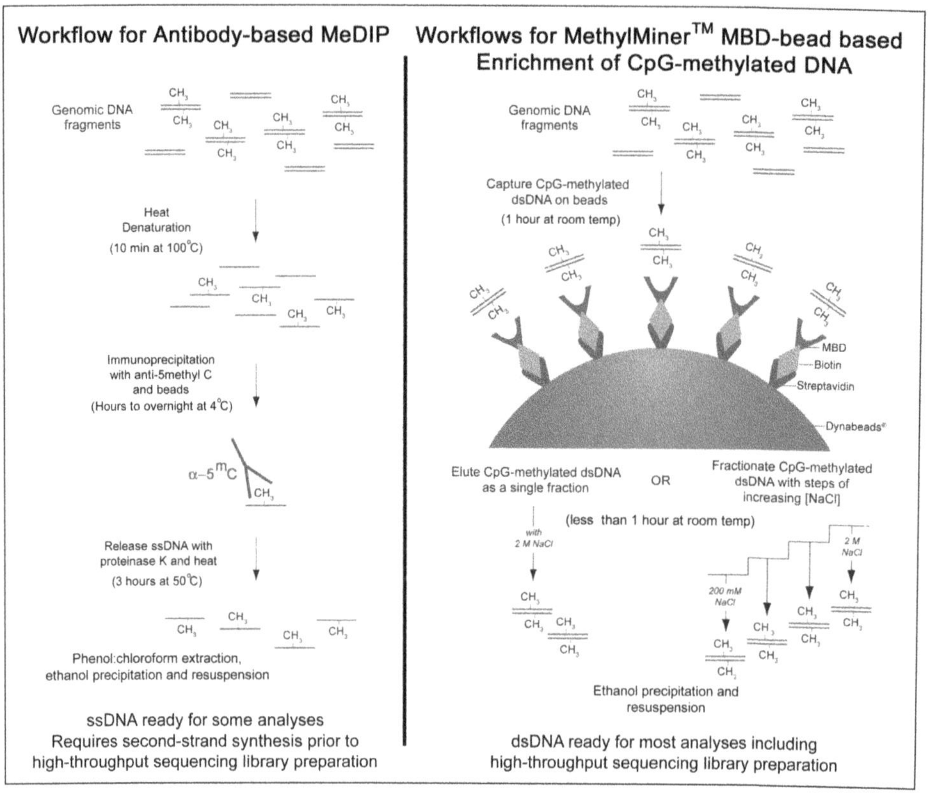

Figure 9.1 MeDIP and MethylMiner™-based methylated DNA enrichment workflows. In contrast to MeDIP, MBD-based enrichment works on double-stranded DNA, can be accomplished within a few hours, can gently yield multiple fractions of DNA with differing degrees of methylation and is directly compatible with high-throughput sequencing library construction methods.

purification, such as by extraction with phenol–chloroform followed by precipitation with ethanol. As a result, MeDIP only partitions sample DNA into two fractions, a methylation-depleted fraction and a methylation-enriched fraction.

Advantages of MBD-based over antibody-based affinity methods

In contrast, use of MBD-based capture, as with the MethylMiner™ system, is fast, specific for methylated CpGs in double-stranded DNA, and permits elution of the captured DNA into one or several fractions. Since MBD2 naturally binds methylated CpG sites in native genomic DNA with high affinity, fragmented sample dsDNA can be directly incubated with the capture media in a binding step that only takes one hour at room temperature (longer times at 4°C are also sufficient). After this brief incubation, the MethylMiner™ beads are ready to be washed and the captured methylated DNA eluted. The elution of double-stranded DNA fragments suitable for direct adaptor ligation is a tremendous advantage for next-generation sequencing library construction. Additionally, the system uses Dynal® paramagnetic beads which provide high-quality and efficient capture kinetics as well as rapid and convenient magnet-based bead capture for washing and changing solutions. Full elution is achieved by mixing the beads with a high-salt buffer. Furthermore, since the affinity of MBDs for methylated DNA can be modulated by ionic strength, fractionation of the captured DNA based on its degree of methylation can be achieved with graded changes in ionic strength (Cross *et al.*, 1994; Gebhard *et al.*, 2006; Schilling *et al.*, 2007). As mentioned above, DNA methylation in various genomic contexts, including regions of low, intermediate, or high CpG density influences gene regulation. Therefore, the ability to fractionate the genome according to the degree of methylation is important for functional studies; thus far, no antibody has been described that can discriminate between DNA fragments with low (but extant) levels of methylation from fragments that are heavily methylated. Finally, only ethanol precipitation is needed to obtain concentrated and purified double-stranded methylated DNA that is ready for further analysis, such as by PCR methods, or by microarray, or by high throughput sequencing (Fig. 9.1).

qPCR to evaluate MBD-based enrichment

Following enrichment, qPCR of control sequences is a valuable tool for monitoring the efficiency of capture prior to global analysis with high-throughput sequencing or with microarrays. Furthermore, when specific loci are of interest, qPCR is useful for determining the characteristic elution patterns of specific regions of interest and for comparison between samples to observe or validate differentially methylated regions (DMRs). In these cases, it is generally sufficient to determine the degree of enrichment within elution fractions relative to a standard curve of input DNA. For absolute measurement of the degree of enrichment, the amount of DNA within each fraction must also be determined. The following sections describe how these measurements can be made and are illustrated with data using the MethylMiner™ methylated DNA enrichment kit (Invitrogen). In short, the enrichment yields salt elution fractions of DNA that are co-precipitated with glycogen from 70% ethanol. The precipitated DNA from each fraction is typically resuspended in 60 µl of nuclease-free water or low ionic strength buffer (TE). At this point the DNA can immediately be used in qPCR. Generally 4 microlitres of the DNA is used as template in each 20 microlitre SYBR-Green-based qPCR.

DNA fragment sizes and qPCR primer design

An example of a typical DMR is the CpG island in the promoter region of the PLAU gene that has been differentially enriched with the MBD-beads (Fig. 9.2). Previous studies with methylation-specific PCR (MSP) methods had been used to determine that this region is methylated in the breast cancer-derived cell-line MCF-7 and relatively non-methylated in the breast cancer-derived cell-line MDA-MB-231 (Guo *et al.*, 2002). Bisulfite cloning and sequencing of DNA samples from these two cell-lines was conducted for the current comparison. As shown in Fig. 9.2, in MCF-7 the CpGs within this single-copy locus are heavily methylated, on average 81% of the time and hypo-methylated, 2% methylated on average, in the MDA-MB-231 cell-line. Also shown are the positions of two qPCR amplicons that have been used to evaluate the separation achieved by affinity purification as will be discussed in following sections. Standard qPCR primer design is sufficient for such analysis but it should be understood that the precision of such measurement depends upon the size of the amplicons and the size distribution of the DNA fragments separated. In this example,

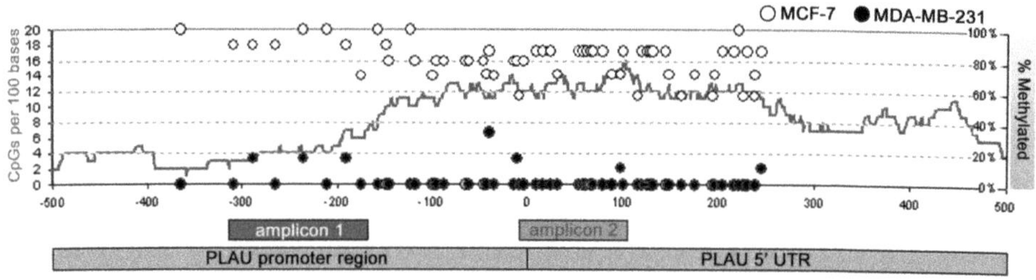

Figure 9.2 Methylation in the promoter and 5′ UTR of the PLAU gene. The frequency of CpG methylation as determined by bisulfite cloning and sequencing across the CpG island in the promoter-region of the PLAU gene for both MCF-7 (light circles) and MDA-MB-231 (dark circles) breast cancer-derived cell-lines is shown. A CpG island (ranging from positions −200 to 500 relative to the transcription start-site) is illustrated by the grey line. The positions of two different qPCR amplicons (blocks) are also indicated. These amplicons are surveyed in figures 4C and 5.

the qPCR amplicons are 145 and 111 bp long, consequently, a significant fraction of the DNA to be enriched should be these sizes or larger after fragmentation in order to be detectable by SYBR-type dye-based qPCR or by TaqMan-type assays. Effective MBD-based enrichment can be achieved with double-stranded fragments ranging from 80 basepairs up to 1 kilobase. However, larger fragments are more likely to carry significant amounts of non-methylated DNA thereby limiting the selectivity of the method for methylated sequences. In practice, fragmentation of the DNA to an average length of 100–200 basepairs yields effective enrichment with the MethylMiner™ workflow.

Creating a standard curve for relative and absolute measures of enrichment

The most reliable qPCR based measures of enrichment depend upon reference to a standard curve generated from non-enriched, 'input', DNA for each primer pair. Since, within the linear range of amplification, the cycle threshold (Ct) is proportional to mass of genomic template or copies of a given target amplicon, the standard curve can be used to determine the equivalent amount of target within an unknown fraction based on its measured Ct value. This calibration curve should be determined on the same plate as unknown fractions are probed and 'no-template' control (NTC) reactions should be run as well. The standard curve is produced from a dilution series of input DNA. For a typical 1 microgram-scale enrichment, from which 4–8 successive salt elution fractions have been obtained, input DNA samples ranging from 64 ng to 62.5 pg (obtained by serial 4-fold dilution and measured in quadruplicate) as template for qPCR are usually sufficient. This range is appropriate since it is convenient to measure $1/15^{th}$ of the elution fraction DNA (e.g. 4 of 60 micolitres) in each qPCR well; 64 ng is approximately the maximal amount of DNA in such a portion from a 1 microgram enrichment experiment. Enrichment is usually carried out in triplicate starting from the same input DNA; the fractions from each replicate separation are measured in duplicate by qPCR. An example of the amplification plot and standard curve (along with unknown data points) from this range is shown in Fig. 9.3. The amplification plot should show well-behaved and appropriately spaced curves and the standard curve plot should

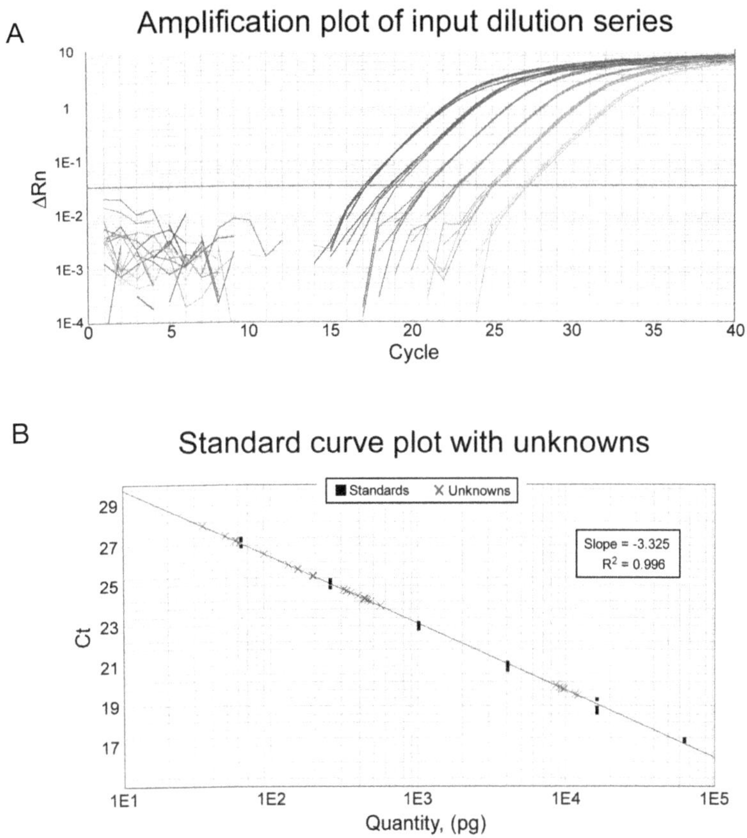

Figure 9.3 qPCR data from non-enriched DNA (input) is used to evaluate amplicon-specific sequence enrichment. (A) Plot of quadruplicate measurements of amplification of a 4-fold dilution series of input DNA (64 ng to 62.5 pg) with an amplicon–specific pair of primers. (B) A linear curve-fit of Ct vs. mass of input DNA (black squares) shows good efficiency (slope = −3.325) and linearity of response (R^2 = 0.996). Cts for unknown samples (portions of enrichment fractions) are plotted as red Xs and an equivalent mass of input is determined by interpolation and limited extrapolation. These values are used to determine relative enrichment and, when the mass of the DNA in enrichment fractions is measured independently, absolute enrichment (see text).

have a slope indicating good amplification efficiency and a high R^2 value indicating that the data lies within a linear range of amplification. For greatest accuracy, the Ct values for experimental fractions should be bracketed by measurements on the standard curve. For the qPCR run, it is useful to include a dissociation stage at the end of the cycling protocol so that the melt-curves can be inspected during data analysis. A single sharp peak in the melt-curves indicates specific amplification of a unique target amplicon. In general, Ct analysis is effective with automated Ct and baseline determination.

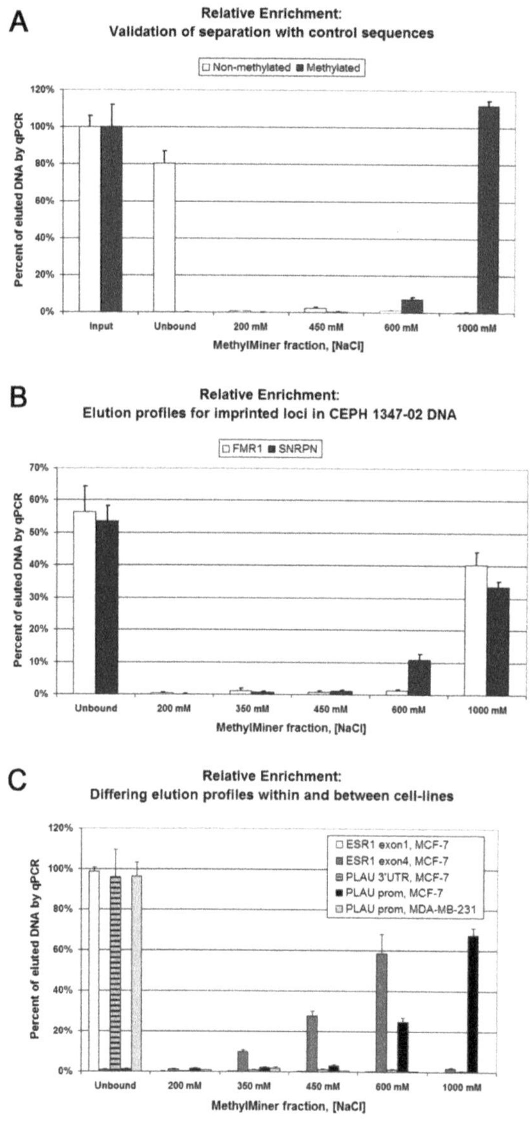

Figure 9.4 Relative enrichment analyses of MethylMiner™-based separation of non-methylated and methylated sequences. (A) Elution profiles of the alien non-methylated and methylated synthetic control duplex DNAs provided in the MethylMiner™ kit. In triplicate, 10 pg of each duplex was spiked into a background of 1 µg of fragmented K-562 DNA prior to enrichment. Values are determined by reference to a standard curve of the Cts of input DNA determined by qPCR; signal magnitude for 10 pg of each duplex (input) is 100%. (B) Relative enrichment profiles from CEPH 1347-02 cell DNA for two imprinted loci were measured in triplicate. As expected, approximately 50% of the amplifiable DNA for each locus is enriched (methylated alleles) and ~50% fails to bind (non-methylated alleles). (C) Distinct relative enrichment profiles for several different genomic loci were measured in triplicate. In MCF-7 cells, the ESR1 exon1 locus is evidently non-methylated while the ESR1 exon4 locus is apparently partially methylated. Also in MCF-7 DNA the 3′ UTR locus shows no evidence of methylation. Matching the data shown in Fig. 9.2, the PLAU locus (amplicon 1) is heavily methylated in MCF-7 cells and poorly methylated in MDA-MB-231 cells.

Assessing relative enrichment

In cases where the precise degree of enrichment is not needed such as when qualitatively testing reagents and procedures or when screening for distinct patterns of enrichment, determining relative enrichment between experimental fractions based on a qPCR standard curve is advised. Examples of this kind of analysis are shown in Fig. 9.4. Fig. 9.4A illustrates qPCR-based analysis of enrichment done with control sequences to validate the MethylMiner™ reagents and procedure. These fractions were obtained from an input sample consisting of 10 pg each of an 80-bp non-methylated control sequence and an 80-bp methylated control sequence mixed into 1 μg of fragmented genomic DNA from the K-562 leukaemia-derived cell-line. The control sequences were designed to be uniquely amplifiable and carry 9 non-methylated CpGs and 8 methylated CpGs respectively and are provided as part of the MethylMiner™ system. A standard curve of the mixture prior to fractionation was generated for each primer set as described above. In addition, 'input' qPCR reactions with 10 pg of either control template were also measured. As can be seen, distinct separation of the non-methylated and methylated control sequences has been achieved. Further, for each control sequence the total amount of amplifiable DNA recovered in the various elution fractions (unbound through to 1000 mM NaCl) closely approximates the 10 pg amount that went into the separation. This indicates a high efficiency of recovery of the DNA.

Fig. 9.4B illustrates relative enrichment profiles for two imprinted loci from genomic DNA from the female donor-derived CEPH 1347-02 cell-line. The FMR1 locus occurs on the X chromosome and is differentially methylated between the active and inactive copies of this chromosome in female human cells (Chhiber *et al.*, 2008). The SNRPN locus resides on chromosome 15 and is subject to maternal allele-specific methylation, i.e. imprinting (Gebhard *et al.*, 2006). As expected, both of these loci show an approximate 50:50 split between copies of the target sequence remaining in the unbound fraction and copies that tightly bind to the MBD-coated capture beads. Such physical separation of differing alleles based on their degree of methylation may permit the discovery of new imprinted loci in future studies.

Fig. 9.4C demonstrates how qPCR-based measures of relative enrichment can be used to distinguish differing patterns of elution and by inference degree of methylation, between loci within a genome and between genomes. Within MCF-7 genomic DNA, distinct methylation is evident for differing portions of two different genes, ESR1, the gene for estrogen receptor alpha and PLAU, the gene for urokinase plasminogen activator. Previous work has shown that the CpG island associated with exon 1 of ESR1 is relatively hypomethylated in MCF-7 cells and, as expected, no MBD-based capture of this DNA could be detected (Gitan *et al.*, 2002). In contrast, DNA corresponding to exon 4 of the same gene from the same cell line is almost entirely captured and elutes with 600 mM NaCl. This result suggests that this locus is subject to partial methylation on both alleles. For the PLAU gene, an amplicon designed against a portion of the 3′ UTR that does not contain any CpG dinucleotides, serves as an endogenous negative control. In contrast and as illustrated in Fig. 9.2 above, DNA that lies immediately ahead of the CpG island in the PLAU promoter (amplicon 1 in Fig. 9.2) is hypermethylated and requires 1000 mM NaCl for complete elution. In contrast and as also depicted in Fig. 9.2, this same PLAU promoter DNA is hypomethylated in MDA-MB-231 cells and also fails to be enriched in the elution fractions.

Assessing absolute enrichment

While determining relative enrichment is broadly useful, it is sometimes important to obtain a more quantitative measure. This is particularly valuable as a means to validate high signal hits from microarray and high-throughput sequencing survey and discovery experiments since qPCR quantitation is generally robust and has a large dynamic range. To obtain such absolute measures of enrichment accurate measures of the mass of DNA within each fraction are needed. In many cases, especially when starting with small amounts of DNA, it is not practical to use spectrophotometry to measure the DNA, but fluorescence-based assays do provide the necessary sensitivity. One such assay system, PicoGreen®, has good specificity for double-stranded DNA and has sensitivity down to 250 pg/ml in fluorescence microplate readers. The mass of input DNA, in picograms, that corresponds to the qPCR signal (Ct) for a given sample reaction can be read directly from the standard curve based on the observed Ct (see Fig. 9.3B). It is important to remember that this value is not the actual mass of DNA used as template in the reaction but is instead the amount of input DNA that would have been required to yield the same Ct. For typical reactions as described above, this value must be multiplied by 15 to obtain the corresponding input mass for the entire fraction. This value, in units of picograms, must be divided by the experimentally determined mass, also in units of picograms, in the fraction (as with a PicoGreen® assay) to yield the unitless quantity 'fold enrichment'. An example of the degree of methylated DNA enrichment that can be achieved with affinity purification is shown in Fig. 9.5. In this case,

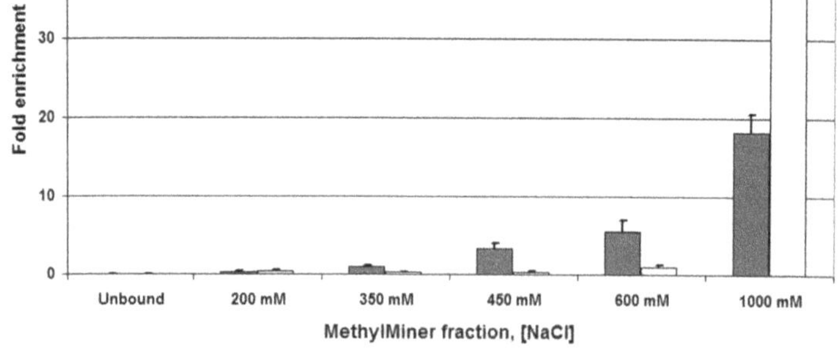

Figure 9.5 Highly methylated DNA can be enriched greater than 40-fold. 2.5 μg of MCF-7 DNA was enriched and eluted in triplicate with the MethylMiner™ system. qPCR was used to assess the enrichment of the two PLAU amplicons shown in Fig. 9.2. Mass of eluted DNA was measured for each fraction with a microplate-based PicoGreen® assay. Absolute fold enrichment was determined as described in the text. Amplicon 1 (dark bars) was enriched in several fractions, possibly due to variable methylation in this locus or neighbouring sequences. In contrast, amplicon 2 (white bars) was almost exclusively enriched, to greater than 40-fold, in the highest salt fraction.

both of the amplicons shown in Fig. 9.2 were enriched from MCF-7 DNA. As can be seen, there is differential enrichment depending upon the position of the amplicons relative to the focal point of highest methylation, the CpG island in this promoter. Amplicon 1, which abuts the leading edge of the CpG island, spans a region of less densely methylated DNA and starts to elute at 600 mM NaCl. In contrast, amplicon 2 spans the portion of the CpG island with the highest methylated CpG density and elutes at the highest ionic strength tested and with greatest degree of enrichment (>40-fold).

Summary

Methyl-CpG binding domain (MBD)-based affinity capture methods are superior for the enrichment of methylated DNA because they are sensitive, specific for CpG-methylation and permit recovery of double-stranded DNA. Further, the MethylMiner™ system in particular combines the utility of the tight binding of biotinylated DNA to streptavidin-coated magnetizable beads with the versatility of salt elution to enable rapid, efficient and reproducible fractionation based on the degree of methylation within individual DNA fragments. Quantitative PCR is a vital tool for assessing the quality and degree of enrichment obtained. Straightforward measures of relative enrichment permit one to evaluate controls, find evidence for differential methylation within and between samples and to distinguish biallelic from monoallelic methylation (i.e. detect imprinting). Obtaining quantitative measures of enrichment necessitates making an accurate measure of DNA mass within samples; however, contemporary fluorescence-based dye-binding assays make this practical. Forty-fold (40×) enrichment can be obtained by these methods.

PCR amplification bias upstream of next-generation sequencing

Bisulfite conversion, cloning and sequencing is considered the gold standard for methylation analysis, but it can be labour intensive and costly. However, the introduction of Next-Generation sequencing technologies promises to revolutionize this approach as it enables whole-genome bisulfite sequencing to achieve nucleotide resolution mapping of methylation patterns. Although the costs associated with these studies remain high and out of reach for many labs, the level of resolution is such that it may eventually become the application of choice.

Following bisulfite treatment and DNA clean-up, the converted DNA is amplified by PCR and either sequenced directly or cloned in a vector for clonal separation and amplification by propagation in bacteria. With these amplification steps, the uridine bases are replicated as thymidine bases, thus bisulfite conversion effectively converts Cs to Ts and transforms normal sequences to AT-rich ones. For a given CpG containing target, copies that are non-methylated become relatively AT-rich as compared to copies that are methylated. As a result, PCR bias in favour of non-methylated targets, due to differences in amplification efficiencies, has been observed and addressed by various strategies (Voss *et al.*, 1998; Wojdacz *et al.*, 2006; Shen *et al.*, 2007). One way to eliminate such bias, reported by Chhibber and Schroeder, is to dilute the DNA after conversion to the point where single-molecule PCR (smPCR) can be conducted (Chhibber *et al.*, 2008). In this method each PCR reaction has a single version of the target thus eliminating the possibility that PCR bias within a reaction-well can occur.

Other results suggest that dilution to the single-molecule level may not be necessary. In a simplified assay system, the PCR bias for non-methylated templates can be readily demonstrated and dilution from millions to thousands of copies per PCR reaction is sufficient to eliminate observable bias. In this experiment, a plasmid carrying a copy of a methylated amplicon from the promoter region of the PLAU gene was obtained from MCF-7 cells and its high degree of methylation (37 of 42 CpGs in 395 bp preserved through bisulfite conversion) was confirmed by cloning and sequencing. Similarly, a copy of the same amplicon (all 42 CpGs converted to TpG and CpA in complementary strands) was obtained from MDA-MB-231 cells and independently cloned into the same vector. These otherwise identical vectors were mixed at the fixed mass ratios of 10:90, 25:75, 50:50, 75:25 and 90:10 to serve as templates for PCR with primers designed against sites containing no CpG dinucleotides. Semi-quantitative melt-curve analysis, based on peak heights of the derivative of the melt curves, was used to determine the relative amount of PCR product

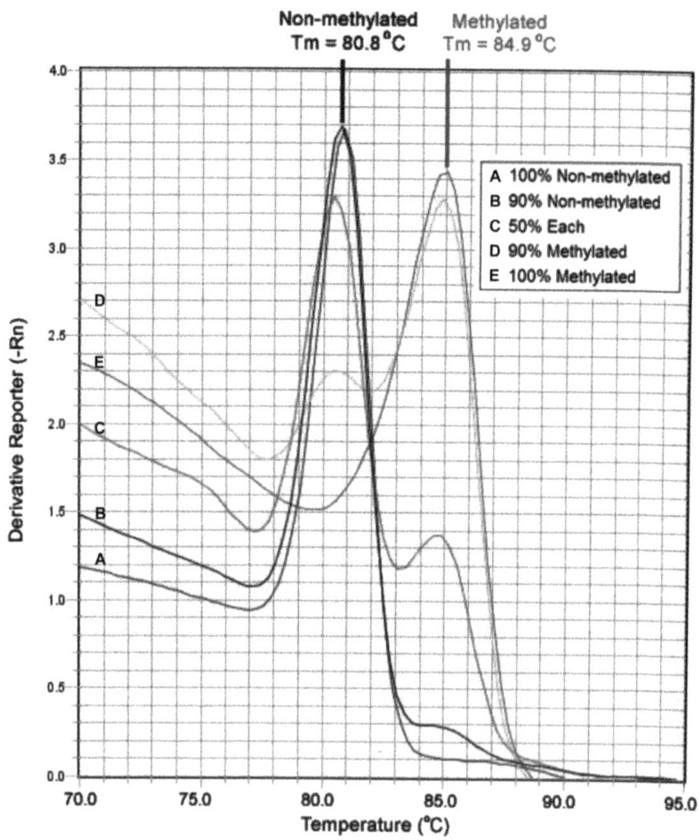

Figure 9.6 Melt-curves of amplified DNA targets. Semi-quantitative estimates of the amount of target DNA (non-methylated vs. methylated) present after amplification can be determined by comparing the derivative peak heights, relative to local background signal amplitude, of the non-methylated amplicon which melts at 80.8°C and its methylated cognate which melts at 84.9°C. Shown are representative control melt curves of mixtures of independently amplified non-methylated and methylated versions of the same locus after bisulfite conversion.

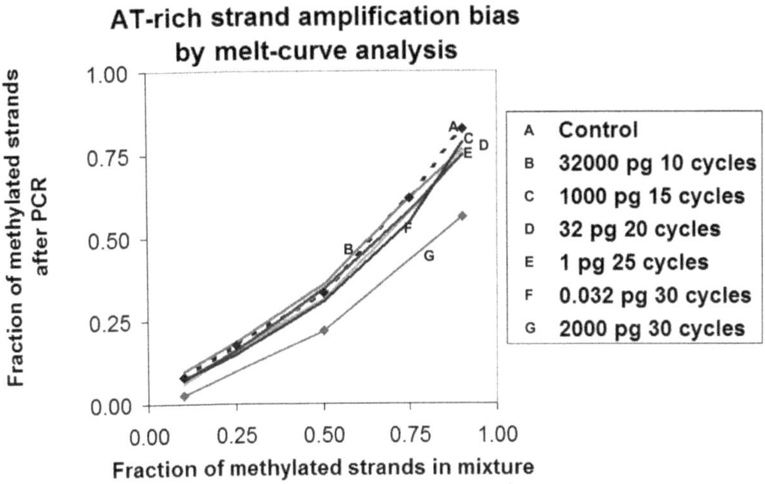

Figure 9.7 Amplification bias and minimization in a model system. Plasmid DNA containing an insert corresponding to either a methylated amplicon after bisulfite conversion or a non-methylated, relatively AT-rich, copy of the same amplicon was mixed after separate PCR amplification at fixed ratios to produce the control data. Samples were composed of mixtures of the plasmid templates at fixed mass ratios. Template mass going into PCR and number of cycles were varied yielding a family of curves that were indistinguishable from the control suggesting minimal PCR bias. In contrast, curve G shows a bias toward the amplification of the AT-rich template at high concentration (2 ng or ~4.3 million target copies/reaction) of template and with 30 cycles of PCR.

derived from either template. As a control, sequences were amplified from pure plasmids prior to mixing and melt-curve analysis (Fig. 9.6). As shown in Fig. 9.7, starting with two nanograms of mixed template and proceeding for 30 cycles yielded underestimates of the degree of methylated template present in all mixtures tested, indicating PCR bias in favour of the AT-rich template. In contrast, using fewer cycles of PCR or starting with less template yielded estimates of the degree of methylation that matched the control. For thirty cycles of PCR, there was no detectable bias when the amount of starting template was reduced to 32 fg (~6800 copies) by dilution.

This model experiment indicates that a practical trade-off can be made between the number of PCR cycles used and the degree of dilution needed to minimize PCR bias when amplifying sequences after bisulfite conversion and suggests that dilution to the extent needed for smPCR may not be necessary. Better guidance on the degree of dilution necessary will come from experiments that replicate the complexity found in actual samples of bisulfite-converted genomic DNA. This model system has only plasmid DNA in approximate 10-fold excess as a background of sequences that compete for primer binding during PCR. In the case where a single allele is the target in a sample of human genomic DNA, the background sequences outweigh the target by approximately 10 million-fold.

High-resolution melting (HRM) applied to methylation analysis

High-resolution melting (HRM) analysis is a relatively new post-PCR analysis method used for identifying genetic variation in nucleic acid sequences that has been successfully used to study locus-specific methylation densities. This simple and rapid method is based on PCR melting (dissociation) curve techniques and is enabled by the recent availability of improved double-stranded DNA (dsDNA)–binding dyes along with next-generation real-time PCR instrumentation and analysis software. HRM analysis can discriminate DNA sequences based on their composition, length, GC content, or strand complementarities. Methylated DNA and unmethylated DNA acquire different sequences after bisulfite treatment resulting in PCR products with markedly different melting profiles. Thus, HRM can be used as a sensitive and specific method for the detection of methylation densities. HRM for methylation analysis starts with bisulfite conversion of the genomic DNA followed by PCR amplification of the region of interest in the presence of a dsDNA-binding dye such as SYTO® 9 (Invitrogen, Carlsbad, CA, USA). This binding dye has high fluorescence when bound to dsDNA and low fluorescence in the unbound state. Amplification is followed by a high-resolution melting step using instrumentation capable of capturing a large number of fluorescence data points per change in temperature, with high precision. When the dsDNA dissociates (or melts) into single strands, the dye is released, causing a change in fluorescence. The result is a melt curve profile characteristic of the amplicon. The melting profile is then compared with a reference amplicon to determine whether the unknown sample contains any sequence variation.

Advantages

High-resolution melt (HRM) is a recent enhancement to traditional melting analyses that significantly increases the detail and information that can be captured. The HRM screening method is sensitive enough to detect single nucleotide differences in several hundred basepairs of sequence. Traditional methods of DNA variant screening using properties of DNA melting such as dHPLC, SSCP and DGGE are more costly and time-consuming than the HRM method. Other advantages of HRM include closed tube reaction for less chance of contamination as well as utilizing existing real-time PCR instruments. HRM screening

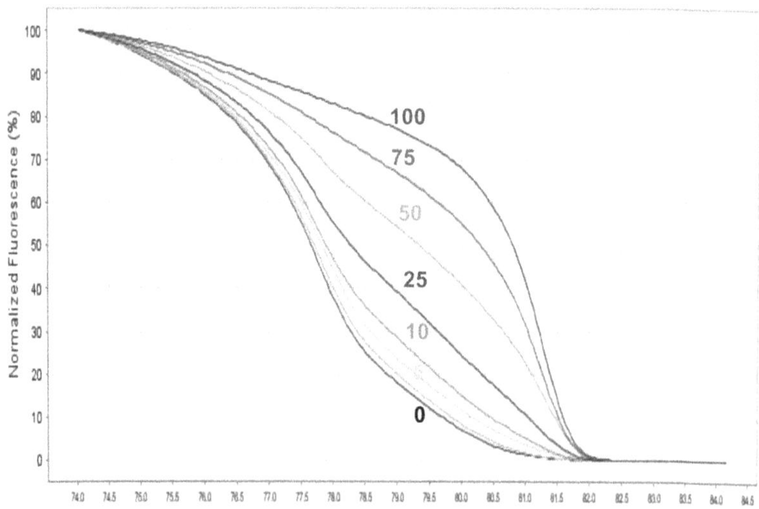

Figure 9.8 Bisulfite converted DNA titration curve from 0% to 100% methylation. The amplicon is 260 basepairs long with 13 CpG sites. The primers used do not contain CpG sequences.

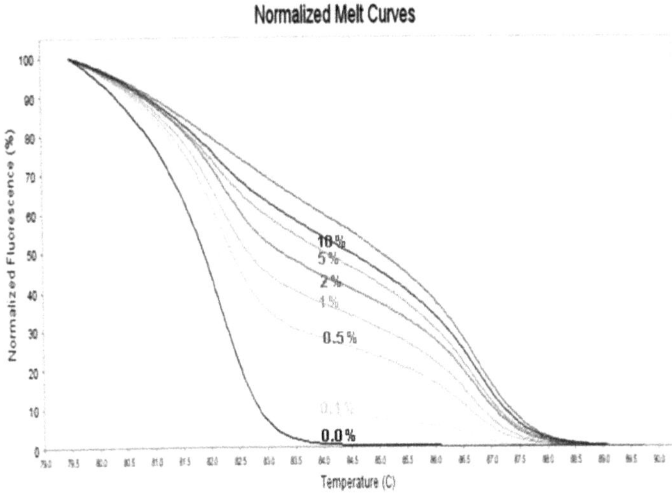

Figure 9.9 Bisulfite converted DNA titration curve from 0% to 100% methylation. The amplicon is 260 basepairs long with 13 CpG sites. The primers contain CpG sequences.

identifies samples containing sequence variation relative to control samples. Furthermore, a titration standard curve can be established using known methylated and unmethylated templates (see example data in Figs. 9.8 and 9.9). The standards can be use to determine quantity of methylation within the amplicon. Using HRM for quantification of methylation is much more sensitive than other techniques such as Sanger sequencing or pyrosequencing. Furthermore, the HRM standard curve can be biased to detect lower methylation levels by manipulating the PCR primer design. By excluding any CpG dinucleotides in the PCR primer binding sites, both methylated and unmethylated DNA templates (bisulfite converted) will amplify in parallel yielding a standard curve favouring a higher percentage of methylation (Fig. 9.8). In contrast, including CpG sites in the primer designs where the primer sequences match the original sequences (before bisulfite conversion) will bias the standard curve toward the lower level of methylation (Fig. 9.9). Using the biased standard curve, the detection of methylation as low as 0.1% is then possible.

Disadvantages

HRM is a target-specific method where one amplicon is screened at a time, as opposed to genome wide screening solutions such as microarrays or next-generation sequencing approaches. However, HRM does have cost and ease of use advantages when the methylation status of only few targets are desired. Furthermore, HRM does not provide nucleotide resolution, i.e. methylation status of individual CpG sites within the target amplicon; bisulfite sequencing is required to gain this level of resolution. However, since HRM is a post-PCR analysis technique, the product of HRM PCR can be used directly in a Sanger sequencing reaction after a simple dilution. The direct sequencing approach reduces processing time by eliminating the additional round of PCR to generate template for sequencing. If the dilution factor of HRM reaction is sufficiently high then there is no need to purify the amplicon prior to use in the sequencing reaction. The fluorescent dye used for HRM does not interfere with the cycle sequencing reaction. Universal M13 tags can be added to the 5′ end of the gene-specific primers allowing the HRM amplicons to be easily sequenced in high throughput fashion.

Chromatin immunoprecipitation (ChIP) and PCR

Chromatin and transcriptional processes are among the most intensively studied fields of biology today. Investigating and understanding epigenetic mechanisms is crucial for vital cellular functions including gene transcription, DNA replication and recombination, repair, chromosomal stability, cell cycle progression and epigenetic silencing. Epigenetic control is an intricate interplay between various molecular mechanisms, e.g. DNA methylation and histone modifications (reviewed in Jenuwein and Allis, 2001; Turner, 2002; Tariq and Paszkowski, 2004; Fuks, 2005)). The role of histone modifications (such as lysine acetylation, lysine and arginine methylation, phosphorylation, ubiquitination) has only been fully appreciated for about 10 years (Fuks, 2005). The impetus was partly driven by histone code hypothesis which predicts that the post-translational modifications of histones, alone or in combination, function to direct specific and distinct DNA-templated programs (Strahl and Allis, 2000). Since then the number of papers on new histone modifications and their possible functions has exploded. Some specific examples such as phosphorylation of serine residues 10 and 28 on histone H3 is a marker for chromosomal condensation. Similarly, the combination of phosphorylation of serine residue 10 and acetylation of a lysine residue 14 on histone H3 is a tell-tale sign of active transcription.

To understand the role of chromatin-associated proteins in chromosomal functions, it is essential to clarify where these regulatory proteins bind to the chromatin *in vivo*. To date, the most widely used and powerful method to identify regions of the genome associated with specific proteins is the Chromatin Immunoprecipitation (ChIP) assay. The ChIP assay has been widely used to study both histones and non-histone proteins, such as transcription factors, within the context of the cell (Orlando *et al.*, 1997; Wells and Farnham, 2002; Weinmann and Farnham, 2002; Das *et al.*, 2004; Nightingale *et al.*, 2006). Transcription factors and other DNA binding proteins have a weaker affinity than histones, which are generally tightly associated within the chromatin complex. To avoid dissociation of non-histone proteins from the chromatin binding site, it is necessary to incorporate a cross-linking step (Nightingale *et al.*, 2006). Importantly, antibodies that recognize a protein of interest can be used to determine the relative association of that antigen in the context of chromatin at one or more loci in the genome. Overall, ChIP has proven to yield very valuable information on chromatin-associated processes in eukaryotes, including plants and humans.

Standard ChIP workflow

In the ChIP technique, live cells are treated with formaldehyde to generate protein–protein and protein–DNA cross-links between molecules in close proximity on the chromatin template *in vivo*. Subsequently, the cells are lysed and the chromatin is released from the nuclei and sheared by sonication to reduce average DNA fragments sizes to ~500bp. The resolution obtained by the ChIP procedure is determined by the size of the chromatin fragments used as input material. Two methods are commonly used to fragment chromatin, sonication (hydrodynamic shearing) and micrococcal nuclease (MNase) digestion. Both methods can show preferential fragmentation of certain chromosomal regions. However when using formaldehyde cross-linking, sonication is the preferred method, as cross-linking may restrict the access of MNase to chromatin. Specific antibodies are then used to immunoprecipitate the protein of interest cross-linked to the chromatin complex. This antibody/chromatin complex is isolated typically using Protein A or G agarose, Sepharose or, more recently, magnetic beads. Reversal of the formaldehyde cross-linking by heating permits the recovery and quantitative analysis of the immunoprecipitated DNA (ChIP'd DNA). This material is now ready for downstream analyses such as end-point PCR and quantitative PCR (qPCR), or genome-wide analyses using promoter-tiling arrays, or massively parallel sequencing. Typically for PCR analysis, primers are designed spanning the desired DNA sequence of interest and data demonstrate whether the specific protein of interest is associated *in vivo* to that DNA region. Quantitative real-time PCR (QPCR) is presented as best practice to analyse the chromatin-immunoprecipitated material.

MAGnify™ ChIP system

The MAGnify™ Chromatin Immunoprecipitation System (Invitrogen Corp) provides a streamlined, optimized assay for the enrichment of chromatin/protein complexes and DNA recovery using magnetic bead capture technology (Fig. 9.10). The isolated DNA is ready for downstream analysis by methods such as PCR- or qPCR-based assays, or massive parallel DNA sequencing. This kit enables researchers to start with lower cell numbers than traditional ChIP workflows, thereby preserving precious samples such as primary cells, stem cells and biopsies. In addition, the protocol can be completed in a single day, compared with 2–3 days for a traditional ChIP assay. In this section we will discuss in detail

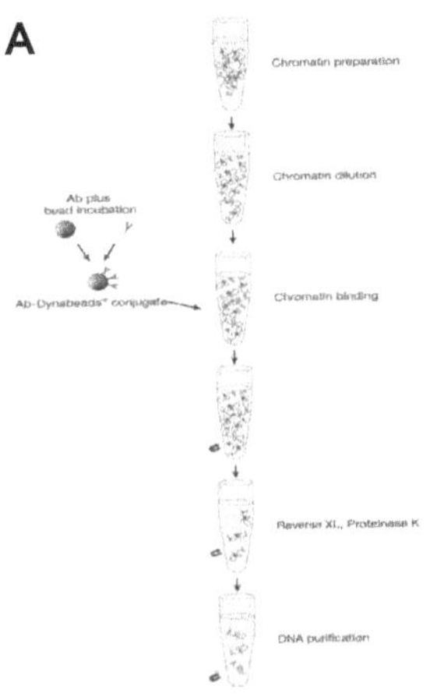

B

Workflow Step	MAGnify™ ChIP Timeline	Conventional ChIP Timeline
Pre-clearing	N/A	1-2 hrs
Antibody/Chromatin Incubation	2 hrs	Overnight
Bead Pulldown	1 hr	2 hrs
Washes	30 min (2 buffers)	1 – 3 hours (4 buffers)
Reverse Crosslinking	1.5 hrs	Overnight
Proteinase K Digestion		2 hrs
DNA Elution from Beads		15 – 30 min
DNA Purification		2 hrs - Overnight
Average Time	5 hrs	36 – 48 hrs

Figure 9.10 Overview of MAGnify™ ChIP workflow vs. conventional ChIP protocols.

the various steps of the ChIP workflow as well as the levels of improvements addressed by the MAGnify™ ChIP System and will describe a method based on the real-time PCR technique to analyse chromatin immunoprecipitation products.

Basic principles of the ChIP workflow

Cross-linking and chromatin shearing

The initial step in the ChIP workflow is to cross-link DNA to the associated proteins by treating the cells with formaldehyde, which ensures that the chromatin structure is preserved during the isolation and ChIP procedure. The MAGnify™ kit provides methods for fixing mammalian cells with formaldehyde prior to lysis and ChIP analysis. Too much cross-linking can lead to less protein bound to the DNA and fewer epitopes or changes in epitopes available for antibody binding. It is important to keep the formaldehyde incubation time and method consistent between samples that you want to compare, to maintain consistency and reproducibility of results. We recommend a 10-min cross-linking step using formaldehyde at a 1% final concentration. However, you may choose to perform a time course experiment to optimize cross-linking conditions.

Two methods are commonly used to fragment chromatin: sonication and micrococcal nuclease (MNase) digestion. Both methods can show preferential fragmentation of certain chromosomal regions. However, sonication is the preferred method when using formaldehyde cross-linking, as cross-linking restricts the access of MNase to chromatin. We have optimized the ratio of chromatin and MAGnify™ lysis buffer for optimal chromatin shearing using the Bioruptor™ 200 and Covaris™ S2 instruments. Shearing the chromatin into 200–500 bp fragments is essential to ensure good resolution for ChIP efficiency but

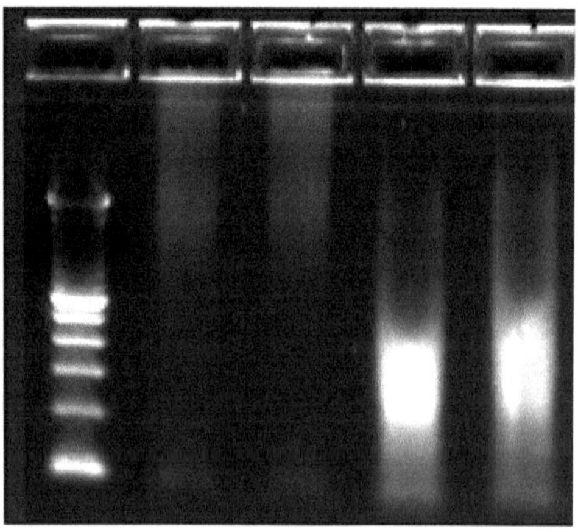

Figure 9.11 Three million Human 293Gt cells were fixed with 1% formaldehyde for 10 min at room temperature. Subsequently, cells were lysed in 150 µl of MAGnify™ Lysis Buffer (10 min on ice) prior to chromatin shearing. Samples were either unsonicated (lanes 2 and 3) or sonicated (lanes 3 and 4) for 12 cycles using Bioruptor™ from Diagenode. 10 µl of DNA was analysed on a 2% E-gel. Lane 1, 100-bp ladder; lane 2, unsonicated DNA; lane 3, unsonicated DNA; lane 4, sonicated DNA from 1 million 293Gt cells; lane 5, sonicated DNA from 1 million 293Gt cells.

fragmenting the DNA to lower sizes may be more suitable for applications such as next generation sequencing. If the average fragment size is greater than 1000 bp, you could be pulling down DNA that contains your target sequence for PCR, but the protein of interest may be a significant distance from your target. If using a probe sonicator, it is important to keep the cell lysate cooled on ice during sonication, as heat released by the sonication probe can reverse the cross-links. If using probe sonication, place the sample on ice between cycles.

We recommend testing various sonication conditions on your cells of interest and then isolating the DNA and running it on a 1.5–2.0% agarose gel (e.g. a 2% E-Gel®) with a 100 bp ladder to determine fragment length (Fig. 9.11).

In general, when starting with 1×10^6 cells per 50 µl, use ~10 µl of isolated DNA per well on a 2% E-Gel®. If desired, you can also add 1 µl of Proteinase K to the 10 µl and incubate at 55°C for 20 min prior to running on a gel. *Note: sonication must be optimized based on cells and equipment used.

Antibody specificity and immunoprecipitation of chromatin

Antibodies are the most important factor for a successful ChIP experiment. It is crucial to choose an antibody carefully. Antibodies can be available as polyclonal or monoclonal. Monoclonal antibodies have a high specificity compared to polyclonal sera, but the polyclonal sera may recognize several epitopes of the target, increasing signal levels of low-abundance templates. Different antibody preparations have distinct properties, which can affect the ChIP results, affecting the resulting signal levels. In addition, the relation between the availability of epitopes and antibody binding may not be linear. Some antibodies are sensitive to inhibitory factors present in the input chromatin sample, resulting in a decrease in binding efficiency of the antibody in the presence of excess chromatin. Additionally, the successful use of a specific antibody in experiments such as Western blotting or immunoprecipitation/Western blotting is not a strong predictor for ChIP compatibility. Standard conventional ChIP protocols are labour-intensive and typically require high number of starting cell numbers per ChIP. The MAGnify™ ChIP System provides the flexibility to test various permutations of antibody–chromatin ratio and determine the optimal ChIP efficiency for the antibody and chromatin of their interest.

To date, protein A/G agarose or Sepharose beads have been widely used to isolate the antibody/chromatin complex. The MAGnify™ ChIP system utilizes the power of magnetic Protein A and Protein G Dynabeads which provide a significant enhancement over agarose and sepharose beads at multiple steps in the workflow. First, Dynabeads® provide a higher capacity of protein A and G to allow for quicker antibody incubation times. Second, Dynabeads® have much lower background than competitor magnetic beads. Thirdly, use of Dynabeads® minimizes manipulations by replacing spin steps with a rapid magnetic separator for easier washing and handling. Overall, magnetic handling is fast, efficient and all steps take place in a single tube, with no centrifugations helping to reduce sample-to-sample variability. However, Dynabeads® Protein A and Dynabeads® Protein G may bind with different affinities depending on the immunoglobulin type and species. The MAGnify™ChIP system provides a proprietary blend of both Dynabeads® Protein A and Dynabeads® Protein G in a single vial allowing for the use of a broader range of antibodies and reducing customer uncertainty regarding matching antibody type with bead type. The MAGnify™ ChIP system utilizes a novel magnet, DynaMag™ PCR, that is compatible with 0.2-ml PCR strip tubes. This magnet is critical for enabling the use of reduced starting cell numbers and volumes, immunoprecipitation and washing steps. This novel magnetic separator allows the researcher to achieve results quickly and easily while offering the freedom to simultaneously perform multiple ChIP assays in PCR tubes using a multichannel pipettor.

Reverse cross-linking and DNA recovery

The MAGnify™ ChIP system includes key buffers that enable the customer to have a streamlined workflow from cell lysis/chromatin prep, immunoprecipitation, washes, reverse cross-link and DNA recovery. In the standard ChIP assay, the most widely used cross-linking reagent is formaldehyde. To reverse formaldehyde cross-linking, the fragments are incubated overnight at 65°C with TES [buffer of 50 mM Tris-HCl (pH 8.0), 10 mM EDTA and 1% SDS]. Then the released DNA fragments have to be purified prior to PCR by another lengthy procedure that includes incubation with proteinase K, repeated phenol–chloroform extraction and ethanol precipitation. This purification step alone consumes at least one whole day and combined with the overnight incubation for cross-link reversal makes the standard ChIP assay both laborious and time-consuming. Recently, commercialized ChIP kits have been developed which have omitted the overnight incubation and phenol–chloroform extraction steps by using Chelex-100 resin to extract DNA from the immunoprecipitates. However, this procedure includes several steps such as tube transfers, repeated boiling step for Chelex extraction, a separate step for the addition of proteinase K digestion. In comparison, the MAGnify™ ChIP system is a simpler and more time-efficient method that simplifies the reversal cross-link and DNA recovery steps and provides more consistency by combining reverse cross-link and proteinase K digestion steps. The proprietary reverse cross-link buffer eliminates the need for multiple buffers and overnight incubations allowing the simultaneous elution of DNA from the beads while reversing the cross-links in less than 1 h. This is followed by an essential DNA recovery/clean up step using a magnetic bead-based approach. (Note: the magnetic beads used for DNA recovery are distinct from the Dynabeads used for immunoprecipitation.) This approach provides higher quality of enriched DNA and results in greater yield of DNA from limiting starting cell number vs. other methods such as phenol-chloroform. Also, this system is simple to remove buffers without disturbing the bead pellet and is now possible to perform multiple ChIPs in PCR strips using a multichannel pipettor.

Controls for ChIP specificity

There are two main controls that are essential to include in your ChIP workflow in order to determine the specificity of your ChIP experiment, the first of which is the Input control sample. This is an aliquot of chromatin that has not been immunoprecipitated but is processed in parallel with ChIP samples from the reverse cross-linking step through DNA purification and is a critical control for PCR efficiency as well as data normalization. Overall, calculating the percent input is very useful in normalizing the starting material and reducing false positives due to variable amounts of DNA in one sample to another or when the experimental design is comparing chromatin association of the protein of interest in various cell lines. The second control is to perform a ChIP with an irrelevant antibody such as Rabbit IgG and Mouse IgG. These antibodies are provided in the MAGnify™ kit for use as negative controls to measure non-specific binding. This negative control is processed through the entire ChIP workflow in parallel with your antibody of interest. The use of the input and negative controls can be used for normalization by the per cent input or fold enrichment method discussed in the following section. In addition to the above-mentioned controls. The use of an antibody that consistently binds chromatin-associated proteins under a wide variety of cellular conditions is helpful to test that your ChIP method is working efficiently. For example, we observe consistent enrichment of heterochromatin markers such as histone H3 lysine 9 trimethylation (H3-K9Me3) at the at the satellite repeat locus (SAT-2). Additional controls may be needed depending on the application; such as the use of RNAi to deplete the protein of interest, mutant cells lacking the protein of interest, or including untreated samples to compare to samples treated under various stimuli.

Analysis of ChIP DNA with real-time quantitative PCR

In order to obtain high-quality qPCR ChIP data, specific criteria must be met regarding primer pair design, optimization and use of primer controls as described below. Success in obtaining high-quality data is critically dependent on good primer design. In general, primers should be 20 to 30 bases long with a T_m of 55–60°C. The design of good primers is greatly facilitated by commercially available software packages such as Oligo 6.6 (see http://www.oligo.net) or Primer Express 3.0 (http://www.appliedbiosystems.com). Most primers require no purification or special treatment prior to PCR. Amplification products should range between 75 and 350 bp; longer products should be avoided, as the amplification efficiency is substantially lower. A final primer concentration of 1 μM works well for most primers, but in some instances, improved product specificity may be obtained by lowering the final primer concentration 5- to 10-fold. High-quality primer pairs should result in ~1.9-fold amplification/cycle (this can be determined from quantitative analysis of raw fluorescence data for each cycle, which is generally available on commercial instruments). Amplified material at the completion of the qPCR should contain only one band which could be assayed on high-percentage agarose or polyacrylamide gels. Specificity information may also be obtained by running dissociation curves on reactions following the conclusion of the PCR run. Typically, samples are melted for 15 min at 95°C, cooled to 60°C and then slowly heated back up to 95°C over a period of 20 min. Some instrument-specific software packages have built-in modules for dissociation curve analysis

Importantly, primer design is very dependent on the overall experimental design and the type of protein being analysed. We advise designing primers within specific elements such as a promoter, intron or other regulatory sequences. For example, specific chromatin states may be associated with specific histone modifications. Transcriptionally active chromatin is generally characterized by gene active regions that are associated with histone H3 lysine acetylation (such as H3-K9 acetylation). In contrast, transcriptionally repressive regions are typically associated with histone H3-K9 trimethylation. Depending your experimental design and protein of interest, it is important to design negative and positive control PCR primers. Negative control primers are designed for a sequence that is not enriched by your ChIP procedure while positive control primers should include the sequence of interest.

For primer pair optimization, the amplification efficiency and melting curve analysis of a given primer pair set should be examined in order to determine the amplification is within a linear range and to eliminate inaccurate quantification if possible. qPCR controls should always include a no template control as well as Input DNA. For each primer pair examined the input DNA samples should be run alongside the immunoprecipitated samples. This is critical in determining the performance and amplification efficiencies among different primer pairs. Different primer pairs may vary slightly on a per-cycle basis, but those slight variations in efficiency translate into substantially different amounts of amplified material in the cycle range used for analysis. Precise quantification of relative binding cannot be accurately performed without primer pair–specific input signal. Below is a description of the use of Input DNA in determining optimal qPCR primer pair conditions.

Determining amplification efficiency and specificity of primer pairs using the input control

We recommend determining the amplification efficiency of your qPCR reaction using 10-fold serial dilutions of the Input Control DNA or genomic DNA (e.g. 1:1 to 1:1000). An acceptable efficiency range is ~1.9–2.1, which corresponds to qPCR standard curve slopes of −2.9 to −3.3. Amplification efficiency (AE) is calculated by the formula $AE = 10^{(-1/\text{slope})}$. The specificity of the product is demonstrated by the presence of one peak in a melting curve and spot-checking by gel electrophoresis. To quantitate the amount of product DNA from the threshold cycle number, a standard curve for each primer pair may be generated for each PCR set from serial dilutions of Input DNA. This curve is used to determine the slope of the standard curve for each test reaction. Below is a representation of the standard curve plot, dissociation and amplification plot for the SAT2 primer pairs used as a positive control for the ChIP workflow. Quantification is performed by measuring the number of cycles necessary to reach a threshold of fluorescence value for each sample. This value has to be in the exponential phase of the reaction as displayed in the amplification plot (Fig. 9.12A). A dissociation curve obtained by progressively increasing the temperature from 60 to 94°C allows measurement of the double-stranded DNA denaturation kinetics (Fig. 9.12B). The kinetics for a specific PCR fragment is dependent on nucleotide composition and size. Therefore, only one peak will be detected if the PCR generates a single clean product (Fig. 9.12B). A standard curve plot generated by the ABI software generating the slope needed to calculate the amplification efficiency of the given primer pair (Fig. 9.12C).

Template set-up for qPCR and data analysis

In general, individual samples should be run in triplicate. Obvious outliers occur with some frequency, generally at <5%. Triplicate analysis of samples permits removal of those outliers while still allowing for inclusion of two accurate measurements for each sample. While this reduces the number of different samples that can be run at any given time, the resulting data is much more reliable and accurate.

Data normalization using the input control

The choice of method for normalization depends on the experimental set-up and the quality and availability of specific controls. As best practice, we normalize ChIP qPCR data by per cent input. In our experience, this method can be reliably used with high quality measurements of at least three replicates of a ChIP QPCR experiment. The five steps below outline the steps taken to retrieve the qPCR data through data analysis.

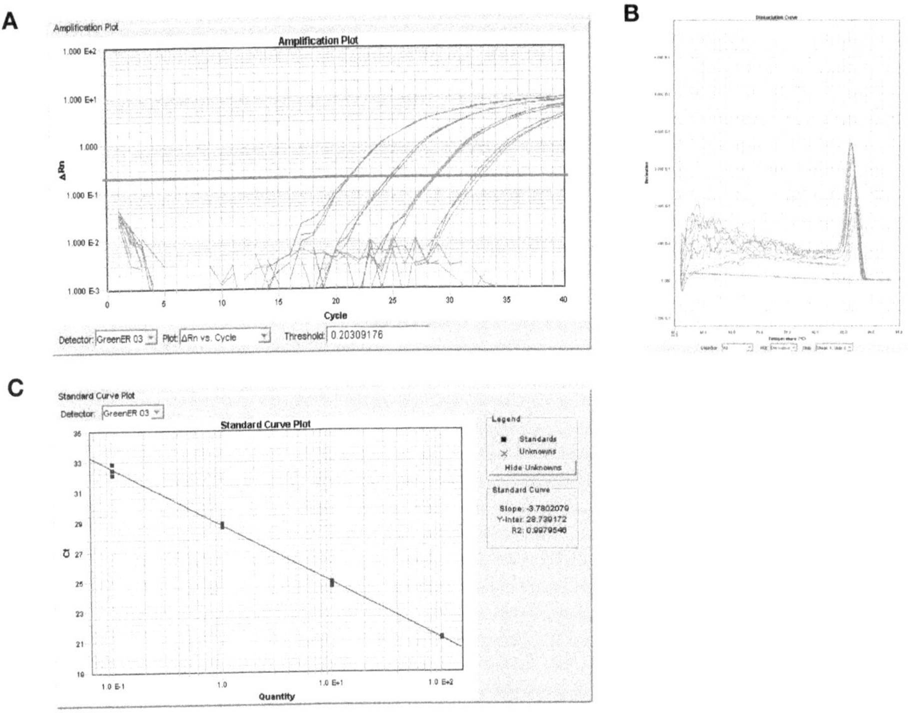

Figure 9.12 (A) Serial dilutions of input DNA were amplified using SYBR GreenER PCR reagents, the ABI Prism 7700 sequence detection system (Applied Biosystems) and chicken lysozyme promoter-specific primers. (B) An example of a dissociation curve indicating the presence of only one product. (C) Standard curve plot generating slope value.

1. Export the qPCR data to a spreadsheet program such as Microsoft Excel by using built in filters. The file should not contain omitted wells and should be in a column format containing well positions, descriptors and Ct values for each selected well.
2. Open the exported file. Average the replicate measurements for each IP reaction in a new column (AVERAGE Ct ChIP).
3. For each primer pair, calculate the adjusted Ct for the input controls. For an input control that was 10% of the IP reaction, then the dilution factor (DF) is 10 and subtract 3.32 cycles (i.e. log2 of 10) from the Ct value of the Input Control. Then average the Input Control replicates (AVERAGE CT INPUT).
4. Subtract the AVERAGECT INPUT from AVERAGECT ChIP in a new column. This number is the ΔCT. This value represents the difference in cycles between the immunoprecipitated sample and the input DNA.
5. Per cent input is calculated by: $100 \times AE^{(\text{AVERAGE Ct INPUT} - \text{AVERAGE Ct ChIP})}$, where AE is the amplification efficiency as calculated above.

Alternatively, a standard curve obtained with known amounts of input DNA will help in converting data expressed as number of cycles into nanograms of DNA. ChIP results

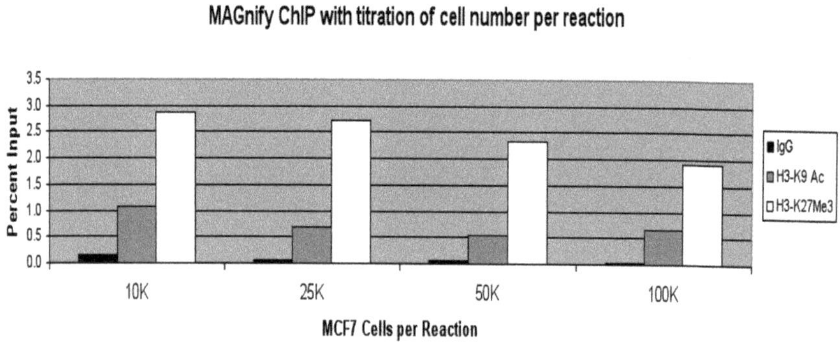

Figure 9.13 Results of typical ChIP experiment. Enrichment for the SAT2 locus shown.

are then expressed as a percentage of the input material and compared with ChIP experiments using a non-specific antibody as a control to determine the background level of the experiment.

Fold enrichment as calculated by signal over background

With this method, signals from the IP reactions are divided by the signals from the negative antibody control reaction (i.e. the rabbit IgG or mouse IgG antibody provided in the kit). This represents the IP signal as the fold increase in signal relative to the background signal. The assumption of this method is that the level of background signal is reproducible between different primer sets, samples and replicate experiments, even though background signal levels can vary due to these factors.

Anticipated results

In the example below we examined the association of two histone modifications, H3-K9 acetylation and H3-K27 trimethylation, with the immunoprecipitated DNA from three or four independent ChIPs and analysed by qPCR. Sheared chromatin from MCF7 cells was prepared from 1 million cells in 50ul lysis buffer and diluted to the indicated number of cells per ChIP according to the MAGnify™ protocol. A fraction (1%) of the sonicated chromatin was set aside as input control and the remaining chromatin was then immunoprecipitated using antibodies specific for H3-K9Ac, H3K27Me3, or rabbit-IgG (1 µg antibody was used per ChIP). Optimized qPCR primers were used to amplify the SAT2 locus. Percent input was calculated by $100 \times 2^{(\text{Ct Adjusted Input} - \text{Ct Enriched})}$. Input DNA Ct was adjusted from 10% to 100% equivalent by subtracting 3.3 Cts; error bars indicate the standard error (Fig. 9.13).

Histone H3-K27Me3 was enriched at the heterchromatin SAT2 locus compared with a significant decrease in the active mark, H3-K9 acetylation. In addition to histones, ChIP is a powerful approach that allows the analysis of the interaction of transcriptional factors and chromatin–regulator complexes in living cells such as RNA polymerase. The immediate early genes such as c-fos, c-jun and c-myc are characterized by their rapid inducibility in response to a variety of signals, including mitogenic stimulation (Clayton, 2003). *c-fos* has been the most extensively studied of this group of genes relative to induction of gene

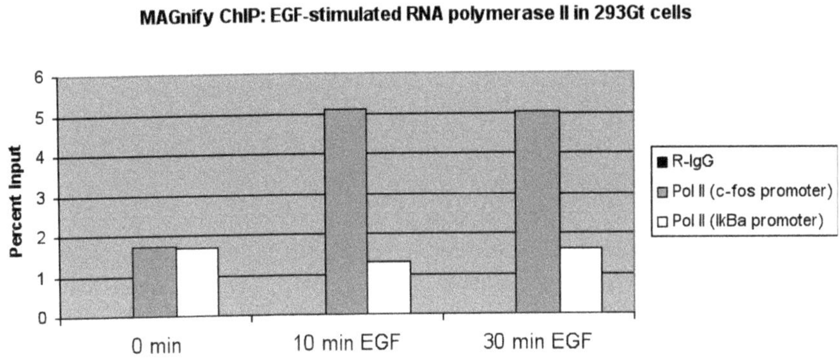

Figure 9.14 Results of typical ChIP experiment.

expression, with evidence demonstrating the involvement of sequence specific transcription factors and associated cofactors controlling transcription induction in a manner dependent on the activation of upstream signal transduction cascades (Jankrecht *et al.*, 1993; Gille *et al.*, 1995; Treisman, 1995). ChIP assays were performed in 293Gt cells that were serum-starved for 12 hours followed by epidermal growth factor, EGF treatment (50 ng/ml), for 10 min. Cells were counted and diluted to 50,000 cells per ChIP. 3 µl of RNA polymerase II antibody was used per ChIP. A fraction (1%) of the sonicated chromatin was set aside as input control. Optimized qPCR primers were used to amplify to different loci *IkBα* or *c-fos* promoter. Percent input was calculated by $100 \times 2^{(\text{Ct Adjusted Input} - \text{Ct Enriched})}$. Input DNA Ct was adjusted from 10% to 100% equivalent by subtracting 3.3 Cts; error bars indicate the standard error (Fig. 9.14). *IkBα* gene expression is not strongly induced by EGF as shown by the overall consistent levels of RNA polII association at the *IkBα* promoter by ChIP assay. In contrast, c-fos is strongly responsive to EGF as the levels of RNA poll I enrichment are enhanced at the c-fos promoter in response to EGF after 10 and 30 min.

RT-qPCR methods for microRNA analysis

MicroRNAs are short, non-coding RNAs that play an important role in regulating gene expression in animals and plants by targeting mRNAs for translational repression or cleavage (Bartel *et al.*, 2004; Kim *et al.*, 2006). The history of the miRNA discovery is a success story of open scientific communications and close collaborations. The lin-4 locus was initially discovered in the Sydney Brenner lab in 1970s. The remarkable developmental defects of lin-4 were first described in the Bob Horvitz lab. Later, Dr Victor Ambros showed that the line-4 was the upstream for the gene lin-14 and identified apparent null alleles of lin-14 which all had developmental timing defects. Victor and his team started the lin-4 project in 1986. After almost seven years of the painstaking efforts, Victor's team together with Dr Gary Ruvkun's discovered the first miRNA, lin-4 in 1993 (Lee *et al.*, 1993; Lee *et al.*, 2004; Ruvkun *et al.*, 2004) It was believed to be an anomaly until 2000 when a second microRNA, let-7, was discovered from Ruvkun lab. Unlike lin-4, the sequence of let-7 was found to be highly conserved in a wide range of organisms (Reinhart, 2000). It was soon realized that similar sequences, first termed miRNAs by Ambros in 2001, were widespread in the genomes of eukaryotes (Lagos-Quintana *et al.*, 2001). To date, there are 706 human microRNA genes listed in the Sanger's miRBase database (http://microrna.sanger.ac.uk/sequences/) (Griffiths-Jones *et al.*, 2008). The total number of miRNA genes is expected to increase to >1000 in human.

Why are miRNAs important?

MicroRNAs are widely recognized as one of the major regulatory gene families in eukaryotic cells. Most miRNAs can potentially target tens to hundreds of mRNA genes. Furthermore, many mRNA genes may have multiple miRNA binding sites that represent targets for one or more miRNAs (Bartel *et al.*, 2004, Kim *et al.*, 2006). MicroRNAs play important roles in a broad range of biological processes including development, cellular differentiation, proliferation and apoptosis. Emerging evidence also implicates miRNAs in the pathogenesis of human diseases such as cancer, metabolic diseases, neurological disorders, infectious diseases and other illnesses (Reinhart *et al.*, 2000; Lu *et al.*, 2005; Triboulet *et al.*, 2007; Ma *et al.*, 2007; Rodriguez *et al.*, 2007; Tavazoie *et al.*, 2008;Yu *et al.*, 2008).

RT-qPCR methods used to quantify miRNAs

MicroRNA molecules are too short to design a conventional qPCR assays which typically include forward, reverse primers and a TaqMan® probe. Use of the novel stem-looped RT primer has made it possible for real-time RT-qPCR. Stem-looped RT based TaqMan® MicroRNA Assays have become one popular method of choice for miRNA quantification (Chen *et al.*, 2005). Another popular method of miRNA quantitation is the universal primer-based SYBR approaches that employ polyA-tailing of the short mature miRNA target prior to first-strand cDNA synthesis (Shi *et al.*, 2005; Raymond *et al.*, 2005). Additional alternatives include Invader assays (Allawi *et al.*, 2004), Padlock probe-based assays (Jonstrup *et al.*, 2006) and even microarrays (Krichevsky *et al.*, 2003; Calin *et al.*, 2004).

NCode™ universal miRNA qRT-PCR

Quantitative RT-PCR (RT-qPCR) is the standard for validating microarray data and is an invaluable tool for highly sensitive and accurate profiling of miRNA population subsets. Most commercially available miRNA RT-qPCR systems use proprietary, pre-designed, miRNA-specific primers for real-time PCR and/or reverse transcription. Unfortunately, such primers require that the miRNA sequence be publicly available and a commercial RT-qPCR assay be developed for that specific sequence, limiting the availability of RT-qPCR assays for many model organisms, recently discovered miRNAs or non-coding RNAs, or proprietary miRNA sequences. The universal primer approach (Shi, 2005), like the one employed by the NCode™ miRNA RT-qPCR kit, overcomes these limitations by combining a carefully optimized RNA polyadenylation and reverse transcription reaction incorporating a universal primer sequence into the 3′ end of the cDNA complementary to the target miRNA. cDNA-specific amplification occurs during the PCR reaction, in which the sequence of the miRNA of interest is used as the target-specific qPCR primer (Fig. 9.15).

Other features of the universal primer method for miRNA RT-qPCR include the requirement for minimal total RNA input with no need for additional small RNA enrichment prior to polyA-tailing and cDNA synthesis (Fig. 9.16). This provides the benefit of being able to conserve precious samples for additional testing. Furthermore, the universal primer tagging approach means that one cDNA reaction can be used to perform qPCR for multiple

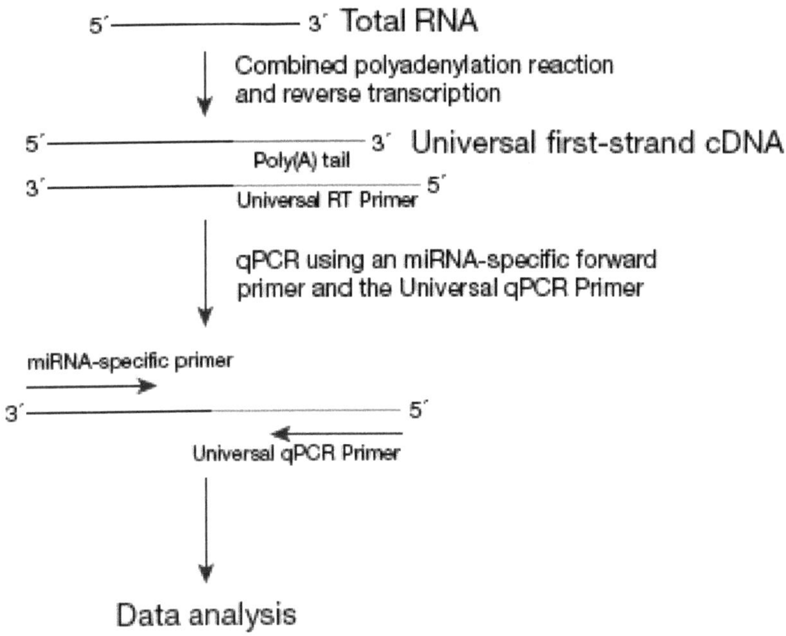

Figure 9.15 The NCode™ miRNA Universal qRT-PCR workflows. miRNAs are polyadenylated in a carefully optimized, combined reverse transcription step using the SuperScript® VILO™ enzyme formulation and a universal primer. The universal first-strand cDNAs are complementary to miRNAs, other non-coding RNAs and mRNAs which can all be analysed subsequently.

miRNA targets. This removes any variability in cDNA synthesis from contributing to variability in the qPCR assays. This also permits the detection of miRNA, other ncRNAs (e.g. piwiRNA, piRNA, or snoRNA) or even mRNA sequences from the same universal cDNA sample. Additionally, it allows one to archive the cDNA samples for analysis at a much later time, for example following the discovery of new miRNA sequences.

Advantages

Some of the other advantages of the universal method are the ability to obtain excellent sensitivity over a broad dynamic range of miRNA abundance (5 logs), the sensitivity to profile closely related miRNAs with single-nucleotide discrimination and the flexibility to use multiple real-time PCR instrument platforms, including standard and fast-cycling modes.

Disadvantages

The primary disadvantage of the universal approach is that often times, design and optimization of the target-specific primer is the end user's responsibility. This can be a difficult and often frustrating exercise since there is very little target landscape to afford primer design optimization. Another perceived disadvantage is the potential for cross reactivity with precursors molecules or variants that have additional nucleotides on the 3′-end.

microRNA	Total RNA	Enriched RNA
26a	16.53	16.53
195	17.19	17.28
1	24.97	25.36

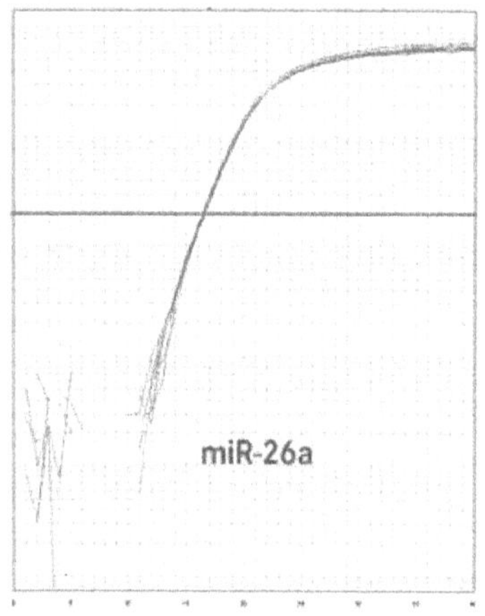

Figure 9.16 miRNA detection from total RNA or enriched miRNA starting materials. cDNA was generated from brain reference RNA or an equivalent volume of enriched miRNA. Average Ct values for 10 ng/µl total RNA or equivalent per qPCR reaction are shown.

TaqMan® microRNA assays

Overview of assay design

TaqMan® MicroRNA Assays are RT-qPCR assays. Each assay includes a miRNA-specific stem-looped RT primer, forward primer, reverse primer and TaqMan® probe. The stem-looped RT primer extends the 3′ end of the miRNA molecule in the presence of reverse transcriptase to produce a template that can be used in the standard TaqMan® PCR. Also, the stem-looped structure in the tail of the RT primer confers a key advantage to these assays: specific detection of the mature miRNA.

How to perform TaqMan® microRNA assays

TaqMan® MicroRNA Assays include RT and real-time PCR two steps (Fig. 9.17). The RT reactions are performed using a single miRNA-specific stem–looped RT primer. Total RNA should be extracted using the mirVana™ kit or other compatible method. Then, 1–300 ng of total RNA is combined in the 15 µl RT reaction with: 1.5 µl 10× RT buffer, 1 µl of 50 U/µl MultiScribe RT enzyme, 0.15 µl 100 mM dNTP mix, 0.19 µl RNase-inhibitor (20 U/µl) and 3 µl 5× specific RT-primer. The RT reactions were incubated on a GeneAmp PCR System 9700 thermocyler by incubating at 16°C for 30 min, 42°C for 30 min, 85°C for 5 min and the 4°C on hold. Each individual RT reaction has a unique set of PCR primers and TaqMan® probe. For a 10 µl reaction, combine 5 µl 2× Universal Mastermix without UNG, 0.66 µl of the RT product, 0.5 µl 20× TaqMan® Assay and 3.84 µl nuclease-free water. PCR

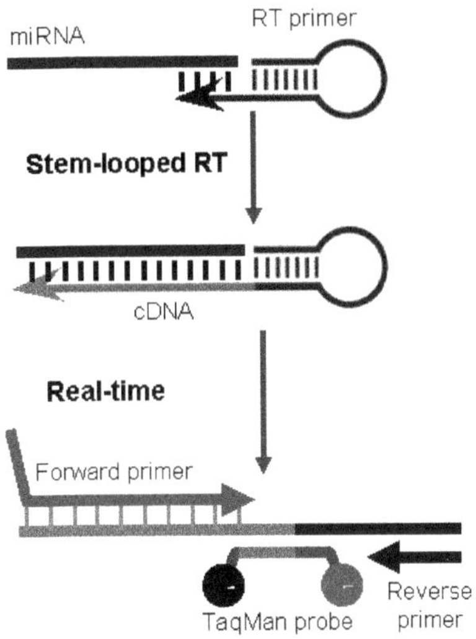

Figure 9.17 Illustration of stem–loop Rt-qPCR method. TaqMan-based real-time quantification of miRNAs involves two steps, stem-loop RT and real-time PCR. Stem–loop RT primers bind to the 3′ end of the targeted miRNA through annealing and are reverse transcribed to make first-strand cDNA which includes the miRNA and reverse primer binding site. The purpose of the tailed forward primer is to increase the melting temperature if needed, depending on the sequence composition of the miRNA.

reactions are incubated in a 384-well plate at 95°C for 10 min, followed by 40 cycles of 95°C for 15 s and 60°C for 1 min.

Advantages

A primary advantage of TaqMan assays is their high degree of specificity and the fact that they guarantee the ability to quantify only the mature miRNAs with no cross-reactivity to precursors. They also provide single-base discrimination of closely related family members and a high level of sensitivity, requiring only 1–10 ng of total RNA or equivalent, which with the use of a pre-amplification protocol, will enable quantitation of miRNAs from single cells. Another benefit is that they are simple, scalable and compatible with the Cell-to-CT™ kit (Applied Biosystems) for direct quantitation of miRNAs from cell lysates.

TaqMan® microRNA arrays

TaqMan® MicroRNA Arrays provide all the advantages of TaqMan® MicroRNA Assays in a convenient, pre-configured micro fluidic card minimizing experimental variability and effort required to run 384 TaqMan® MicroRNA Assays in parallel. Megaplex™ RT Primers are highly multiplexed RT primers designed to convert up to 384 miRNAs and control genes to cDNA prior to real-time analysis and simplify the TaqMan® MicroRNA Array. When sensitivity is of the utmost importance and/or RNA sample is limiting, Megaplex™ PreAmp Primers provide an optional preamplification step. TaqMan® MicroRNA Arrays v2.0 and Megaplex™ Primer Pools enable a comprehensive expression profile consistent with Sanger miRBase v10 within 5 hours, providing the ideal miRNA profiling solution.

Megaplex™ RT Primers are stem-looped RT primer pools that streamline the profiling of hundreds of miRNA targets in a single experiment and reduce the number of RT reactions and the amount of total RNA required to generate a comprehensive miRNA expression profile. Optional Megaplex™ PreAmp Primers significantly enhance the ability to detect low expressed miRNAs enabling the generation of a comprehensive expression profile from as little as single cells.

Expression profiling of miRNAs from single mouse ES cells

Megaplex™ RT and PreAmp TaqMan® MicroRNA Assays have been used for simultaneously quantifying all 237 mouse miRNAs and 21 stem-cell related mRNA genes from single individual cells (Fig. 9.18). Results suggest that assays are highly specific, reproducible and quantitative for at least four logs. Single cell expression signatures could classify mouse individual ES, embryoid body (EB), or somatic cells (Fig. 9.19). Furthermore, there are

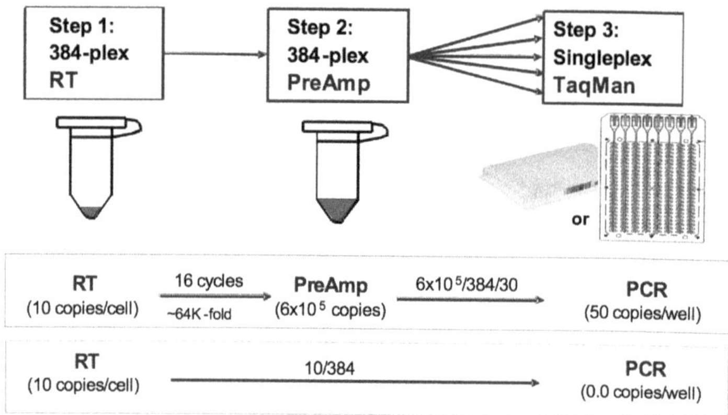

Figure 9.18 Schematic of Megaplex Preamp TaqMan® MicroRNA Assays for single cell gene expression profiling.

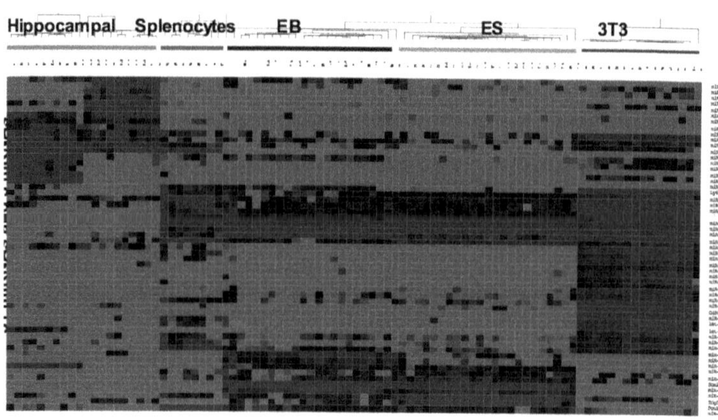

Figure 9.19 Single cell miRNA and mRNA expression profile in 90 individual mES, mEB and somatic cells.

significant variations in miRNA expression from cells to cells within or between ES cell lines, suggesting the existence of spontaneous differentiation and heterogeneity of ES cells. Based on combined expression signatures of miRNAs and mRNAs, we could distinguish between differentiated and undifferentiated ES cells. The results reveal new insight of both miRNA and mRNA expression from a single cell level.

References

Affymetrix E.N.C.O.D.E Transcriptome Project Cold Spring Harbor Laboratory E.N.C.O.D.E Transcriptome (2009). Post-transcriptional processing generates a diversity of 5′-modified long and short RNAs. Nature *457*, 1028–1032.

Allawi, H.T., Dahlberg, J.E., Olson, S., Lund, E., Olson, M., Ma, W.-P., Takova, T., Neri, B.P., and Lyamichev, V.I. (2004). Quantitation of microRNAs using a modified Invader assay. RNA *10*, 1153–1161.

Bartel, D.P. (2004). MicroRNAs: genomics, biogenesis, mechanism and function. Cell *116*, 281–297.

Boyd, V.L., Moody, K.I., Karger, A.E., Livak, K.J., Zon, G., and Burns, J.W. (2006). Methylation-dependent fragment separation: direct detection of DNA methylation by capillary electrophoresis of PCR products from bisulfite-converted genomic DNA. Anal. Biochem. *354*, 266–273.

Calin, G.A., Liu, C.G., Sevignani, C., Ferracin, M., Felli, N., Dumitru, C.D., Shimizu, M., Cimmino, A., Zupo, S., Dono, M., *et al.* (2004). MicroRNA profiling reveals distinct signatures in B cell chronic lymphocytic leukemias. Proc. Natl. Acad. Sci. U.S.A. *101*, 11755–11760.

Chen, C., Ridzon, D.A., Broomer, A.J., Zhou, Z., Lee, D.H., Nguyen, J.T., Barbisin, M., Xu, N.L., Mahuvakar Andersen, V.R., M.R., *et al.* (2005). Real-time quantification of microRNAs by stem-loop RT-PCR. Nucleic Acids Res. *33*, e179.

Chibber, A., and Schroeder, B.G. (2008). Single-molecule polymerase chain reaction reduces bias: application to DNA methylation analysis by bisulfite sequencing. Anal. Biochem. *377*, 46–54.

Clayton, A., and Mahadevan, L. (2003). MAP kinase-mediated phosphoacetylation of histone H3 and inducible gene regulation FEBS Lett. *1*, 51.

Cottrell, S.E., Distler, J., Goodman, N.S., Mooney, S.H., Kluth, A., Olek, A., Schwope, I., Tetzner, R., Ziebarth, H., and Berlin, K. (2004). A real-time PCR assay for DNA-methylation using methylation-specific blockers. Nucleic Acids Res. *1332*, e10.

Cross, S.H., Charlton, J.A., Nan, X., and Bird, A.P. (1994). Purification of CpG islands using a methylated DNA binding column. Nat. Genet. *6*, 236–244.

Das, P.M., Ramachandran, K., van Wert, J., and Singal, R. (2004). Chromatin immunoprecipitation assay. BioTechniques *6*, 961–969.

Eads, C.A., Danenberg, K.D., Kawakami, K., Saltz, L.B., Blake, C., Shibata, D., Danenberg, P.V., and Laird, P.W. (2000). MethyLight: a high-throughput assay to measure DNA methylation. Nucleic Acids Res. *28*, E32.

Fraga, M.F., Ballestar, E., Montoya, G., Taysavang, P., Wade, P.A., and Esteller, M. (2003). The affinity of different MBD proteins for a specific methylated locus depends on their intrinsic binding properties. Nucleic Acids Res. *31*, 1765–1774.

Frommer, M., McDonald, L.E., Millar, D.S., Collis, C.M., Watt, F., Grigg, G.W., Molloy, P.L., and Paul, C.L. (1992). A genomic sequencing protocol that yields a positive display of 5-methylcytosine residues in individual DNA strands. Proc. Natl. Acad. Sci. U.S.A. *89*, 1827–1831.

Fuks, F. (2005). DNA methylation and histone modifications: teaming up to silence genes. Curr. Opin. Genet. Dev. *15*, 490–495.

Gebhard, C., Schwarzfischer, L., Pham, T.H., Schilling, E., Klug, Andreesen, M., R., and Rehli, M. (2006). Genome-wide profiling of CpG methylation identifies novel targets of aberrant hypermethylation in myeloid leukemia. Cancer Res. *66*, 6118–6128.

Gille, H., Kortenjann, M., Thomae, O., Moomaw, C., Slaughter, C., Cobb, M., and Shaw, P. (1995). EMBO J. *14*, 951–962.

Gitan, R.S., Shi, H., Chen, C.M., Yan, P.S., and Huang, T.H. (2002). Methylation-specific oligonucleotide microarray: a new potential for high-throughput methylation analysis. Genome Res. *12*, 158–164.

Gonzalgo, M.L., and Jones, P.A. (2002). Quantitative methylation analysis using methylation-sensitive single-nucleotide primer extension (Ms-SNuPE). Methods 27, 128–133.

Griffiths-Jones, S., Saini, H.K., van Dongen, S., and Enright, A.J. (2008). miRBase: tools for microRNA genomics. Nucleic Acids Res. *36*, D154–158.

Guo, Y., Pakneshan, P., Gladu, J., Slack, A., Szyf, M., and Rabbani, S.A. (2002). Regulation of DNA methylation in human breast cancer. Effect on the urokinase-type plasminogen activator gene production and tumor invasion. J. Biol. Chem. *277*, 41571–41579.

Hawkins, P.G., Santoso, S., Adams, C., Anest, V., and Morris, K.V. (2009). Promoter targeted small RNAs induce long-term transcriptional gene silencing in human cells. Nucleic Acids Res. *37*, 2984–2995.

Hendrich, B., and Bird, A. (1998). Identification and characterization of a family of mammalian methyl-CpG binding proteins. Mol. Cell Biol. *18*, 6538–6547.

Huang, T.H., Laux, D.E., Hamlin, B.C., Tran, P., Tran, H., and Lubahn, D.E. (1997). Identification of DNA methylation markers for human breast carcinomas using the methylation-sensitive restriction fingerprinting technique.. Cancer Res. *57*, 1030–1034.

Irizarry, R.A., Ladd-Acosta, C., Wen, B., Wu, Z., Montano, C., Onyango, P., Cui, H., Gabo, K., Rongione, M., Webster, M., Ji, H., Potash, J.B., Sabunciyan, S., and Feinberg, A.P. (2009). The human colon cancer methylome shows similar hypo- and hypermethylation at conserved tissue-specific CpG island shores. Nat. Genet. *41*, 178–186.

Janknecht, R., Ernst, W., Pingoud, V., and Nordheim, A. (1993). Activation of ternary complex factor Elk-1 by MAP kinases. EMBO J. *12*, 5097–5104.

Jenuwein, T., and Allis, C.D. (2001). Translating the histone code. Science *293*, 1074–1080.

Jonstrup, S.P., Koch, J., and Kjems, J.A. (2006). microRNA detection system based on padlock probes and rolling circle amplification. RNA *12*, 1747–1752.

Kangaspeska, S., Stride, B., Métivier, R., Polycarpou-Schwarz, M., Ibberson, D., Carmouche, R.P., Benes, V., Gannon, F., and Reid, G. (2008). Transient cyclical methylation of promoter DNA. Nature *452*, 112–115.

Kim, V.N., and Nam, J.W. (2006). Genomics of microRNAs. Trends Genet. *22*, 165–173.

Krichevsky, A.M., King, K.S., Donahue, C.P., Khrapko, K., and Kosik, K.S. (2003). A microRNA array reveals extensive regulation of microRNAs during brain development. RNA *9*, 1274–1281.

Kuo, M.H., and Allis, C.D. (1999). *In vivo* cross-linking and immunoprecipitation for studying dynamic Protein: DNA associations in a chromatin environment. Methods *3*, 425–433.

Lagos-Quintana, M., Rauhut, R., Lendeckel, W., and Tuschl, T. (2001). Identification of novel genes coding for small expressed RNAs. Science *294*, 853–858.

Lee, R., Feinbaum, R., and Ambros, V. (2004). A short history of short RNA. Cell *116*, S89–92; p. 81 following S96.

Lee, R.C., Feinbaum, R.L., and Ambros, V. (2004). The *C. elegans* heterochronic gene lin-4 encodes small RNAs with antisense complementarity to lin-14. Cell *75*, 843–854.

Leu, Y.W., Yan, P.S., Fan, M., Jin, V.X., Liu, J.C., Curran, E.M., Welshons, W.V., Wei, S.H., Davuluri, R.V., Plass, C., Nephew, K.P., and Huang, T.H. (2004). Loss of estrogen receptor signaling triggers epigenetic silencing of downstream targets in breast cancer. Cancer Res. *22*, 8184–8192.

Lu, J., Getz, G., Miska, E.A., Alvarez-Saavedram E., Lamb, J., Peck, D., Sweet-Cordero, A., Ebert, B.L., Mak, R.H., Ferrando, A.A., *et al.* (2005). MicroRNA expression profiles classify human cancers. Nature *435*, 834–838.

Ma, L., Teruya-Feldstein, J., and Weinberg, R.A. (2008). Tumour invasion and metastasis initiated by microRNA-10b in breast cancer. Nature *449*, 682–688.

Métivier, R., Gallais, R., Tiffoche, C., Le Péron, C., Jurkowska, R.Z., Carmouche, R.P., Ibberson, D., Barath, P., Demay, F., Reid, G., Benes, V., Jeltsch, A., Gannon, F., and Salbert, G. (2008). Cyclical DNA methylation of a transcriptionally active promoter. Nature *452*, 45–50.

Nightingale, K.P., O'Neill, L.P., and Turner, B.M. (2006). Histone modifications: signaling receptors and potential elements of a heritable epigenetic code. Curr. Opin. Genet. Dev. *16*, 125–136.

Orlando, V., Strutt, H., and Paro, R. (1997). Analysis of chromatin structure by *in vivo* formaldehyde cross-linking. Methods *11*, 205–214.

Rauch, T., and Pfeifer, G.P. (2005). Methylated-CpG island recovery assay: a new technique for the rapid detection of methylated-CpG islands in cancer. Lab. Invest. *85*, 1172–1180.

Raymond, C.K., Roberts, B.S., Garrett-Engele, P., Lim, L.P., and Johnson, J.M. (2005). Simple, quantitative primer-extension PCR assay for direct monitoring of microRNAs and short-interfering RNAs. RNA *11*, 1737–1744.

Reinhart, B.J., Slack, F.J., Basson, M., Pasquinelli, A.E., Bettinger, J.C., Rougvie, A.E., Horvitz, H.R., and Ruvkun, G. (2000). The 21-nucleotide let-7 RNA regulates developmental timing in *Caenorhabditis elegans*. Nature *403*, 901–906.

Reynaud, C., Bruno, C., Boullanger, P., Grange, J., Barbesti, S., and Niveleau, A. (1992). Monitoring of urinary excretion of modified nucleosides in cancer patients using a set of six monoclonal antibodies. Cancer Lett. *61*, 255–262.

Rodriguez, A., Vigorito, E., Clare, S., Warren, M.V., Couttet, P., Soond, D.R., van Dongen, S., Grocock, R.J., Das, P.P., Miska, E.A., *et al.* (2007). Requirement of bic/microRNA-155 for normal immune function. Science *316*, 608–611.

Ruvkun, G., Wightman, B., and Ha, I. (2004). The 20 years it took to recognize the importance of tiny RNAs. Cell *116*, S93–96; p. 92 following S96.

Schilling, E., and Rehli, M. (2007). Global, comparative analysis of tissue-specific promoter CpG methylation. Genomics *90*, 314–323.

Schmidl C., Klug, M., Boeld, T.J. Andreesen R., Hoffmann, P., Edinger, M., and Rehli, M. (2009). Lineage-specific DNA methylation in T cells correlates with histone methylation and enhancer activity. Genome Res. *19*, 1165–1174.

Sharbati-Tehrani, S., Kutz-Lohroff, B., Bergbauer, R., Scholven, J., and Einspanier, R. (2008). miR-Q: a novel quantitative RT-PCR approach for the expression profiling of small RNA molecules such as miRNAs in a complex sample. BMC Mol. Biol. *9*, 34.

Shen, L., Gou, Y., Chen, X., Ahmed, S., and Issa, J.P. (2007). Optimizing annealing temperature overcomes bias in bisulfite PCR methylation analysis. BioTechniques 42:48–52.

Shi, R., and Chiang,V.L. (2005). Facile means for quantifying microRNA expression by real-time PCR. BioTechniques *39*, 519–525.

Strahl, B.D., and Allis, C.D. (2000). The language of covalent histone modifications. Nature *403*, 41–45.

Swift-Scanlan, T., Blackford, A., Argani, P., Sukumar, S., and Fackler, M.J. (2006). Two-color quantitative multiplex methylation-specific PCR. BioTechniques *40*, 210–219.

Taft, R.J., Kaplan, C.D., Simons, C., and Mattick, J.S. (2009). Evolution, biogenesis and function of promoter-associated RNAs. Cell cycle Aug 2:8(15).

Tariq, M., and Paszkowski, J. (2004). DNA and histone methylation in plants. Trends Genet. *20*, 244–251.

Tavazoie, S.F., Alarcon, C., Oskarsson, T., Padua, D., Wang, Q., Bos, P.D., Gerald, W.L., Massague,J. (2008). Endogenous human microRNAs that suppress breast cancer metastasis. Nature *451*, 147–152.

Thomassin, H., Kress, C., and Grange, T. (2004). MethylQuant: a sensitive method for quantifying methylation of specific cytosines within the genome. NAR 32, e168:

Treisman, R. (1995). Journey to the surface of the cell: Fos regulation and the SRE. EMBO J. *14*, 4905–4913.

Triboulet, R., Mari, B., Lin, Y.L., Chable-Bessia, C., Bennasser, Y., Lebrigand, K., Cardinaud, B., Maurin, T., Barbry, P., Baillat, V., *et al.* (2007). Suppression of microRNA-silencing pathway by HIV-1 during virus replication. Science *315*, 1579–1582.

Turner, B.M. (2002). Cellular memory and the histone code. Cell *111*, 285–291.

Voss, K.O., Roos, K.P., Nonay, R.L., and Dovichi, N.J. (1998). Combating PCR bias in bisulfite-based cytosine methylation analysis. betaine-modified cytosine deamination PCR. Anal. Chem. *70*, 3818–3823.

Weber, M., Davies, J.J., Wittig, D., Oakeley, E.J., Haase, M., Lam, W.L., and Schübeler, D. (2005). Chromosome-wide and promoter-specific analyses identify sites of differential DNA methylation in normal and transformed human cells. Nat. Genet. *37*, 853–862.

Weinmann, A.S., and Farnham, P.J. (2002). Identification of unknown target genes of human transcription factors using chromatin immunoprecipitation. Methods *26*, 37–47.

Wells, J., and Farnham, P.J. (2002). Characterizing transcription factor binding sites using formaldehyde cross-linking and immunoprecipitation. Methods, *1*, 48–56.

Wojdacz, T.K., and Hansen, L.L. (2006). Reversal of PCR bias for improved sensitivity of the DNA methylation melting curve assay. BioTechniques *41*, 274–278.

Xiong, Z., and Laird, P.W. (1997). COBRA: a sensitive and quantitative DNA methylation assay. Nucleic Acids Res. *25*, 2532–2534.

Yegnasubramanian, S., Lin, X., Haffner, M.C., DeMarzo, A.M., and Nelson, W.G. (2006). Combination of methylated-DNA precipitation and methylation-sensitive restriction enzymes (COMPARE-MS) for the rapid, sensitive and quantitative detection of DNA methylation. Nucleic Acids Res. *34*, e19.

Yu, S.L., Chen, H.Y., Chang, G.C., Chen, C.Y., Chen, H.W., Singh, S., Cheng, C.L., Yu, C.J., Lee, Y.C., Chen, H.S., *et al.* (2008). MicroRNA signature predicts survival and relapse in lung cancer. Cancer Cell *13*, 48–57.

High-Resolution Melting Analysis 10

John F. Mackay and Carl T. Wittwer

Abstract

Real-time qPCR using SYBR Green and melting curve analysis to verify specific product amplification has become a standard laboratory technique for rapid, high throughput gene quantification. An extension of this melting curve method – High-resolution melting analysis (HRMA)– is now doing the same for the analysis of sequence variation, allowing rapid cost-effective discrimination of sequences to SNP level in an automated closed-tube method. Two PCR primers are typically required as with SYBR Green quantification but HRMA differs in its requirement for the use of a saturating dye, precise reaction temperature control and software algorithms to cluster the melting curves. Originally described for SNP analysis (and still the leading application), HRMA is now being used in a wider context-HLA comparisons, microsatellite genotyping and methylation status of DNA sequences. New developments such as unlabelled probes and snapback elements on the PCR primers allow the simultaneous genotyping of a desired SNP with the scanning of the whole amplicon for other sequence variation. This chapter covers some of these developments and provides a guide to those wishing to establish this technique, as well as troubleshooting advice for those already underway.

Introduction

The melting and reassociation of DNA has been used as a tool for decades – studying and characterizing genome structures and repetitive sequences. It therefore seems appropriate that this fundamental property of nucleic acids continues to develop and be refined in combination with amplification technologies so prevalent today. When PCR joined the on-line revolution of the 1990s with the advent of real-time qPCR, the fluorescent detection of DNA lent itself to the rapid monitoring of this melting behaviour. The first use was in characterizing amplicons generated in the presence of SYBR Green I (Ririe *et al.*, 1997). This chemistry – the first commercial alternative to hydrolysis probes – coupled with melting curve analysis still remains the most popular research methodology for real-time PCR today. Around the same time, the first report of SNP analysis via fluorescent probe melting coupled to real-time PCR was described (Lay and Wittwer, 1997). Once again, it therefore seems appropriate that this same SNP – the Leiden thrombophilia variant factor V – was the first clinical genetic test to pass FDA approval in 2003, using the by-then standard dual hybridization probe technology (Fig. 10.1).

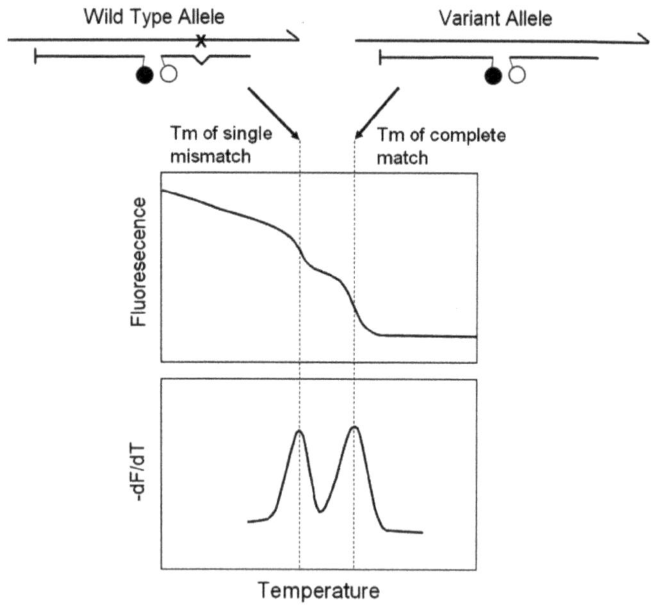

Figure 10.1 Dual hybridization probe (HybProbe) technology for SNP genotyping. Two probes, each with one fluorophore are designed to hybridize next to each other on the target so that they are in resonance energy transfer. One probe is blocked at the 3′-end and hybridizes to invariant sequence (anchor probe) while the other hybridizes over the variable locus (sensor probe). After PCR, the solution is heated and fluorescence monitored. A melting curve for a heterozygote is shown, along with the corresponding negative derivative plot.

Melting curve analysis of an amplicon using SYBR Green I is generally not reliable enough to detect heterozygous genotypes, although success has been reported using post-PCR procedures (e.g. Lipsky *et al.*, 2001). Accurate SYBR Green I genotyping is possible by allele-specific amplification that results in different sized amplicons – either by addition of a GC tail to one allele (Germer and Higuchi, 1999; Wang *et al.*, 2005) or bidirectional amplification of the respective alleles (Waterfall and Cobb, 2002).

Despite the sensitivity of dual hybridization probes for SNP analysis and their continued popularity, they necessitate the use of two individually labelled (and additionally in one case, 3′ blocked) oligonucleotides, adding to the cost and complexity of the reaction.

The use of a single hybridization probe for melting analysis first appeared using guanosine as a fluorescent quencher (Crockett and Wittwer, 2001), followed by the commercial release of SimpleProbe chemistry. While resulting in lower fluorescent yields than dual hybridization probes, they permitted similarly efficient genotyping. The use of these single-labelled probes gave rise to the first use of mutation scanning (in contrast to known variant genotyping) using melting curves, analysing exons 5 to 8 of the p53 gene (Millward *et al.*, 2002).

High-resolution melting analysis (HRMA)

As with real-time PCR, HRMA for mutation scanning has travelled through a number of chemistry developments. Initially, a fluorescein-labelled primer in the PCR was used to analyse SNPs within the amplicons (Gundry *et al.*, 2003). However, as with other single probe genotyping methods, a specific labelled oligonucleotide was required for each reaction (unless using a generic indicator oligonucleotide homologous to a 5′ tail of the assay primer (Gundry *et al.*, 2003)).

HRMA, as it is commonly used today, is based on monitoring the melting of whole PCR amplicons (amplicon HRMA). This method was enabled by three developments. The first was the introduction of 'saturating' dyes – dyes that can be used at high enough concentrations that heteroduplexes are detected. Due to its inhibitory effects on PCR, SYBR Green I is used at low concentrations and its translocation during melting curve acquisition can cause product identification issues (Giglio *et al.*, 2003; Monis *et al.*, 2005). The first saturating dye introduced was LC Green I and was described together with the first high-resolution melting instrument, HR-1 (both from Idaho Technology Inc.). Many HRMA applications were initially developed using this dye and instrument, along with software methods and normalization steps for analysis (Wittwer *et al.*, 2003).

Dyes for HRMA

Since the introduction of LC Green I, LC Green PLUS (Idaho Technology Inc.) was introduced as a brighter dye better adapted to microtiter plates as opposed to capillaries. A number of other dyes have since been reported for use as dyes in HRMA. These include SYTO9 and SYBR GreenER (not to be confused with SYBR Green I) from Invitrogen (Monis *et al.*, 2005; Krypuy *et al.*, 2006; Wolff *et al.*, 2008), EvaGreen from Biotium (Mao *et al.*, 2007), Chromofy and BEBO from TATAA Biocenter (Mader *et al.*, 2008) and Resolight from Roche Applied Science (Schütz and von Ahsen, 2009). The cost per reaction, optimal excitation wavelengths, and performance of these dyes varies considerably. Some are only available as PCR mastermixes while others are available as free dye. LC Green PLUS and Resolight have been compared in terms of brightness, clustering variability, and effect on amplification Cq (Grievink and Stowell, 2008). LC Green PLUS, SYTO9 and EvaGreen have been compared for their performance in heterozygote detection (Farrar *et al.*, 2010). SYBR Green I can be used for some applications where detection of heteroduplexes is not necessary or desirable (Stephans *et al.*, 2008). These include genotyping of monoploid organisms, methylation analysis, and repeat typing. However, for all applications that depend on heterozygote detection, a saturating DNA dye with high heteroduplex sensitivity is crucial.

Instruments for HRMA

The first of the instruments capable of high-resolution melting and with the appropriate software algorithms was the HR-1 (Idaho Technology Inc.). This instrument is designed to work with LightCycler capillaries amplified in either the capillary LightCycler instruments (Roche Applied Science) or the RapidCycler 2 (Idaho Technology Inc.) before being transferred to the HR-1. Samples are analysed one at a time so that testing large numbers of samples is laborious and more prone to human error (Vandersteen *et al.*, 2007). This instrument was followed by the LightScanner (Idaho Technology Inc.); this time using 96-well or 384-well plates with PCR also performed on a separate instrument. The first of the instruments to combine real-time PCR amplification and high-resolution melting analysis were the Corbett (now Qiagen) Rotorgene 6000 and the LightCycler 480 (Roche Applied Science). The combination of real-time PCR and HRMA provides a useful quality check of

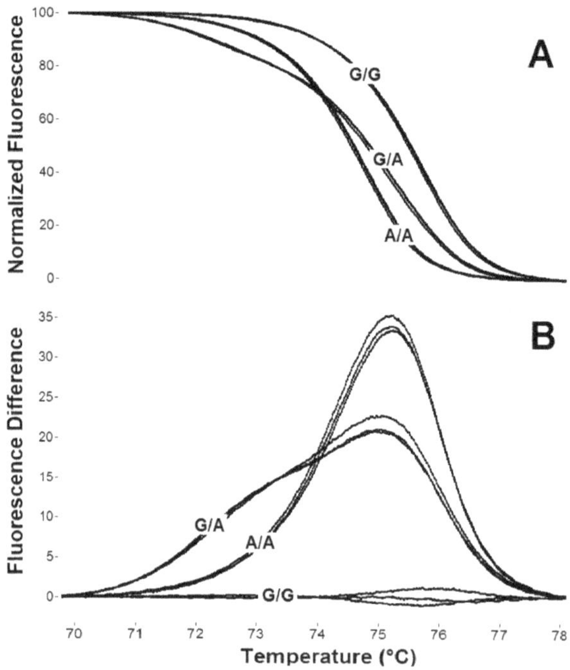

Figure 10.2 Basic software analysis for HRMA genotyping. Fluorescent normalization of melting curves before and after the amplicon melting transition is shown in (A) with difference plots of the same data shown in (B). Three samples of each genotype are analysed using a 42 bp product (Liew *et al.*, 2004) bracketing the FV Leiden locus (1691G>A).

the PCR that can be important in interpreting HRMA plots (Grievink and Stowell, 2008; Kristensen and Dobrovic, 2008). It also allows large deletion detection (by real time Cq differences) to be combined with SNP detection (by subsequent HRMA) (Rouleau *et al.*, 2009). To date, a number of other real-time PCR instruments claim HRMA in their offering, including the ABI 7500 and 7900HT instruments, the BioRad CFX instruments and also a multiple capillary instrument (LightScanner 32, Idaho Technology Inc.). It seems certain that additional systems will adopt this technology. However, different instruments vary widely in their resolution with a direct effect on their sensitivity and specificity (Herrmann *et al.*, 2006, 2007a,b). The choice regarding a suitable instrument is also affected by other factors such as budget, need for integrated real time PCR, turnaround time, batch size, and throughput requirements.

Software algorithms

The third facet required for HRMA is the appropriate software algorithms in the analysis software. The raw data looks similar to SYBR Green I melting curves. The first analysis step is to exclude samples that do not amplify, such as no template controls. Then, the fluorescence is normalized between 100% before melting and 0% after melting (Fig. 10.2A).

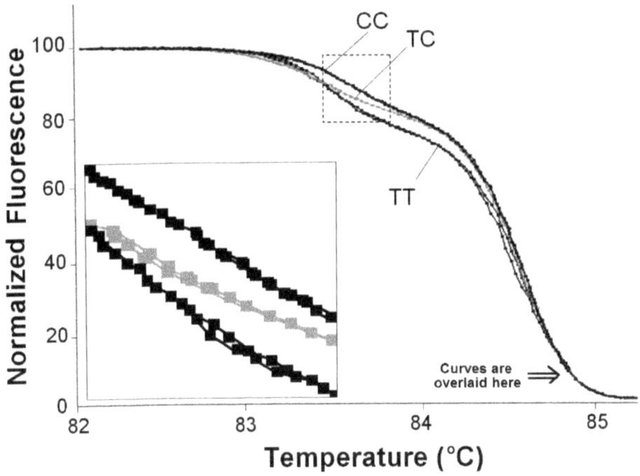

Figure 10.3 Advanced software analysis for HRMA heterozygote scanning. When many samples are analysed and multiple domains or alleles are present, curve overlay or 'temperature shifting' increases accuracy. Curve shapes are compared by shifting the temperature axis to superimpose the curves, usually over a low fluorescence segment. Samples should also be analysed without curve overlay for homozygote detection. Shown is a 544 bp PCR product that melts in two domains, with an SNP revealed in the lower temperature domain (Wittwer *et al.*, 2003).

Normalization adjusts for volume, concentration, and optical path differences between samples that may obscure any sequence differences. For small amplicon genotyping with a single melting domain, no further analysis is necessary or desired, although difference plots where the curves are shown relative to each other are popular (Fig. 10.2B). However, for heterozygote scanning, particularly when many samples are analysed and multiple alleles and/or domains are present, sensitivity is increased by curve overlay, also known as 'temperature shifting' to focus on shape differences between curves (Fig. 10.3). Curve overlay not only corrects for any well-to-well temperature difference between samples, but more importantly corrects unavoidable sample-to-sample chemistry differences including ionic strength and PCR product concentration. In effect, this provides normalization in both the fluorescence and temperature axes (Fig. 10.3). Advanced instruments include curve overlay, and the data are analysed both before (homozygote detection) and after (heterozygote detection) curve overlay. Difference plots may also be used after curve overlay.

Clustering of different genotypes may be trained based on known genotypes or untrained using unbiased hierarchal clustering when variant information is unknown. Simple normalization is adequate for single locus genotyping, but exponential background removal or deviation analysis is superior when multiple loci are analysed (Seipp *et al*, 2008), or for simultaneous analysis of amplicon and probe transitions (Zhou *et al.*, 2005). Details of mathematical algorithms for high-resolution melting have recently been published (Palais and Wittwer, 2009).

HRMA chemistries

As with labelled probes in melting curve analysis, several adaptations of HRMA beyond amplicon melting have developed. The use of an unlabelled probe covering variants of interest allows the combination of genotyping for a known variant with scanning of the entire amplicons (Zhou *et al.*, 2005; Fig. 10.4). One issue with these probes (both dual hybridization probes and unlabelled probes) is the requirement for a 3′-end label block to prevent extension during PCR. Traditionally this blocking agent has been a phosphate group, but concern over its stability led Cradic *et al.* (2004) to use a C3 carbon spacer instead. More recently, a number of blocking elements have been investigated but all showed some breakdown under harsh conditions (Dames *et al.*, 2007). A more recent development that overcomes this 3′ blocking requirement is snapback genotyping – a method that attaches the probe element to the 5′-end of one of the primers (Zhou *et al.*, 2008; Fig. 10.5). The probe element folds back in a similar manner to scorpion primers and binds to the variable region of interest, as with unlabelled probes. While this chemistry offers the specificity of probe-based melting with the simplicity of only two primers, there is more to be considered in the design of snapback primer design compared to unlabelled probes. Both the unlabelled probe format and snapback genotyping require asymmetric primer concentrations and 40–60 cycles to generate strong probe signals. Therefore careful attention to primer design is needed for specific amplification.

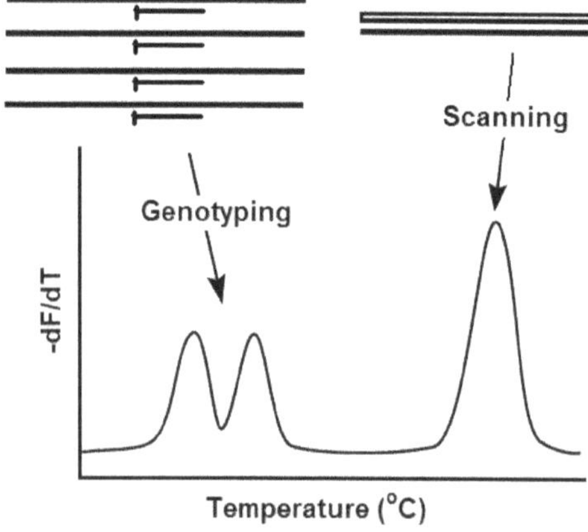

Figure 10.4 Unlabelled probe chemistry. Asymmetric PCR is performed in the presence of an unlabelled probe that is blocked at its 3′-end. Both probe/target and full length duplexes are produced, resulting in two melting regions – one for the SNP region covered by the probe and the other for the amplicon. Shown is a derivative plot after exponential background subtraction.

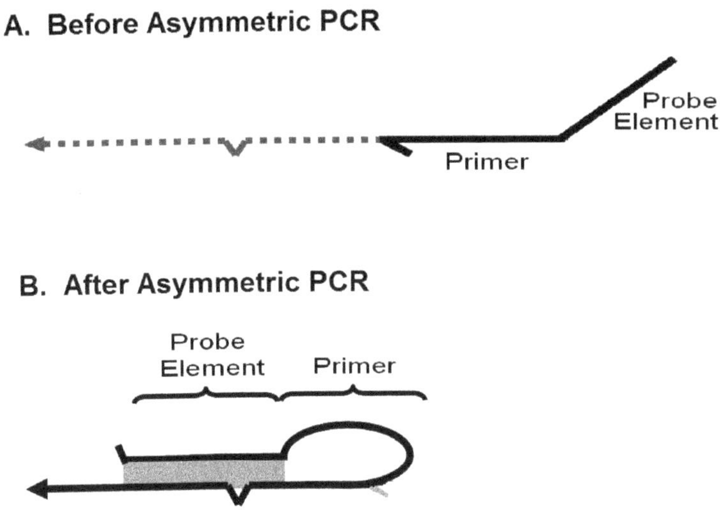

Figure 10.5 Snapback primer chemistry: The probe element is attached the 5′-end of one primer (A). After asymmetric amplification, the probe element folds (snaps back) over the SNP of interest. As with the unlabelled probe approach, both probe and product transitions can be analysed.

Applications

HRMA is primarily used to (1) scan for unknown variants and (2) genotype known variants, both by rapid amplicon melting. A wide range of applications have been published – from clinical diagnostics (Erali *et al.*, 2008; Reed *et al.*, 2007) to research (Vossen *et al.*, 2009; Smith *et al.*, 2009). This breadth of HRMA applications arises from the ability to rapidly generate new assays and the fact that the reagent costs for a new assay are negligible if HRMA is already performed in the laboratory. Applications range from simple SNP genotyping to rapid scanning of complex genes such as BRCA1 (van der Stoep *et al.*, 2009). Up to four amplicons/SNPs have been multiplexed for simultaneous genotyping in a single colour by using the entire temperature spectrum (Seipp *et al.*, 2008, 2009). Another growing application is the detection and quantification of methylated DNA in oncology and diseases such as Prader–Willi and Angelman syndromes (White *et al.*, 2007). After bisulfite conversion of unmethylated C to U (leaving methylated C unaffected), the relative levels of C and T in the amplicon are determined by HRMA.

Although not as widely adopted as amplicon melting, probe chemistries enable both rapid detection of common known variants and scanning for disease-causing SNPs with only a dye and unlabelled oligonucleotides (Nguyen-Dumont *et al.*, 2009). Probe formats also allow SNP genotyping on standard real-time PCR instruments that are not otherwise set up for HRMA (Liew *et al.*, 2007; Habalovà *et al.*, 2009).

In contrast to establishing the exact sequence by genotyping, HRMA can also assess variable regions when the actual sequence is unimportant. In this case, HRMA is used as a efficient confirmation of 'like' vs 'different'. The first example of this was for HLA typing where genotypes between potential donors could be rapidly identified as a match for a recipient at a particular HLA locus (Zhou *et al.*, 2004b). Other examples include grapevine and olive cultivar genotyping using HRMA of microsatellites as an alternative to polyacrylamide gels or capillary sequencers (Mackay *et al.*, 2008; Muleo *et al.*, 2009). Using a reference sample, suspected varieties can be rapidly confirmed as 'true to type'. This method can also be used in mapping applications for the establishment of codominant markers (Mader *et al.*, 2008).

Additional mapping applications using SNP markers include diploid species such as apple (Chagné *et al.*, 2008), barley (Lehmensiek *et al.*, 2008), lupin (Croxford *et al.*, 2008) and almond (Wu *et al.*, 2008, 2009), and polyploid species such as tetraploid hops (D. Chagné and E. Buck, pers. comm.) and potato (De Koyer *et al.*, 2009).

Assay approaches

One of the features of HRMA is that only assay-specific PCR primers are required for different targets and applications; performing an assay is very similar to standard SYBR Green I PCR and melting. However the user should be clear as to the purpose of the assay – whether it is to scan for heterozygous sequence variants or to genotype a known SNP. One difference between these assays is in the size of the amplicons; while heterozygous scanning can be performed successfully in amplicons over 500 bp (Wittwer *et al.*, 2003), smaller amplicons are better for genotyping known SNPs. Short amplicons (<100 bp) generate a greater T_m differential between wild-type and homozygous variants and 100 bp has been recommended as an upper limit on certain instruments (Kristensen and Dobrovic, 2008; Vossen *et al.*, 2009). However, some homozygous SNPs are not resolved because 'basepair neutrality' results in little or no change to the amplicon melting temperature. In these instances, one of the probe approaches (unlabelled probe or snapback primers) may be required. Temperature calibrators (double-stranded oligonucleotides used to normalize melting curve temperatures) can resolve some of these basepair neutral SNPs (Gundry *et al.*, 2008) – however these methods are currently only available on the LightScanner instrument. Another alternative is to mix in DNA of known genotype for quantitative heteroduplex analysis, resulting in different curve shapes for each genotype (Palais *et al.*, 2005). Mixing unknowns with wild type DNA will also convert homozygous and hemizygous (X-linked) variants into easily detected heterozygotes (Laurie *et al.*, 2007). An elegant plate-based mixing scheme to detect homozygotes after initial detection of heterozygotes is provided by Vossen *et al.*, 2009.

Nucleic acid extraction

The quality of DNA extraction is important for HRMA. The melting of DNA is dependent on salt and co-factor concentrations. While the particular DNA diluent or elution buffer (for column-based systems) may depend on preference, the same buffer should be used for all samples if possible. Extraction methods that generate crude templates tend to introduce variability into the melting curves. Although DNA extracted by different methods has been analysed concurrently in clinical samples (Van der Stoep *et al.*, 2009), plant and other matrixes containing large amounts of polysaccharides and phenolic compounds are more problematic. For grapevines and other 'difficult' species a modified CTAB protocol has been used with good success and column-based methods are also possible (Mader *et al.*, 2008; Mackay *et al.*, 2008). For highest accuracy, samples extracted with different methods should be analysed separately. While template concentrations varying by up to 10-fold may be used for HRMA (Montgomery *et al.*, 2007; Kristensen and Dobrovic, 2008), sample normalization to a stock concentration (e.g. 10 ng/µl or higher depending on genome complexity) improves reproducibility and sensitivity and helps minimize fluorescent differences due to initial template concentration variances (Wittwer *et al.*, 2003; Mackay *et al.*, 2008; Nguyen-Dumont *et al.*, 2009).

Assay design

As described in the excellent review by Vossen *et al.* (2009), assay design for amplicon HRMA is straight forward. As discussed earlier, the size of the amplicon generated depends on the application in question. Mutation scanning for unknown variants can use larger amplicons whereas SNP genotyping should use as small an amplicon as possible. Given the cost of standard primers, it is worthwhile to design multiple primers around a given SNP and use the best-performing assay. Primer dimers that may melt near the amplicon and interfere with fluorescence normalization must be avoided. Secondary structure prediction programs such as mfold (see Websites at end) are also useful to ensure primers do not terminate in loop structures that affect assay performance (Fig. 10.6).

For unlabelled probe assays, the SNP should be near the centre of a 20–30 bp probe and 3′-end blocked to prevent extension by the polymerase. Preferable blocking agents are the C3 spacer and amino C6 modifications (Cradic *et al.*, 2004; Dames *et al.*, 2007). The probe should generally be at least 10% of the amplicon length in order to provide a good probe signal. Modifications to the probe sequence include masking bases over common variants

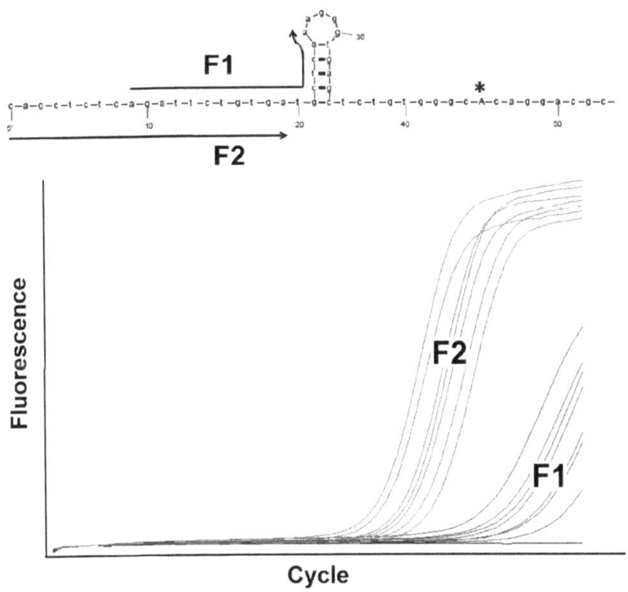

Figure 10.6 The effect of secondary structure at the 3′ end of a primer. On top, two alternative primers (F1 and F2) are shown with identical Tms of 59°C. The F1 primer region forms a loop structure while the F2 region has no secondary structure as predicted by mfold at 55°C, the assay annealing temperature. The reverse primer was the same in both cases and (*) indicates the targeted SNP. Below, the real time growth curves of several samples are shown, amplified with either F1or F2 forward primers.

(bases in the probes designed to mismatch both allele sequences) (Margraf *et al.*, 2007) and LNA bases to permit greater differentiation between genotypes (Chou *et al.*, 2005). Snapback probes can be similarly modified. As an intramolecular hybridization, higher melting temperatures are observed over those calculated for linear probes (Zhou *et al.*, 2008). Smaller amplicons are useful to prevent dominating amplicon peaks. In addition, the first two bases should be non-complementary to the amplicon sequence to prevent extension of the minor strand.

If multiple PCR products are being amplified, then it is convenient to design the primers with similar melting temperatures for simultaneous analysis on the same instrument run. In order to predict the melting curves of the amplicons (perhaps while waiting for oligonucleotide delivery), software such as POLAND can be employed (See websites at end) (Rasmussen *et al.*, 2007).

Like any other fluorescent PCR application, a number of options now exist for the amplification. The saturating dyes may be purchased individually and added to favourite amplification reagents or commercial mastermixes may be used. If the user is new to the technique, it may be worthwhile to start with a commercial mastermix irrespective of whether the plan is to use home-brew reagents later on.

Assay optimization and performance

Often (but not always) the PCR may require Mg^{2+} and temperature optimization. All saturating dyes affect primer Tm – often raising the T_m by 2–3°C at recommended concentrations. Therefore, optimization of new assay primers is best performed with HRMA reagents. Optimal annealing temperatures can be found using a PCR machine with gradient capability (conventional PCR and gel analysis or real-time qPCR with melting analysis). Touchdown PCR can improve specificity (Evrard *et al.*, 2007) and allow assays with varying primer T_ms in the same run (Mackay *et al.*, 2008). Optimal primer concentrations (typically 300–400 nM) are a compromise between specificity and robust amplification. For unlabelled probe and snapback assays, ratios of 5:1 to 10:1 are typical with a high concentration primer of 0.5–1.0 µM. The optimized protocol should allow robust and reproducible amplification of all samples. Real-time PCR may be useful if HRMA is planned on a melting-only instrument. Samples that amplify late (higher Cq) are often those that show non-specific amplification products or poorly clustered melting curves (Grievink and Stowell, 2008; Kristensen and Dobrovic, 2008; Fig. 10.6). If mfold indicates secondary structures or POLAND predicts a high amplicon Tm (e.g. above 90°C) then destabilizing additives may be beneficial (Tindall *et al.*, 2009), such as 0.5–1M betaine, 2% formamide or 5% DMSO (Montgomery *et al.*, 2007). When amplicons include a high temperature melting domain or probes cover GC-rich regions (Habalovà *et al.*, 2009), such destabilizing reagents may be necessary, even though they may lower PCR efficiency and delay amplification.

After PCR, samples are usually denatured at 90–95°C, followed by cooling to randomly associate opposite strands. Heteroduplex detection is maximized by rapid cooling and rapid heating (Gundry *et al.*, 2003), although these rates may be limited by the instrumentation used. For example, the easiest way for manufacturers to attain 'higher' resolution melting is to slow down the heating rate, allowing more time for fluorescence and temperature acquisition and integration. The best instruments allow rapid heating at 0.1–0.3°C so that a 30°C temperature range can be acquired in 1.7–5.0 min. Melting for amplicon HRMA

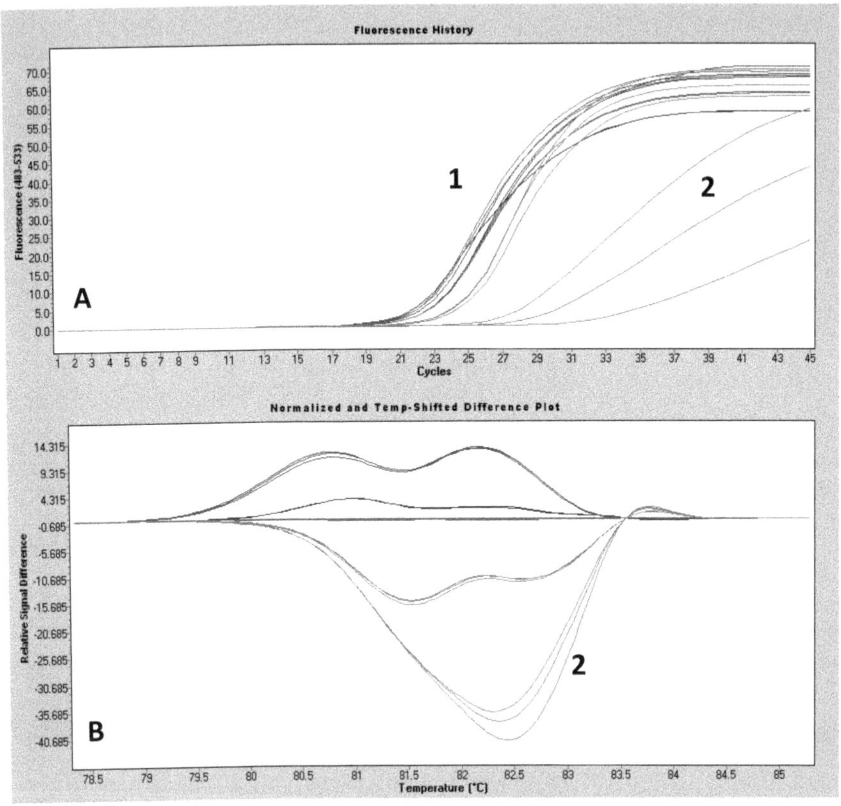

Figure 10.7 The effect of varying template concentration and purity, observed by late amplification and poor reproducibility of melting curves. Replicate DNA extracts from several grapevine varieties were diluted to 10 ng/μl and amplified for microsatellite analysis. In (A), most of the samples amplified early (1) while three samples of one variety were delayed (2), suggesting that poor purity led to overestimation of DNA concentration and/or inhibited amplification, leading to higher Cqs. Late and poor reproducibility amplification in real time analysis resulted in lower melting curve reproducibility as shown in the difference plot (B2).

usually starts around 65–70°C, while unlabelled probes and snapback primers should begin around 40°C. If the melting curve is previously known or predicted, narrower melting ranges may be used. Further optimization and troubleshooting advice that will prove useful in establishing a robust assay is given in Table 10.1.

Future directions

Although first published in 2003, HRMA is still described as a new technology. In our opinion – and to use a popular phrase from a Malcolm Gladwell book – HRMA has hit its 'tipping point' at this time (late 2009) and the rate of new applications will only increase. HRMA provides very high sequence specificity confirmation for amplicon identity and one can envisage instruments becoming smaller with higher resolution. Miniaturization will not only permit wider use of nucleic acid HRMA but will also permit faster turn-around times.

Table 10.1 Troubleshooting

Issue	Remedy
Non-specific amplification	Decrease Mg^{2+} concentration and/or increase the annealing temperature Decrease the primer concentrations Use a gradient cycler to determine the optimal annealing temperature Try touchdown PCR Select new primers
Late cycle amplification	Increase Mg^{2+} concentration and/or lower the annealing temperature Increase the primer concentrations Check secondary structure at or near primer sequences that may limit primer annealing.
Poor melting reproducibility among identical genotypes	Re-purify DNA templates Dilute all samples to the same initial concentration, using the same diluent (e.g. 10–25 ng/μl) Add a high-salt buffer to the reactions and re-melt (Vossen *et al.*, 2009)
Poor probe signal (unlabelled probes or snapback primers)	Design smaller amplicon Use longer probe sequence Increase asymmetry in primer concentrations to generate more probe-binding template strands
Poor resolution of homozygous variants by amplicon HRMA	Mix with wild type DNA to form heterozygotes and remelt (e.g. mixing scheme of Vossen *et al.*, 2009. Try temperature calibrators Mix with known genotypes pre-PCR for quantitative heteroduplex analysis (Palais *et al.*, 2005) Use a probe-based method (unlabelled probe or snapback primers)
Presence of common variants confusing interpretation	Include probes to cover common variants (Nguyen-Dumont *et al.*, 2009)
High T_m amplicons with lower reproducibility and/or secondary structure	Addition of PCR additive such as betaine or DMSO (Tindall *et al.*, 2009; Montgomery *et al.*, 2007)

Another application that will prove useful is the generation of databases for HRMA curve data, already considered by us and others (Mackay *et al.*, 2008; Mader *et al.*, 2008). This would be most useful in 'like vs different' scenarios and for species and type identification, i.e., in bacteria (Cheng *et al.*, 2006; Yang *et al.*, 2009). Stored melting curves could obviate the need for large numbers of standards. However, because absolute temperature measurements vary across instruments and some variation in PCR product, salt, and dye concentrations may be unavoidable over time, internal temperature controls or curve overlay to compare curve shapes may provide the bests comparisons. Even if some variation in curve shape occurs, potential matches may be narrowed considerably so that a limited number of likely candidates could be used as references or mixing controls in a confirmation run.

Summary

High-resolution melting analysis is a highly flexible technology with applications that continue to expand rapidly. While most applications scan for unknown SNPs and/or genotype known variants, HRMA also offers cost-effective and rapid alternatives for HLA matching, varietal certification, phage display clone identification and even the simple gel electrophoresis of PCR products. As the melting resolution of instruments continues to improve, this range of applications will only increase as PCR does itself: exponentially!

Useful websites for HRMA

mfold for predicting secondary structures of amplicons (http://dinamelt.bioinfo.rpi.edu/twostate-fold.php)

POLAND melting simulation software for predicting melting domains of amplicons (http://www.biophys.uni-duesseldorf.de/local/POLAND/poland.html).

References

Chagné, D., Gasic, K., Crowhurst, R.N., Han, Y., Bassett, H.C., Bowatte, D.R., Lawrence, T.J., Rikkerink, E.H., Gardiner, S.E., and Korban, S.S. (2008). Development of a set of SNP markers present in expressed genes of the apple. Genomics 92, 353–358.

Cheng, J.C., Huang, C.L., Lin, C.C., Chen, C.C., Chang, Y.C., Chang, S.S., and Tseng, C.P. (2006). Rapid detection and identification of Clin. ically important bacteria by high-resolution melting analysis after broad-range ribosomal RNA real-time PCR. Clin. Chem. *52*, 1997–2004.

Chou, L.S., Meadows, C., Wittwer, C.T., and Lyon, E. (2005). Unlabeled oligonucleotide probes modified with locked nucleic acids for improved mismatch discrimination in genotyping by melting analysis. BioTechniques 39, 644–648.

Cradic, K.W., Wells, J.E., Allen, L., Kruckeberg, K.E., Singh, R.J., and Grebe, S.K. (2004). Substitution of 3′-phosphate cap with a carbon-based blocker reduces the possibility of fluorescence resonance energy transfer probe failure in real-time PCR assays. Clin. Chem. *50*, 1080–1082.

Crockett, A.O., and Wittwer, C.T. (2001). Fluorescein-labeled oligonucleotides for real-time PCR: Using the inherent quenching of deoxyguanosine nucleotides. Anal. Biochem. *290*, 89–97.

Croxford, A.E., Rogers, T., Caligari, P.D., and Wilkinson, M.J. (2008). High-resolution melt analysis to identify and map sequence-tagged site anchor points onto linkage maps: a white lupin (*Lupinus albus*) map as an exemplar. New Phytol. *180*, 594–607.

De Koeyer, D., Douglass, K., Murphy, A., Whitney, S., Nolan, L., Song, Y., and De Jong W. (2009). Application of high-resolution DNA melting for genotyping and variant scanning of diploid and autotetraploid potato. Mol. Breeding 25, 67–90.

Evrard, A., Raynall, C., Boyer, J.-C., Le Gallic L., and Lumbroso, S. (2007). Mutation scanning of the cytidine deaminase gene by high-resolution melting curve analysis using the LightCycler® 480 System. Biochemica 3, 2007.

Farrar, J.S., Reed, G.H., and Wittwer, C.T. (2010). High resolution melting curve analysis for molecular diagnostics. In Molecular Diagnostics, 2nd edn, Patrinos, G.P., Ansorge, W., eds. (Burlington: Elsevier) pp. 229–245.

Germer, S., and Higuchi, R. (1999). Single-tube genotyping without oligonucleotide probes. Genome Res. 9, 72–78.

Giglio, S., Monis, P.T., and Saint, C.P. (2003). Demonstration of preferential binding of SYBR Green I to specific DNA fragments in real-time multiplex PCR. Nucleic Acids Res. *31*, e136.

Gundry, C.N., Vandersteen, J.G., Reed, G.H., Pryor, R.J., Chen, J., and Wittwer, C.T. (2003). Amplicon melting analysis with labeled primers: a closed-tube method for differentiating homozygotes and heterozygotes. Clin. Chem. *49*, 396–406.

Gundry, C.N., Dobrowolski, S.F., Martin, Y.R., Robbins, T.C., Nay, L.M., Boyd, N., Coyne, T., Wall, M.D., Wittwer, C.T., and Teng, D.H. (2008). Basepair neutral homozygotes can be discriminated by calibrated high-resolution melting of small amplicons. Nucleic Acids Res. *36*, 3401–3408.

Grievink, H., and Stowell, K.M. (2008). Identification of ryanodine receptor 1 single nucleotide polymorphisms by high-resolution melting using the LightCycler 480 system. Anal. Biochem. *374*, 396–404.

Habalová, V., Klimčáková, L., Židzik, J., and Tkáč, I. (2008). Rapid and cost effective genotyping method for polymorphisms in PPARG, PPARGC1 and TCF7L2 genes. Mol. Cell. Probes *23*, 52–54.

Herrmann, M.G., Durtschi, J.D., Bromley, L.K., Wittwer, C.T., and Voelkerding, K.V. (2006). Amplicon DNA melting analysis for mutation scanning and genotyping: cross-platform comparison of instruments and dyes. Clin. Chem. *52*, 494–503.

Herrmann, M.G., Durtschi, J.D., Bromley, L.K., Wittwer, C.T., and Voelkerding, K.V. (2007a). Instrument comparison for heterozygote scanning of single and double heterozygotes: a correction and extension of Herrmann *et al.*, Clin. Chem. 2006, *52*, 494–503. Clin. Chem. *53, 150–152.*

Herrmann, M.G., Durtschi, J.D., Wittwer, C.T., and Voelkerding, K.V. (2007b). Expanded instrument comparison of amplicon DNA melting analysis for mutation scanning and genotyping. Clin. Chem. *53*, 1544–1548.

Kristensen, L.S., and Alexander Dobrovic A. (2008). Direct genotyping of single nucleotide polymorphisms in methyl metabolism genes using probe-free high-resolution melting analysis. Cancer Epidemiol. Biomarkers Prev. *17*, 1240–1247.

Krypuy, M., Newnham, G.M., Thomas, D.M., Conron, M., and Dobrovic, A. (2006). High resolution melting analysis for the rapid and sensitive detection of mutations in clinical samples: KRAS codon 12 and 13 mutations in non-small cell lung cancer. BMC Cancer *6*, 295.

Laurie, A.D., Smith, M.P., and George, P.M. (2007). Detection of factor VIII gene mutations by high-resolution melting analysis. Clin. Chem. *53*, 2211–2214.

Lay, M.J., and Wittwer, C.T. (1997). Real-time fluorescence genotyping of factor V Leiden during rapid-cycle PCR. Clin. Chem. *43*, 2262–2267.

Lehmensiek, A., Sutherland, M.W., McNamara R.B. (2008). The use of high resolution melting (HRM) to map single nucleotide polymorphism markers linked to a covered smut resistance gene in barley. Theor. Appl. Genet. *117*, 721–728.

Liew, M., Seipp, M., Durtschi, J., Margraf, R.L., Dames, S., Erali, M., Voelkerding, K., and Wittwer, C. (2007). Closed-tube SNP genotyping without labeled probes/a comparison between unlabeled probe and amplicon melting. Am. Clin. J. Pathol. *127*, 341–348.

Lipsky, R.H., Mazzanti, C.M., Rudolph, J.G., Xu, K., Vyas, G., Bozak, D., Radel, M.Q., and Goldman, D. (2001). DNA melting analysis for detection of single nucleotide polymorphisms. Clin. Chem. *47*, 635–644.

Mackay, J.F., Wright, C.D., and Bonfiglioli, R.G. (2008). A new approach to varietal identification in plants by microsatellite high resolution melting analysis: application to the verification of grapevine and olive cultivars. Plant Methods *19*, 4:8.

Millward, H., Samowitz, W., Wittwer, C.T., and Bernard, P.S. (2002). Homogeneous amplification and mutation scanning of the p53 gene using fluorescent melting curves. Clin. Chem. *48*, 1321–1328.

Monis, P.T., Giglio, S., and Saint, C.P. (2005). Comparison of SYTO9 and SYBR Green I for real-time polymerase chain reaction and investigation of the effect of dye concentration on amplification and DNA melting curve analysis. Anal. Biochem. *340*, 24–34.

Muleo, R., Colao, M.C., Miano, D., Cirilli, M., Intrieri, M.C., Baldoni And, Rugini, L. E. (2009). Mutation scanning and genotyping by high-resolution DNA melting analysis in olive germplasm. Genome *52*, 252–260.

Palais, R.A., Liew, M.A., and Wittwer, C.T. (2005). Quantitative heteroduplex analysis for single nucleotide polymorphism genotyping. Anal. Biochem. *346*, 167–75.

Palais, R., and Wittwer, C.T. (2009). Mathematical algorithms for high-resolution DNA melting analysis. Methods Enzymol. *454*, 323–343.

Rasmussen, J.P., Saint, C.P., and Monis, P.T. (2007). Use of DNA melting simulation software for in silico diagnostic assay design: targeting regions with complex melting curves and confirmation by real-time PCR using intercalating dyes. BMC Bioinformatics *8*, 107.

Reed, G.H., Kent, J.O., and Wittwer, C.T. (2007). High-resolution DNA melting analysis for simple and efficient molecular diagnostics. Pharmacogenomics *8*, 597–608.

Ririe, K.M., Rasmussen, R.P., and Wittwer, C.T. (1997). Product differentiation by analysis of DNA melting curves during the polymerase chain reaction. Anal. Biochem. *245*, 154–160.

Rouleau, E., Lefol, C., Bourdon, V., Coulet, F., Noguchi, T., Soubrier, F., Bieche, I., Olschwang, S., Sobol, H., Lidereau, R. 2009. qPCR HRM, a new approach to screen simultaneously point mutations and large rearrangements – application to MLH1 germline mutations in Lynch Syndrome. Hum. Mutat. *30*, 867–875.

Schütz, E and von Ahsen N. (2009). Influencing factors of dsDNA dye (high-resolution) melting curves and improved genotype call based on thermodynamic considerations. Anal. Biochem. *385*, 143–152.

Seipp, M.T., Pattison, D., Durtschi, J.D., Jama, M., Voelkerding, K.V., and Wittwer, C.T. (2008). Quadruplex genotyping of F5, F2, and MTHFR variants in a single closed tube by high-resolution amplicon melting. Clin. Chem. *54*, 108–115.

Seipp, M.T., Durtschi, J.D., Voelkerding, K.V., and Wittwer, C.T. (2009). Multiplex amplicon genotyping by high-resolution melting. Biomol. Tech. J. *20*, 160–164.

Smith, B.L., Lu, C.-P., and Alvarado Bremer J.R. (2009). High-resolution melting analysis (HRMA): a highly sensitive inexpensive genotyping alternative for population studies. Mol. Ecol. Res. 1–4.

Stephens, A., Inman-Bamber J., Giffard, P., and Huygens, F. (2008). High-Resolution melting analysis of the spa repeat region of *Staphylococcus aureus*. Clin. Chem. *54*, 432–436.

Tindall, E.A., Petersen, D.C., Woodbridge, P., Schipany, K., and Hayes, V.M. (2009). Assessing high-resolution melt curve analysis for accurate detection of gene variants in complex DNA fragments. Hum. Mutat. *30*, 876–883.

Vandersteen, J.G., Bayrak-Toydemir P., Palais, R.A., and Wittwer, C.T. (2007). Identifying common genetic variants by high-resolution melting. Clin. Chem. *53*, 1191–1198.

van der Stoep, N., van Paridon C.D.M., Janssens, T., Krenkova, P., Stambergova, A., Macek, M., Matthijs, G., and Bakker, E. (2009). Diagnostic guidelines for high resolution melting curve analysis: an inter-laboratory validation of BRCA1 mutation scanning using the 96- well LightScanner (IT). Hum. Mutat. *30*, 899–909.

Wang, J., Chuang, K., Ahluwalia, M., Patel, S., Umblas, N., Mirel, D., Higuchi, R., and Germer, S. (2005). High-throughput SNP genotyping by single-tube PCR with Tm-shift primers. BioTechniques *39*, 885–893.

Waterfall, C.M., and Cobb, B.D. (2002). SNP genotyping using single-tube fluorescent bidirectional PCR. BioTechniques *33*, 80–84.

White, H.E., Hall, V.J., and Cross, N.C. (2007). Methylation-sensitive high-resolution melting-curve analysis of the SNRPN gene as a diagnostic screen for Prader-Willi and Angelman syndromes. Clin. Chem. *53*, 1960–1962.

Wittwer, C.T., Reed, G.H., Gundry, C.N., Vandersteen, J.G., and Pryor, R.J. (2003). High-resolution genotyping by amplicon melting analysis using LCGreen. Clin. Chem. *49*, 853–860.

Wolff, B., Thacker, W., Schwartz, S., and Winchell, J. (2008). Detection of macrolide resistance in *Mycoplasma pneumoniae* by real time PCR and high-resolution melt analysis. Antimicrob. Agents Chemother. *52*, 3542–3549.

Wu, S.B., Wirthensohn, M.G., Hunt, P., Gibson, J.P., and Sedgley, M. (2008). High resolution melting analysis of almond SNPs derived from ESTs. Theor. Appl. Genet. *118*, 1–14.

Wu, S.B., Tavassolian, I., Rabiei, G., Hunt, P., Wirthensohn, M., Gibson, J.P., Ford, C.M., and Sedgley, M. (2009). Mapping SNP-anchored genes using high-resolution melting analysis in almond. Mol. Genet. Genomics *282*, 273–81.

Yang, S., Ramachandran, P., Rothman, R., Hsieh, Y.H., Hardick, A., Won, H., Kecojevic, A., Jackman, J., and Gaydos, C. (2009). Rapid identification of biothreat and other clinically relevant bacterial species by use of universal PCR coupled with high-resolution melting analysis. Clin. J. Microbiol. *47*, 2252–2255.

Zhou, L., Myers, A.N., Vandersteen, J.G., Wang, L., and Wittwer, C.T. (2004a). Closed-tube genotyping with unlabeled oligonucleotide probes and a saturating DNA dye. Clin. Chem. *50*, 1328–1335.

Zhou, L., Vandersteen, J., Wang, L., Fuller, T., Taylor, M., Palais, B., and Wittwer, C.T. (2004b). High-resolution DNA melting curve analysis to establish HLA genotypic identity. Tissue Antigens *64*, 156–164.

Zhou, L., Wang, L., Palais, R., Pryor, R., and Wittwer, C.T. (2005). High-resolution DNA melting analysis for simultaneous mutation scanning and genotyping in solution. Clin. Chem. *51*, 1770–1777.

Zhou, L., Errigo, R.J., Lu, H., Poritz, M.A., Seipp, M.T., and Wittwer, C.T. (2008). Snapback primer genotyping with saturating DNA dye and melting analysis. Clin. Chem. *54*, 1648–1656.

Microfluidic Emulsion PCR

11

N. Reginald Beer and John H. Leamon

Abstract

PCR has traditionally been performed in microlitre-scale reactions because larger scale volumes are prohibitively expensive and wasteful while the smaller scales (nanolitre and below) are impractical with available sample handling tools and detection systems. At the microlitre scale, samples can contain mutually competitive and distinct targets, introducing amplification bias and competitive inhibition that degrade assay performance. Microfluidic Emulsion PCR has emerged as a technique to resolve these challenges by a combination of two enabling technologies. Emulsion PCR provides the advantages of fluid partitioning, namely elimination of sample bias and the ability to run millions of reactions in discrete volumes, while microfluidics simultaneously reduces the sample volume, introduces a level of control over emulsion parameters, and provides optical observability of the partitioned microreactors. Furthermore, since microfluidic emulsions can be made monodisperse in size, they allow the assumption of an average dilution per reactor to permit the exploitation of Poisson statistics for very accurate titre estimation. Microfluidic emulsions can also be employed to perform solid-phase amplification with bead-based assays, combining yet another useful technique with the sample partitioning benefits of droplets. We expect the advantages of both emulsion PCR and microfluidics will encourage new applications and the integration of these enabling technologies will improve PCR performance.

Emulsion PCR

Traditionally, PCR has been performed in microlitre-scale reactions because larger scale (millilitre and above) is prohibitively expensive and wasteful while the smaller scale (nanolitre and below) is unserved by available sample handling tools. Despite the lack of available tools, however, the advantages of going to smaller sizes – namely less reagent use and the ability to approach single copy analyses, beckoned researchers to investigate non-traditional methods. One of these methods has entered the mainstream of nucleic acids analysis, the use of emulsions in PCR.

Emulsions are mixtures of two or more immiscible liquids. A common example is oil and water, but even mayonnaise is an emulsion (oil, vinegar, egg yolks, etc.). Emulsions are two-phase systems of colloids where both phases are liquid. Because they do not blend, the dispersed phase forms droplet-like compartments within the continuous phase. It is this compartmentalization that partitions an aqueous sample especially well for assays such as PCR. Prior to digital microfluidics and batch-generated emulsions, the only available method for sample discretization down to the nanolitre had been performed by dilution and pipetting a sample into microarrays. In this way individual nucleic acids could be isolated, amplified, and detected without interference from other nucleic acids that may dominate the sample, such as the early detection of cancer in blood-borne DNA from a previously healthy individual (Diehl *et al.*, 2008). Ngai *et al.* developed a microchamber array for picolitre PCR with 86 pl volumes (Nagai *et al.*, 2001a), and Leamon *et al.* demonstrated a massively parallel PicoTiter-Plate with up to 300,000 discrete reactions at volumes down to just below 40 pl (Leamon *et al.*, 2003).

Emulsion thermostability

The overwhelming advantage provided by emulsified reactions is the ability to generate literally millions of discrete, compartmentalized reactions in a single tube (Leamon *et al.*, 2006), permitting entire genomic libraries to be universally amplified in a single reaction(Williams *et al.*, 2006). Retaining droplet integrity during a PCR reaction is complicated by the stress imposed by the cyclical thermal changes, and has sparked research into optimization of oils, surfactants and emulsion processes beyond the 4.5% Span-80, 0.5% Tween-80 in mineral oil originally described by Tawfik and Griffiths (1998). Marguiles *et al.* (2005) replaced the Tawfik and Griffiths oil system with a silicon based system comprising 40% (w/w) DC 5225C Formulation Aid (Dow Chemical Co., Midland, MI, USA), 30% (w/w) DC 749 Fluid (Dow Chemical Co.), and 30% (w/w) AR20 Silicone Oil (Sigma), and generated an emulsion of small, sub-micron droplets dubbed 'microfines' prior to the generation of the PCR reaction containing microreactors. These microfines are reported to increase emulsion stability (Huang *et al.*, 2009; Kumaresan *et al.*, 2008; Margulies *et al.*, 2005), possibly by providing a protective barrier of smaller, more stable droplets around the larger PCR droplets, preventing their aggregation or coalescence, as the thermostability of the emulsion droplets has been shown to decrease with increased diameter (Kumaresan *et al.*, 2008). Additional emulsion formulations have also been described (Li *et al.*, 2006; Tiemann-Boege *et al.*, 2009; Williams *et al.*, 2006; Fig. 11.1) that show improved thermostability and are composed of ABIL EM 90 and mineral oil, differing in their use of either Triton X-100 (Williams *et al.*, 2006) or TEGOSOFT (Li *et al.*, 2006; Tiemann-Boege *et al.*, 2009).

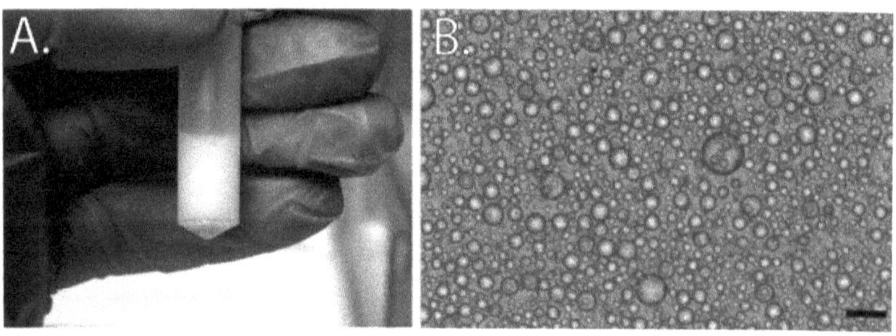

Figure 11.1 Emulsified PCR reactions generated by horizontal agitation of a stainless steel mixing ball as described in (Li *et al.*, 2006). (A) Emulsified reaction in 1.5 ml tube prior to dispensation into PCR tubes. (B) Same emulsion as seen in a disposable haemocytometer under a microscope. The scale bar at the lower right is approximately 10 μm. Some of the droplets can be seen to contain 2.8 μm solid particles.

Ease of use

Despite its many advantages, one of the drawbacks of emulsified reactions has been the emulsification process used in many of the protocols. Bulk emulsions are typically generated through some form of mechanical agitation of an aqueous and an oil phase, with a resulting emulsion that is oily, thick and difficult to pipette. Many different strategies have been used to emulsify reactions, including vortexing (Wetmur *et al.*, 2005), drop-wise addition of the aqueous phase into a stirred cell containing the oil phase (Tawfik and Griffiths, 1998), reciprocal horizontal agitation both with (Li *et al.*, 2006; Tiemann-Boege *et al.*, 2009) and without (Margulies *et al.*, 2005) a stainless steel mixing ball, and sonification of an oil and water mixture (Musyanovych *et al.*, 2005).

In addition to the advantages of amplifying entire genomic libraries, PCR also benefits greatly from the emulsification-induced reactor miniaturization and isolation, especially if the reactors are monodisperse, which ensures consistent reaction rates as well as optical tuning for real-time PCR. Since bulk emulsions are polydisperse, however, unwanted variations in the reaction rate can occur due to a continuous droplet size distribution. This has led researchers to employ higher homogenization speeds with the goal of producing smaller droplets with a narrower size distribution (Miller *et al.*, 2006; Fig. 11.2),but the distribution itself is still Gaussian. Unfortunately these tube-based bulk emulsion PCR methods can only include end point amplification detection(Musyanovych *et al.*, 2005). To perform real-time detection the droplets must be focused into a channel so that background fluorescence from droplets above or below the focal depth does not affect the fluorescence intensity measurement. Otherwise only an average fluorescent intensity measurement is taken, and the optical interrogation of individual droplets or reactors cannot be performed. Finally, batch processes lack the requisite controls for individual droplet addressability provided by on-chip architectures. This latter disadvantage has been overcome by utilizing solid phase amplification in droplets which we discuss below.

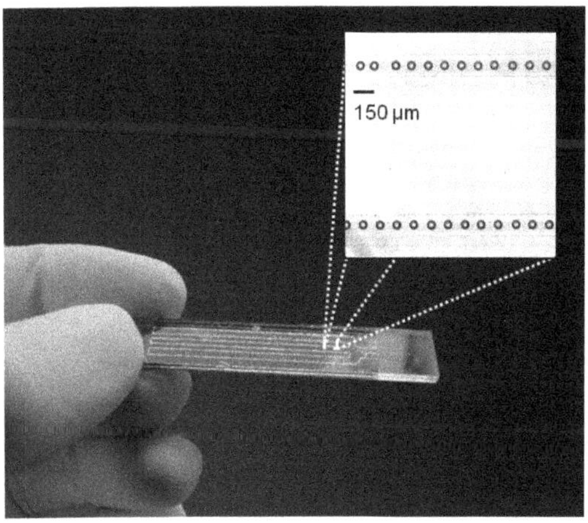

Figure 11.2 Picolitre-scale droplets on-chip prior to performing real-time reverse-transcription PCR.

Microfluidic approaches to emPCR

The burgeoning field of microfluidics takes advantage of semiconductor manufacturing processes to generate extremely small (by benchtop standards) fluidic compartments, allowing complex laboratory assays like PCR to be performed on semiconductor or polymer 'Chips'. These devices exhibit feature sizes on the micron scale, hence the term 'micro' in microfluidics. Reduction of sample processing volumes to the microchannel, when coupled to massively parallel high-throughput analyses, implies the potential for faster and more efficient reactions, as well as the ability to isolate individual cells, proteins, virions, or nucleic acids. These devices can either use architectural compartmentalization to isolate individual targets, or employ emulsification on the microfluidic scale to exploit the benefits of emPCR described in above. This has been the driver for the implementation of PCR on the microchip.

The microfluidics approach dramatically reduced the reaction volumes for performing biochemical reactions. 'Microfluidics' can perform on-chip discretization of an aqueous sample into small and addressable volumes on the micron scale. This can be done through droplet generation at a flow-focusing device (Anna, 2003) or at a T-junction (Garstecki *et al.*, 2006; Tan *et al.*, 2004; Thorsen *et al.*, 2001). The application of droplet-based microfluidics to the polymerase chain reaction (PCR) represents a paradigm shift from microlitre-scale microarray assays to the picolitre regime, promising fewer cycles with lower durations, reduced reagent consumption, and the ability to detect very low copies of nucleic acid. These advantages arise from the very small heat and molecular diffusion distances Lab-on-a-chip systems provide, which explains the growing interest. As of this writing these systems are still experimental, but given a reduction in their operational complexity and instrument expense, they may eventually find their way to the laboratory benchtop. If they do make the commercial transition, droplets on a chip will offer a level of control over microdroplet compartmentalization not achievable by the 'shake-and-bake' batch-generated emulsion methods previously described. Mainly, droplet size in the emulsion is no longer polydisperse in these experimental systems, as the droplet generation on the chip is capable of producing a monodisperse distribution (one size only), and droplet size can be adjusted by adjusting flow parameters(Tan *et al.*, 2004; Thorsen *et al.*, 2001). Like batch-generated emulsion reactors, however, these droplets are still isolated from each other and the channel walls, but can be individually optically addressed for real-time PCR. Due to their miniscule size and observability, these droplets are ideal for single-copy PCR.

Assay adjustments for microfluidics

Adapting benchtop assays to the microscale necessitates some changes to the assay as well. For instance, introducing BSA helps by preferentially adsorbing to channel and tubing walls, keeping the polymerase and targets available for the reaction. Also, doubling or even tripling the polymerase enzyme concentration has been effective in on-chip PCR(Beer *et al.*, 2007), as it is thought that the enzyme accumulates at the surface of the aqueous droplet near the water/oil interface.

Ease of use

Assuming the microfluidic instrument developer has prepared the system appropriately with accurate valves and pumps for metering the sample and generating a monodisperse emulsion, as well as utilized a substrate with a hydrophobic surface (by either coating or inherent substrate property), all that remains is to thermal cycle and detect template amplification in the droplets. Thermal cycling can be performed on-chip through Peltier or other means, and the droplets can either be optically interrogated with Taqman probes (Beer *et al.*, 2007; Kiss *et al.*, 2008) or intercalating dyes (Mazutis *et al.*, 2009), or the droplets can be run off-chip, the emulsion broken, and the aggregated aqueous sample run on capillary gel electrophoresis (Kiss *et al.*, 2008). Of course, all of these steps require a great deal of commitment to perform repeatably at this time, and the challenge for droplet-based instrument providers will be to reduce the operational complexity of any commercial product while maintaining affordability.

Poisson statistics and small volume qPCR

Perhaps the most significant advantage of small volume PCR is the ability to detect a single copy of target nucleic acid in a complex background, and it has been used to detect a low concentration of mutations in alleles associated with colorectal cancer This groundbreaking research was termed 'digital PCR' by Sykes *et al.* (1992), popularized by Vogelstein and Kinzler (1999) and has demonstrated single molecule sensitivity in a microarray-pipetted sample with a high background (Vogelstein and Kinzler, 1999).

Since the early work for cancer detection the method has evolved with commercial arrays, Taqman-based FRET probes, and smaller sample volumes. For example, single molecule sensitivity has recently been shown in experimental droplet systems with picolitre-scale reaction volumes (Beer *et al.*, 2007; Kiss *et al.*, 2008), but it also can be attained with non-droplet-based microfluidic systems like the Fluidigm (Warren *et al.*, 2006) and BioTrove (Brenan and Morrison, 2005; Sogin *et al.*, 2006) approaches, as well as the micromachined silicon array (Nagai *et al.*, 2001b) and the PicoTiter Plate (Leamon *et al.*, 2003) approaches. In fact all one needs is a method for isolating small volumes of a dilute sample (pipetting onto a multiwell plate will do), and a simple understanding of Poisson statistics.

Since sample discretization at these volumes allows for individual wells, droplets, or equivalently, reactors to hold either one copy of a genomic template or none, the process can be described as a Poisson process. This process models the probability of events that occur at discrete intervals, and has the following discrete probability density function:

$$P(\lambda)=\frac{\lambda^{x}e^{-\lambda}}{x!}$$

where x is an integer and λ is the mean. The probability distribution function is then simply:

$$\sum_{i=0}^{x}\frac{\lambda^{i}e^{-\lambda}}{i!},\lambda>0$$

This distribution describes a process that can only exhibit a discrete non-negative integer number of occurrences, i.e. $x = \{0, 1, 2, 3, \ldots\}$. Such a distribution arises when a well or reactor is either empty or containing only a discrete number of molecules. For DNA dilutions of 50% (one copy for every two wells of solution on a microarray plate, or equivalently one copy for every two droplets), λ = 0.5. Under standard dilution chemistry one might expect half of the wells or droplets to have positive reactions. In fact only ~39.3% will be positive according to Poisson statistics. This is because some wells get zero copies, others will get one copy, a small percentage will get two copies, and so on. This effect is demonstrated in Figs. 11.3 and 11.4 for 1 and 0.5 copies per droplet respectively.

Thus we see an important benefit of digital PCR: we can infer starting copy concentration based on the percentage of droplets that 'go hot' for the Poisson regime and based on Ct values for the greater than one copy per reactor dilutions (see Fig. 11.5). This convenient statistical observation compares favourably to traditional qPCR methods which require the user to do a time-consuming calibration titration series over three or four magnitudes of starting copy concentration with a well-known standard and measure the Ct shift for each. We have shown previously the accuracy of digital PCR in a continuous flow microfluidic platform, which is displayed in below. Note the blue curve is the predicted percentage of droplets that support amplification verses starting genomic template copies per droplet according the Poisson statistics. Tracking the shape of this curve are the experimental results of 16,000 droplets per data point at titration-quantified starting copy numbers. The displayed excellent agreement points to the power of this method.

Emulsified solid-phase amplification

Combining emulsification with solid-phase amplification avoids some bulk amplification limitations described previously, and provides several key advantages. Due to the large number of droplets generated by emulsification, template concentrations can be diluted down to levels where the average droplet contains, on average, less than a single template.

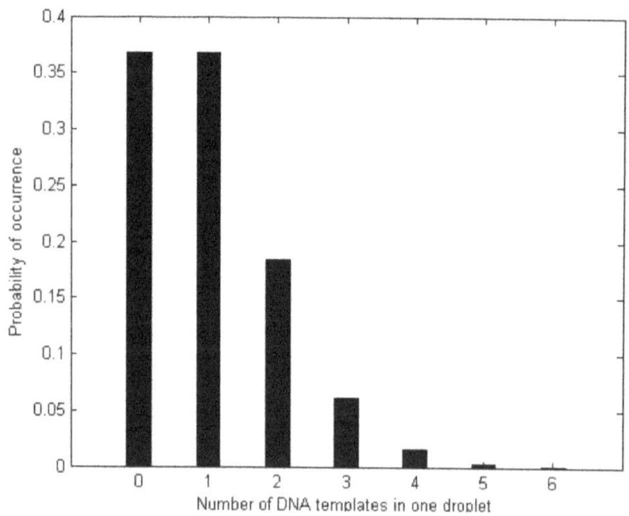

Figure 11.3 Distribution of genomic copies in droplets under a Poisson process for an average dilution of one copy per droplet. Note that approximately 37% of droplets are empty due to some droplets receiving more than one copy.

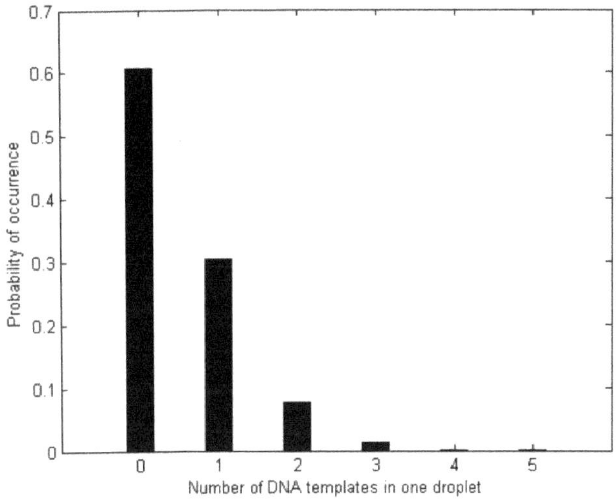

Figure 11.4 Distribution of genomic copies in droplets under a Poisson process for an average dilution of 0.5 copies per droplet. Note that approximately 60% of droplets are empty due to some droplets receiving more than one copy.

clonally amplified template, which then served as a template for sequencing by synthesis (Leamon and Rothberg, 2007; Margulies *et al.*, 2005). These approaches have integrated solid-phase emPCR in commercial high-throughput sequencing strategies (Margulies *et al.*, 2005; Shendure *et al.*, 2005), where solid phase amplification provides clonally amplified template as a substrate for subsequent sequencing, obviating previous requirements for bacterial cloning (Margulies *et al.*, 2005; Shendure *et al.*, 2005), and avoiding bias that can be introduced by the cloning process (Cottrell and Kirchman, 2000). First brought to market by 454 Life Sciences in 2005, solid-phase emPCR has become incorporated in three of the four commercially available 'next-gen' sequencers (454 Life Sciences sequencers, George Church's Polonator, and ABI's SOLiD use some form of emPCR, while Solexa's line of GA sequencers rely on non-emulsified bridge amplification to amplify template molecules to detectable concentrations).

The challenge of amplicon length

One of the persistent challenges facing solid-phase amplification whether in bulk or emulsified conditions, has been the difficulty of amplifying or capturing amplicons above a certain length. Mitra and Church (1999) experienced this effect when generating Polonies (PCR colonies) via solid-phase amplification in an acrylamide matrix with one primer immobilized in matrix, the other free in solution. They reported that polony diameter decreased both with increased acrylamide percentage in the matrix as well as fragment length; as the templates length increased from 100 to 1009 bases, the diameter of the resulting polony decreased both in size and variability. This was attributed to the fact that with one of the primers immobilized, the polony can only grow in size via template diffusion through the

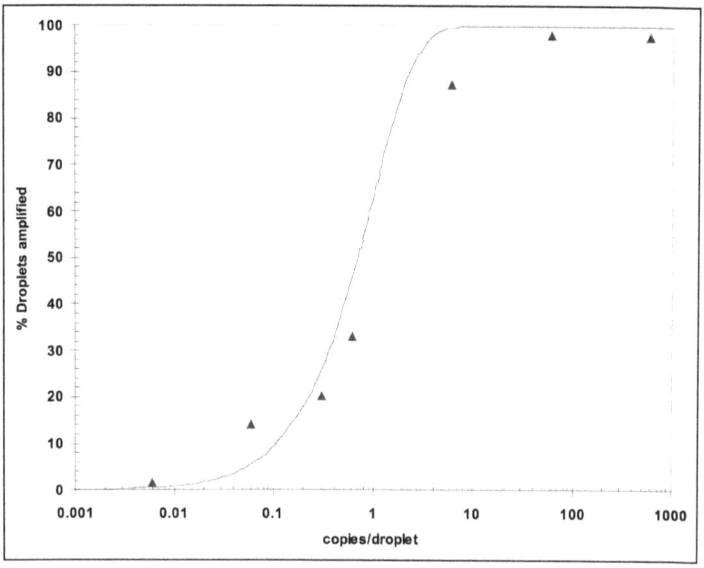

Figure 11.5 Observed % amplification versus theory (*n* ~16,000 droplets per data point) (Kiss *et al.*, 2008).

Limited dilution ensures that templates were able to amplify individually and uniformly (Williams *et al.*, 2006) within discrete droplets, without bias resulting from competition with other template species (Hori *et al.*, 2007; Williams *et al.*, 2006) based on amplification efficiency (Jordan *et al.*, 2002; Kwok *et al.*, 1990), differences in amplicon GC content (Benita *et al.*, 2003), or amplicon reanimating (Suzuki and Giovanni, 1996).

Utility of emulsified DNA capture beads

The utility of bulk emulsified reactions was substantively increased by the incorporation of DNA capture beads in the microreactors, as the solid phase amplification process preserved the clonal nature of the amplified single molecules after the emulsion was broken. For example, solid-phase emPCR combined with cell-free protein synthesis was used to link individual DNA molecules to the specific protein for which they code (Gann *et al.*, 2008). The immobilized DNA also served as effective substrates for enzymatic processes including rolling circle amplification as in the BEAMing process (Diehl *et al.*, 2006; Li *et al.*, 2006), where the copy number of solid-phase emPCR immobilized templates was further increased by 100× or more via rolling circle amplification. This permits the amplicon-covered beads to be flow sorting based on hybridization of gene-specific fluorescent probes, with sensitivity sufficient to detect mutations present in ratios of less than 1 mutant to 10,000 wild-type alleles, enabling such applications as the detection and enumeration of rare mutations (Diehl *et al.*, 2006; Dressman *et al.*, 2003) and transcription factor screening (Kojima *et al.*, 2005). emPCR-immobilized templates have also proven effective templates for DNA ligation (Shendure *et al.*, 2005). Shendure *et al.* (Shendure *et al.*, 2005) used emPCR to amplify genomic fragments onto DNA capture beads for subsequent sequencing operations. Another sequencing strategy also employed emPCR to populate DNA capture beads with acrylamide matrix. As the matrix density and/or the length of the template increases the rate of diffusion is reduced, and so is the size of the polony. Some of the steric factors controlling the growth of PCR colonies during SPA have been extensively modelled (Mercier and Slater, 2005; Mercier *et al.*, 2003).

In solid-phase reactions conducted in emulsified droplets containing 34 μm Sepharose DNA capture beads, Kumersan *et al.* (2008) found amplification yield of immobilized template to be correlated with the length of the respective fragment: 175 amol from a 380-bp fragment, approximately 155 amol from a 642-bp fragment and 10 amols of 1139 bp amplicon (Fig. 11.6). It is interesting to note that the slope of the negative relationship between amplicon length and amplification yield is –1.3, quite similar to the slopes of –1.5 and –1.8 obtained from Mitra and Church's data in Fig. 11.7 (Mitra and Church, 1999). Kumersan *et al.* attributed their success at long fragment amplification to the larger 5 nl droplet size and relatively higher reagent concentration.

McCalla *et al.* (2009) conducted a study of enzymatic activity on immobilized templates. Although not specific to emulsion PCR generated products, but rather enzymatic manipulation of any immobilized substrate, their findings attributed the low efficiency of solid phase processes to steric inhibition imposed by the solid surface itself as well as the density of neighbouring DNA strands.

PCR efficiency

Despite the utility of bead-based emPCR, solid phase amplification suffers from inherent inefficiencies generating active research in process optimization. Although the efficiency of solid phase amplification hasn't been directly calculated for emulsion-based reactions, it is well known for solid-phase amplification in general. Kohsaka and Carson (1994) reported 10-fold lower sensitivity for solid-phase PCR in microwells relative to convention solution-phase PCR and the efficiency of bridge amplification was shown to range from 0.2 to 0.3 per cycle (Bing *et al.*, 1996) as opposed to efficiencies of 0.8–0.9 for typical solution phase. Solid-phase amplification on bulk acrylamide beads in solution displayed amplification efficiencies of 0.3 (Shapero *et al.*, 2001) compared to solution phase (0.8 to 0.9), and Ge *et al.* (2008) described poor surface phase amplification (0.6) relative to typical solution phase efficiencies of approximately 0.8. The discrepancy between solution and solid phase amplification efficiency focused considerable research on leveraging the success of solution phase PCR to improve solid-phase template immobilization. Despite the overall reduction in PCR efficiency, yields of immobilized template are typically enhanced by the excess of single-stranded complementary DNA relative to those obtained with the dsDNA produced by symmetrical PCR (Wei *et al.*, 2004). The practical benefits of this strategy are evident in the fact that asymmetric amplification has been incorporated in many commercial emulsified systems with DNA capture components (Kojima *et al.*, 2005; Margulies *et al.*, 2005).

Improved primer design

PCR primer design is a critical component for the success of any PCR reaction as indicated by the depth of literature surrounding the process. Primer design for solid-phase emPCR is more complicated as the primers must function in solution, interfacial and possibly solid-phase environments. Solution phase primers have benefited from several rounds of optimization, the first resulting in LATE (linear after the exponential) PCR (Sanchez

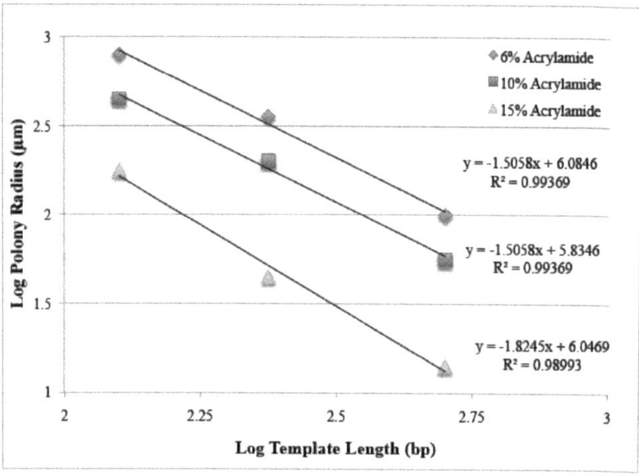

Figure 11.6 Inverse relationship between fragment length and polony diameter amplified by bridge amplification at three acrylamide concentrations. Data from Mitra and Church (1999).

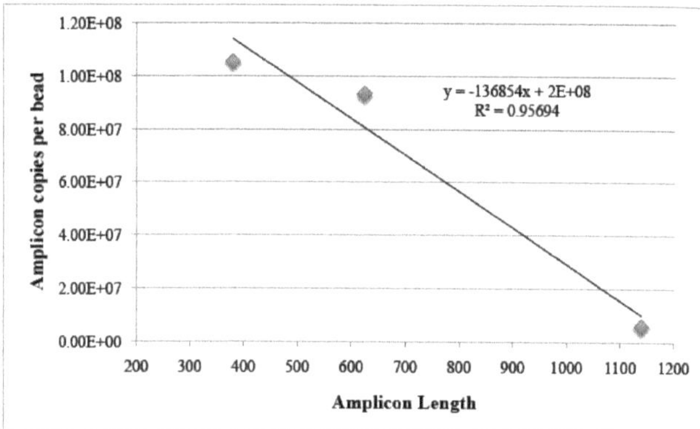

Figure 11.7 Inverse relationship between fragment length and amplification yield on emulsified DNA capture beads. Data from Kumarescan *et al.* (2008).

et al., 2004), successfully used to generate template for pyrosequencing by Salk *et al.* (2006). LATE-PCR improved the efficiency of asymmetric PCR by incorporating in the primer design the fact that primer T_m is concentration dependant. Thus, the limiting primer starts at an effectively lower T_m than the excess primer, resulting in lower hybridization efficiency if the two primers were designed with T_m equivalence based on the typical equimolar assumption. The limiting primer's inefficient hybridization is only exacerbated as it is consumed during the course of exponential amplification and the concentration drops further. LATE-PCR utilized a modified algorithm for primer design that compensated for the varying primer concentrations and was able to obtain efficiencies similar to symmetric PCR with a relative increase in real-time signal strength from 80 to 250%(Sanchez *et al.*, 2004). It must be noted that although LATE-PCR has been used to generate product for pyrosequencing from single cells(Salk *et al.*, 2006), this sequencing reaction was conducted in solution without a solid-phase component, and to date LATE-PCR has not been utilized for solid-phase capture or emPCR.

Enhanced solid-phase PCR (ESP-PCR)(Khan *et al.*, 2008) refers to PCR primer optimization designed specifically for solid-phase amplifications. Khan *et al.* (2008) tested an improved primer design on silica microspheres that increased amplification efficiency and yield by employing nested primers, thereby decoupling the solution and solid phase primer kinetics. The annealing temperatures were also controlled to ensure that limiting primer didn't decrease in efficiency with reduced concentration, with solution phase primers possessing a lower T_m than the solid phase. Using these improvements, Khan *et al.* (2008) were able to improve solid-phase amplification from 1.4- to 9.8-fold across three separate diagnostic targets.

Carmon *et al.* (2002) investigated non-emulsified solid-phase PCR in microwells and found optimal amplification when the primers were distanced from the binding moieties by five to ten HEG (hexaethylene glycol) spacers with an approximate length of 90–180 nucleotides. Under these conditions roughly 20% of the immobilized primers were extended(Carmon *et al.*, 2002), significantly more than the 0.3% extension of oligo-dt(10) primers coupled to glass slides obtained by Adeesi *et al.* (Adeesi *et al.*, 2000; Kohsaka and Carson, 1994). Carmon *et al.* found that the majority of the amplification inefficiencies in amplification were due to steric hindrance of the *Taq* polymerase binding to the immobilized template(Carmon *et al.*, 2002) as opposed to steric hindrance of the binding of the template to the immobilized primer as had been suggested by others (Guo *et al.*, 1994; Shchepinov *et al.*, 1997).

Enrichment

As discussed previously, the massive throughput inherent in emulsified reactions enables the use of limiting dilution to provide single molecule amplification. With millions to billions of reactions generated per tube, DNA concentrations can be lowered to levels where of the droplets containing DNA template, single template droplets are predominant. This allows clonal amplification without competition between templates. However, limiting dilutions yield a large percentage of empty droplets with no DNA template but may contain a DNA capture bead nonetheless. Screening the DNA capture beads from these non-templated droplets would be extremely inefficient or expensive for some applications, such as genomic sequencing.

Several methods have been developed to enrich the bead population for those covered with amplicon and discard the empty, uninformative beads. BEAMing (Dressman *et al.*, 2003) utilized hybridization of fluorescent probes complementary to known allelic variations on the immobilized templates to select amplified beads via FACS. Streptavidin-coated paramagnetic beads tethered to the templates on DNA capture beads via hybridized biotinylated were used by Marguiles *et al.* (2005) to selectively retain the amplicon coated DNA capture beads. Shendure *et al.* (2005) effectively reversed this process by binding large, non-magnetic beads to amplicon-coated positive beads via biotinylated probes. The amplicon beads were then selected by centrifugation down a density gradient, with un-templated beads sinking down the gradient, while DNA beads bound to the larger non-magnetic beads floated at the top. Regardless of the methodology, enrichment of immobilized emulsion PCR products permits the efficient segregation and utilization of clonally amplified templates for a variety of applications.

Continued improvements

Considerable effort has been expended on improving the efficiency and sensitivity of solid-phase amplification, and many of these advances have been incorporated into emPCR. However, it should be noted in advance that most, if not all of the optimization research to date has been conducted in bulk rather than emulsified reactions. This is primarily because of the difficulty inherent in monitoring solid-phase emulsified hybridization and amplification in real time – although possible (Beer *et al.*, 2007; Kiss *et al.*, 2008), requires specialized equipment and is not amenable to traditional real time studies conducted in PCR tubes. As a result, relative amplification efficiency and success of optimization programs is typically determined by endpoint assays such as fluorescent probe hybridization, electrophoretic resolution or even sequencing, as exemplified by the optimization process described by Tiemann-Boege *et al.* (2009). While these methodologies have resulted in optimized systems capable of providing solid-phase amplicons for sequencing (Fig. 11.8) or other downstream processes, the finer details concerning, for example, the relative rates and efficiencies of solution versus solid phase amplification remain unknown, leaving the actual kinetics surrounding solid-phase emPCR a 'black box in many respects, with ample room for future studies.

As the widespread adoption of next-gen sequencing broadens the commercial use of emPCR, the potential to automate the process through microfluidic droplet generators is attracting attention. As described (Diehl *et al.*, 2006; Dressman *et al.*, 2003; Margulies *et al.*, 2005), the actual bulk emulsification and amplification process is a labour intensive, multistep process, with little or no possibility for direct transition to automated laboratory workstations. Conversely, automated microfluidic flow-based emulsification has been successfully demonstrated with (Kumaresan *et al.*, 2008) and without (Kiss *et al.*, 2008) encapsulated DNA capture beads, providing a possible 'hands-free' solution. However; while microfluidic systems enable automated and highly controlled generation of PCR droplets, the serial nature of the droplet generation is a current drawback when compared to the bulk processes' simultaneous generation rates. Whereas the bulk methods can generate 10^7 to 10^9 droplets in a matter of seconds, the flow-focused systems generate droplets at a much lower rate: 5.7 (Kumaresan *et al.*, 2008) and 530 (Kiss *et al.*, 2008) droplets per second. This throughput is further exacerbated if single molecule amplifications are required, where limiting dilution and the Poisson distribution result in only 10 to 25% of the droplets containing a template to amplify (Leamon *et al.*, 2006).

While the current solid-phase throughput limitation may be solved in the near future, possibly through some highly parallel method of droplet formation such as that described by Sugiura *et al.* (2005), the first commercial flow-based emPCR product to reach the market uses microfluidics to generate droplets for solution, not solid-phase, amplification. The RDT-1000 (www.RainDanceTechnologies.com, Lexington, MA, USA) utilizes microfluidics to generate droplets containing genomic DNA and PCR reagents, and then merges those droplets with a secondary stream of droplets containing PCR primer pairs specific to particular genes (Tewhey *et al.*, 2009). While this process allows the genome of interest to be effectively 'enriched' with only specific targets amplified for subsequent sequencing, it should be noted that the RDT-1000 serves only as a preparative process for solution-phase bulk emulsions amplified in traditional thermal cyclers. The amplified regions are then processed by sequencing platform-specific sample preparation and amplification protocols.

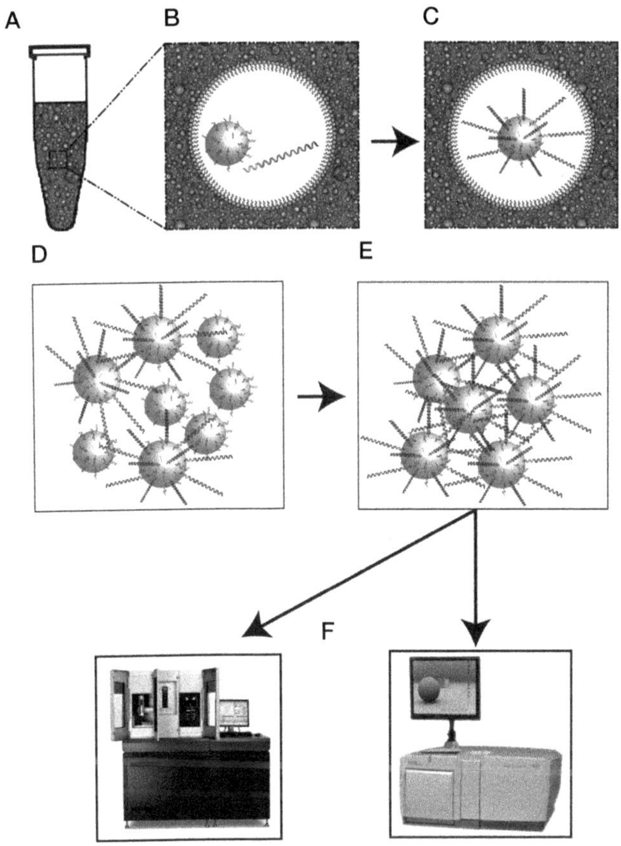

Figure 11.8 Process flow from bulk solid-phase emPCR to next-generation sequencing. (A) Bulk emulsified reaction in tube. (B) Enlargement showing single reaction droplet, containing DNA capture bead, single DNA template and PCR reagents prior to amplification. (C) Depiction of the same droplet following amplification, wherein the DNA capture bead is now covered with immobilized copies of the amplified template. (D) Emulsions are broken; bead population is comprised of both PCR(+) and PCR(–) beads as determined by limiting dilution of template DNA. (E) Enrichment processes select for PCR (+) beads. (F) PCR positive beads can then be sequenced by various Next-Gen sequencing systems (ABI SOLiD and 454 FLX shown).

Conclusion

Microfluidic Emulsion PCR combines the advantages of both emulsion PCR, namely elimination of sample bias and the ability to run millions of reactions in discrete volumes, with the advantages of microfluidics, principally the reduction in sample volume and level of control and optical observability of the emulsion parameters. Furthermore, since monodisperse emulsions allow the assumption of an average dilution per reactor, these microfluidic systems permit the exploitation of Poisson statistics for very accurate titre estimation compared to benchtop qPCR processes. We expect the advantages of both emulsion PCR and microfluidics will encourage continued development and integration of both of these enabling technologies for improved PCR performance.

References

Adeesi, C., Matton, G., Ayala, G., Turcatti, G., Mermod, J.-J., Mayer, P., and Kawashima, E. (2000). Solid phase DNA amplification: characterisation of primer attachment and amplification mechanisms. Nucleic Acids Res. *28*, e87.

Anna, S.L. (2003). Formation of dispersions using 'flow focusing' in microchannels. Appl. Physics Lett. *82*, 364.

Beer, N.R., Hindson, B.J., Wheeler, E.K., Hall, S.B., Rose, K.A., Kennedy, I.M., and Colston, B.W. (2007). On-chip, real-time, single-copy polymerase chain reaction in picoliter droplets. Anal. Chem. *79*, 8471–8475.

Benita, Y., Oosting, R.S., Lok, M.C., Wise, M.J., and Humphery-Smith, I. (2003). Regionalized GC content of template DNA as a predictor of PCR success. Nucleic Acids Res. *31*, e99.

Bing, D.H., Boles, C., Rehman, F.N., Audeh, M., Belmarsh, M., Kelley, B., and Adams, C.P. (1996). Bridge amplification: A solid phase PCR system for the amplification and detection of allelic differences in single copy genes. Paper presented at: Genetic Identity Conference Proceedings, Seventh International Symposium on Human Identification (Phoenix, AZ, Promega Corporation).

Brenan, C., and Morrison, T. (2005). High throughput, nanoliter quantitative PCR. Drug Discov. Today: Technologies 2, 247–253.

Carmon, A., Vision, T.J., Mitchell, S.E., Thannhauser, T.W., Muller, U., and Kresovich, S. (2002). Solid-phase PCR in microwells: effects of linker length and composition on tethering, hybridization, and extension. Biotechniques *32*, 410, 412, 414–418, 420.

Cottrell, M.T., and Kirchman, D.L. (2000). Community composition of marine bacterioplankton determined by 16S rRNA gene clone libraries and fluorescence in situ hybridization. Appl. Environ. Microbiol. *66*, 5116–5122.

Diehl, F., Li, M., He, Y., Kinzler, K.W., Vogelstein, B., and Dressman, D. (2006). BEAMing: single-molecule PCR on microparticles in water-in-oil emulsions. Nat. Methods *3*, 551–559.

Diehl, F., Schmidt, K., Choti, M.A., Romans, K., Goodman, S., Li, M., Thornton, K., Agrawal, N., Sokoll, L., Szabo, S.A., *et al.* (2008). Circulating mutant DNA to assess tumor dynamics. Nat. Med. *14*, 985–990.

Dressman, D., Yan, H., Traverso, G., Kinzler, K.W., and Vogelstein, B. (2003). Transforming single DNA molecules into fluorescent magnetic particles for detection and enumeration of genetic variations. Proc. Natl. Acad. Sci. U.S.A. *100*, 8817–8822.

Gan, R., Yamanaka, Y., Kojima, T., and Nakano, H. (2008). Microbeads display of proteins using emulsion PCR and cell-free protein synthesis. Biotechnol. Prog. *24*, 1107–1114.

Garstecki, P., Fuerstman, M.J., Stone, H.A., and Whitesides, G.M. (2006). Formation of droplets and bubbles in a microfluidic T-junction-scaling and mechanism of break-up. Lab. Chip *6*, 437–446.

Ge, Q., Yu, P., Bai, Y., and Lu, Z. (2008). An emulsion system based on a chip polymerase chain reaction. Molecules *13*, 3057–3068.

Guo, Z., Guilfoyle, R.A., Thiel, A.J., Wang, R., and Smith, L.M. (1994). Direct fluorescence analysis of genetic polymorphisms by hybridization with oligonucleotide arrays on glass supports. Nucleic Acids Res. *22*, 5456–5465.

Hori, M., Fukano, H., and Suzuki, Y. (2007). Uniform amplification of multiple DNAs by emulsion PCR. Biochem. Biophys. Res. Commun. *352*, 323–328.

Huang, H., Qi, Z., Deng, L., Zhou, G., Kajiyama, T., and Kambara, H. (2009). Highly sensitive mutation detection based on digital amplification coupled with hydrogel bead-array. Chem. Commun. (Camb), 4094–4096.

Jordan, B., Charest, A., Dowd, J.F., Blumenstiel, J.P., Yeh, R.F., Osman, A., Housman, D.E., and Landers, J.E. (2002). Genome complexity reduction for SNP genotyping analysis. Proc. Natl. Acad. Sci. U.S.A. *99*, 2942–2947.

Khan, Z., Poetter, K., and Park, D.J. (2008). Enhanced solid phase PCR: mechanisms to increase priming by solid support primers. Anal. Biochem. *375*, 391–393.

Kiss, M.M., Ortoleva-Donnelly, L., Beer, N.R., Warner, J., Bailey, C.G., Colston, B.W., Rothberg, J.M., Link, D.R., and Leamon, J.H. (2008). High-throughput quantitative polymerase chain reaction in picoliter droplets. Anal. Chem. *80*, 8975–8981.

Kohsaka, H., and Carson, D.A. (1994). Solid-phase polymerase chain reaction. J. Clin. Lab. Anal. *8*, 452–455.

Kojima, T., Takei, Y., Ohtsuka, M., Kawarasaki, Y., Yamane, T., and Nakano, H. (2005). PCR amplification from single DNA molecules on magnetic beads in emulsion: application for high-throughput screening of transcription factor targets. Nucleic Acids Res. *33*, e150.

Kumaresan, P., Yang, C.J., Cronier, S.A., Blazej, R.G., and Mathies, R.A. (2008). High-throughput single copy DNA amplification and cell analysis in engineered nanoliter droplets. Anal. Chem. *80*, 3522–3529.

Kwok, S., Kellogg, D.E., Spasic, D., Goda, L., Levenson, C., and Sninsky, J.J. (1990). Effects of primer-template mismatches on the polymerase chain reaction: Human immunodeficiency virus type 1 model studies. Nucleic Acids Res. *18*, 999–1005.

Leamon, J.H., Lee, W.L., Tartaro, K.R., Lanza, J.R., Sarkis, G.J., deWinter, A.D., Berka, J., Weiner, M., Rothberg, J.M., and Lohman, K.L. (2003). A massively parallel PicoTiterPlate based platform for discrete picoliter-scale polymerase chain reactions. Electrophoresis *24*, 3769–3777.

Leamon, J.H., Link, D.R., Egholm, M., and Rothberg, J.M. (2006). Overview: methods and applications for droplet compartmentalization of biology. Nat. Methods *3*, 541–543.

Leamon, J.H., and Rothberg, J.M. (2007). Cramming more sequencing reactions onto microreactor chips. Chem. Rev. *107*, 3367–3376.

Li, M., Diehl, F., Dressman, D., Vogelstein, B., and Kinzler, K.W. (2006). BEAMing up for detection and quantification of rare sequence variants. Nat. Methods *3*, 95–97.

Margulies, M., Egholm, M., Altman, W.E., Attiya, S., Bader, J.S., Bemben, L.A., Berka, J., Braverman, M.S., Chen, Y.J., Chen, Z., *et al.* (2005). Genome sequencing in microfabricated high-density picolitre reactors. Nature *437*, 376–380.

Mazutis, L., Araghi, A.F., Miller, O.J., Baret, J.C., Frenz, L., Janoshazi, A., Taly, V., Miller, B.J., Hutchison, J.B., Link, D., *et al.* (2009). Droplet-based microfluidic systems for high-throughput single DNA molecule isothermal amplification and analysis. Anal. Chem. *81*, 4813–4821.

McCalla, S.E., Luryi, A.L., and Tripathi, A. (2009). Steric effects and mass-transfer limitations surrounding amplification reactions on immobilized long and clinically relevant DNA templates. Langmuir *25*, 6168–6175.

Mercier, J.-F., and Slater, G.W. (2005). Solid phase DNA amplification: a Brownian dynamics study of crowding effects. Biophys. J. *89*, 32–42.

Mercier, J.-F., Slater, G.W., and Mayer, P. (2003). Solid phase DNA amplification: a simple Monte Carlo lattice model. Biophys. J. *85*, 2075–2086.

Miller, O.J., Agresti, J.J., Amitai, G., Kelly, B.T., Mastrobattista, E., Taly, V., Tawfik, D.S., and Griffiths, A.D. (2006). Directed evolution by *in vitro* compartmentalization. Nat. Methods *3*, 561–570.

Mitra, R.D., and Church, G.M. (1999). *In situ* localized amplification and contact replication of many individual DNA molecules. Nucleic Acids Res. *27*, e34.

Musyanovych, A., Mailander, V., and Landfester, K. (2005). Miniemulsion droplets as single molecule nanoreactors for polymerase chain reaction. Biomacromolecules *6*, 1824–1828.

Nagai, H., Murakami, Y., Morita, Y., Yokoyama, K., and Tamiya, E. (2001a). Development of a microchamber array for picoliter PCR. Anal. Chem. *73*, 1043–1047.

Nagai, H., Murakami, Y., Yokoyama, K., and Tamiya, E. (2001b). High-throughput PCR in silicon based microchamber array. Biosens. Bioelectron. *16*, 1015–1019.

Salk, J.J., Sanchez, J.A., Pierce, K.E., Rice, J.E., Soares, K.C., and Wangh, L.J. (2006). Direct amplification of single-stranded DNA for pyrosequencing using linear-after-the-exponential (LATE)-PCR. Anal. Biochem. *353*, 124–132.

Sanchez, J.A., Pierce, K.E., Rice, J.E., and Wangh, L.J. (2004). Linear-After-The-Exponential (LATE)-PCR: An advanced method of asymmetric PCR and its uses in quantitative real-time analysis. Proc. Natl. Acad. Sci. U.S.A. *101*, 1933–1938.

Shapero, M.H., Leuther, K.K., Nguyen, A., Scott, M., and Jones, K.W. (2001). SNP genotyping by multiplexed solid-phase amplification and fluorescent minisequencing. Genome Res. *11*, 1926–1934.

Shchepinov, M.S., Case-Green, S.C., and Southern, E.M. (1997). Steric factors influencing hybridisation of nucleic acids to oligonucleotide arrays. Nucleic Acids Res. *25*, 1155–1161.

Shendure, J., Porreca, G.J., Reppas, N.B., Lin, X., McCutcheon, J.P., Rosenbaum, A.M., Wang, M.D., Zhang, K., Mitra, R.D., and Church, G.M. (2005). Accurate multiplex polony sequencing of an evolved bacterial genome. Science *309*, 1728–1732.

Sogin, M.L., Morrison, H.G., Huber, J.A., Welch, D.M., Huse, S.M., Neal, P.R., Arrieta, J.M., and Herndl, G.J. (2006). Microbial diversity in the deep sea and the underexplored 'rare biosphere'. Proc. Natl. Acad. Sci. U.S.A.*103*, 12115–12120.

Sugiura, S., Oda, T., Izumida, Y., Aoyagi, Y., Satake, M., Ochiai, A., Ohkohchi, N., and Nakajima, M. (2005). Size control of calcium alginate beads containing living cells using micro-nozzle array. Biomaterials *26*, 3327–3331.

Suzuki, M.T., and Giovannoni, S.J. (1996). Bias caused by template annealing in the amplification of mixtures of 16S rRNA genes by PCR. Appl. Environ. Microbiol. *62*, 625–630.

Sykes, P.J., Neoh, S.H., Brisco, M.J., Hughes, E., Condon, J., and Morley, A.A. (1992). Quantitation of targets for PCR by use of limiting dilution. Biotechniques *13*, 444–449.

Tan, Y.C., Fisher, J.S., Lee, A.I., Cristini, V., and Lee, A.P. (2004). Design of microfluidic channel geometries for the control of droplet volume, chemical concentration, and sorting. Lab. Chip *4*, 292–298.

Tawfik, D.S., and Griffiths, A.D. (1998). Man-made cell-like compartments for molecular evolution. Nat. Biotechnol. *16*, 652–656.

Tewhey, R., Warner, J.B., Nakano, M., Libby, B., Medkova, M., David, P.H., Kotsopoulos, S.K., Samuels, M.L., Hutchison, J.B., Larson, J.W., *et al.* (2009). Microdroplet-based PCR enrichment for large-scale targeted sequencing. Nat. Biotechnol. *27*, 1025–1031.

Thorsen, T., Roberts, R.W., Arnold, F.H., and Quake, S.R. (2001). Dynamic pattern formation in a vesicle-generating microfluidic device. Phys. Rev. Lett. *86*, 4163–4166.

Tiemann-Boege, I., Curtis, C., Shinde, D.N., Goodman, D.B., Tavaré, S., and Arnheim, N. (2009). Product length, dye choice, and detection chemistry in the bead-emulsion amplification of millions of single DNA molecules in parallel. Anal. Chem. *81*, 5770–5776.

Vogelstein, B., and Kinzler, K.W. (1999). Digital PCR. Proc. Natl. Acad. Sci. U.S.A. *96*, 9236–9241.

Warren, L., Bryder, D., Weissman, I.L., and Quake, S.R. (2006). Transcription factor profiling in individual hematopoietic progenitors by digital RT-PCR. Proc. Natl. Acad. Sci. U.S.A. *103*, 17807–17812.

Wei, Q., Liu, S., Huang, J., Mao, X., Chu, X., Wang, Y., Qiu, M., Mao, Y., Xie, Y., and Li, Y. (2004). Comparison of hybridization behavior between double and single strands of targets and the application of asymmetric PCR targets in cDNA microarray. J. Biochem. Mol. Biol. 37, 439–444.

Wetmur, J.G., Kumar, M., Zhang, L., Palomeque, C., Wallenstein, S., and Chen, J. (2005). Molecular haplotyping by linking emulsion PCR: analysis of paraoxonase 1 haplotypes and phenotypes. Nucleic Acids Res. 33, 2615–2619.

Williams, R., Peisajovich, S.G., Miller, O.J., Magdassi, S., Tawfik, D.S., and Griffiths, A.D. (2006). Amplification of complex gene libraries by emulsion PCR. Nat. Methods 3, 545–550.

Index

Current Books of Interest

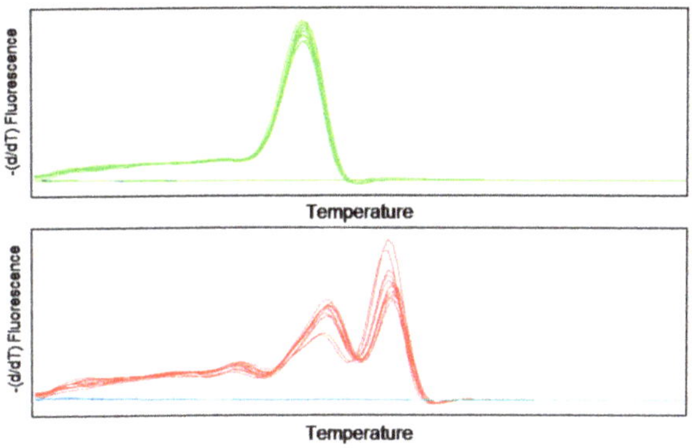

Plate 7.1 Melt curve analysis. The upper plot reveals a unique melt peak, suggestive for specific amplification. The plot at the bottom contains two clearly distinct melt peaks, indicative for non-specific amplification.

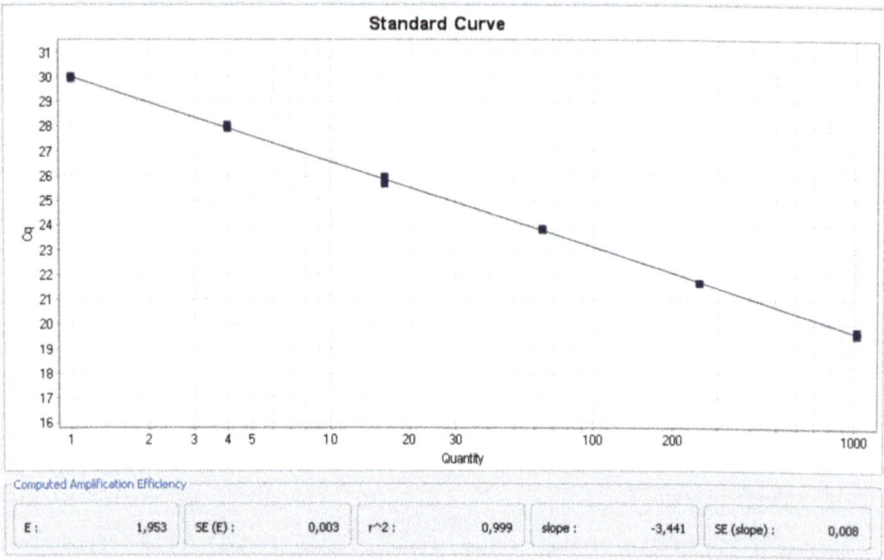

Plate 7.2 Typical standard curve obtained from a 6-point 4-fold dilution series for an assay with a good amplification efficiency (E = 1.953, efficiency = 95.3%) as calculated by qbasePLUS. A high correlation coefficient (r^2 = 0.999) and low error on the estimated efficiency (SE(E) = 0.3%) indicate a high-quality standard curve from which reliable efficiency estimations can be drawn.

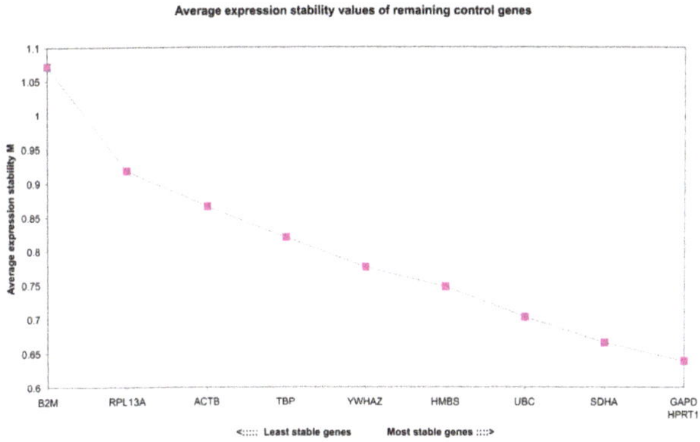

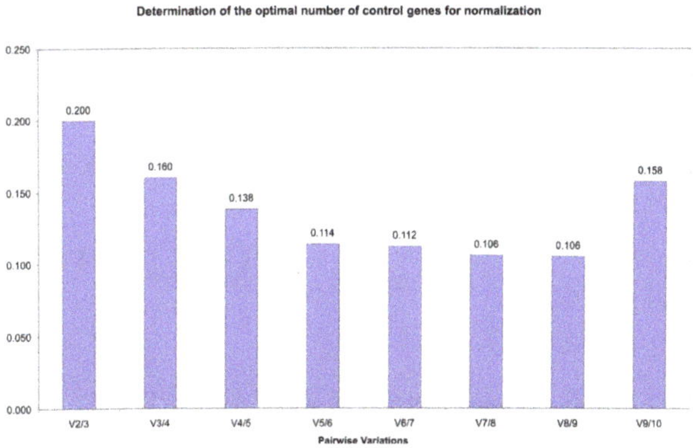

Plate 7.3 Outcome of a geNorm analysis on a heterogeneous cancer biopsy sample set. M values (line graph on top) for candidate reference genes start out rather high (1.35 for B2M at the left), then gradually drop towards the more stably expressed genes at the right (TBP and YWHAZ with an M value of approximately 0.5). The analysis of *V* values (bar chart at the bottom) indicates that 4 reference genes are minimally required for optimal normalization in this heterogeneous sample set (V4/5 is below 0.15, indicating that there is not need to include a 5th gene in the normalization factor).

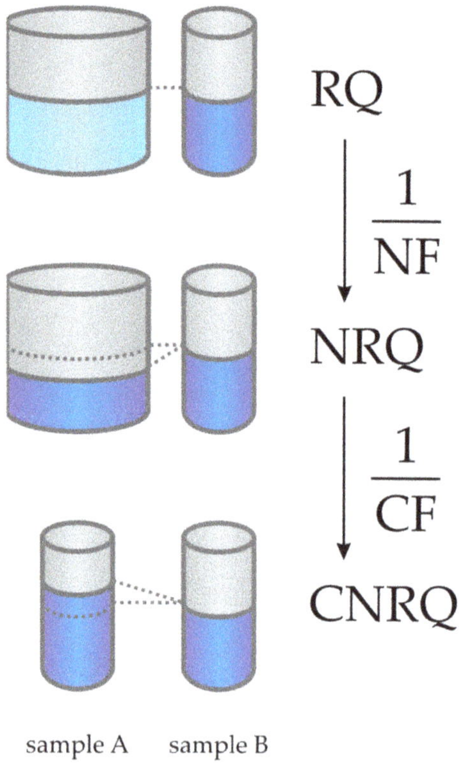

Plate 7.4 Calculation workflow. Identical relative quantities (RQ) of a given target between two samples (A and B) do not necessarily imply equal expression levels of the measured target by the applied assay. Differences in final cDNA concentration (represented by a different shade) or technical run-to-run variation (represented by a variable width) may obscure the picture of true differential expression. The normalization procedure removes the variation resulting from the use of different cDNA concentrations, thus yielding normalized relative quantities (NRQ) in which the expression level is only affected by possible run-to-run variation. This remaining variation can be removed by calibrating NRQs into calibrated normalized relative quantities (CNRQ) that truly reflect the differences in gene expression between the two samples. From a mathematical perspective normalization and inter-run calibration are equivalent procedures and are effected by division by a normalization factor (NF, geometric mean of reference gene NRQ values) or a calibration factor (CF, geometric mean of IRC sample NRQ values), respectively.

www.ingramcontent.com/pod-product-compliance
Lightning Source LLC
LaVergne TN
LVHW051050210926
848486LV00001B/59
9781904455721